Fantastic Realities

49 Mind Journeys and A Trip to Stockholm

Fantastic Realities

49 Mind Journeys and A Trip to Stockholm

FRANK WILCZEK

with a contribution from Betsy Devine

World Scientific

NEW JERSEY • LONDON • SINGAPORE • BEIJING • SHANGHAI • HONG KONG • TAIPEI • CHENNAI

Published by

World Scientific Publishing Co. Pte. Ltd.
5 Toh Tuck Link, Singapore 596224
USA office: 27 Warren Street, Suite 401-402, Hackensack, NJ 07601
UK office: 57 Shelton Street, Covent Garden, London WC2H 9HE

Library of Congress Cataloging-in-Publication Data
Wilczek, Frank, 1951–
　　Fantastic realities : 49 mind journeys and a trip to Stockholm
　　　p. cm.
　　Includes bibliographical references.
　　ISBN 981-256-649-X -- ISBN 981-256-655-4 (pbk.)
　　1. Physics--Miscellanea. I. Wilczek, Frank.

QC75.F36 2006
530--dc22
　　　　　　　　　　　　　　　　　　2005057870

British Library Cataloguing-in-Publication Data
A catalogue record for this book is available from the British Library.

Front cover:　Author's photo courtesy of Justin Knight Photography.
　　　　　　　Nobel Prize award ceremony photo courtesy of Scanpix.

Copyright © 2006 by World Scientific Publishing Co. Pte. Ltd.

All rights reserved.

Printed in Singapore by Mainland Press.

Introduction

This collection is meant to give pleasure. There are some jokes, and poems, and extracts from Betsy Devine's sparkling blog, which should all be easy entertainment. There's also some history, some philosophy, some exposition of frontier science, and some frontier science. I've organized the 49 pieces by style and subject into a dozen sections, and written a little introduction to each section. You can pick and choose and read them in any order you please; although there are many connections among the ideas, there are no dependencies. Betsy's contribution comes last, but you might want to start with it. I would.

I've made only minimal changes to the pieces that have been published previously. I want to thank all the publishers involved for their permission to reprint, and my various editors for their help and advice. My special thanks go to Gloria Lubkin, who solicited and edited each and every one of the 18 Reference Frame articles.

Books with short introductions are well begun.

Contents

Introduction v

Constructing This World, and Others 1
1. The World's Numerical Recipe 3
2. Analysis and Synthesis 1: What Matters for Matter 10
3. Analysis and Synthesis 2: Universal Characteristics 16
4. Analysis and Synthesis 3: Cosmic Groundwork 22
5. Analysis and Synthesis 4: Limits and Supplements 28

Musing on Mechanics 35
6. Whence the Force of $F = ma$? 1: Culture Shock 37
7. Whence the Force of $F = ma$? 2: Rationalization 43
8. Whence the Force of $F = ma$? 3: Cultural Diversity 49

Making Light of Mass 55
9. The Origin of Mass 57
10. Mass without Mass 1: Most of Matter 72
11. Mass without Mass 2: The Medium is the Mass-age 76

QCD Exposed 81
12. QCD Made Simple 83
13. 10^{12} Degrees in the Shade 100
14. Back to Basics at Ultrahigh Temperatures 114

Breathless at the Heights 121
15. Scaling Mount Planck 1: A View from the Bottom 123
16. Scaling Mount Planck 2: Base Camp 128
17. Scaling Mount Planck 3: Is That All There Is? 133

At Sea in the Depths — 137

18. What is Quantum Theory? — 140
19. Total Relativity: Mach 2004 — 144
20. Life's Parameters — 150

Once and Future History — 155

21. The Dirac Equation — 157
22. Fermi and the Elucidation of Matter — 191
23. The Standard Model Transcended — 211
24. Masses and Molasses — 216
25. In Search of Symmetry Lost — 223
26. From 'Not Wrong' to (Maybe) Right — 250
27. Unification of Couplings — 253

Methods of Our Madness — 277

28. The Social Benefit of High Energy Physics — 279
29. When Words Fail — 285
30. Why Are There Analogies between Condensed Matter and Particle Theory? — 288
31. The Persistence of Ether — 293
32. Reaching Bottom, Laying Foundations — 298

Inspired, Irritated, Inspired — 305

33. What Did Bohr Do? — 307
34. Dreams of a Final Theory — 314
35. Shadows of the Mind — 318
36. The Inflationary Universe — 323
37. Is the Sky Made from Pi? — 327

Big Ideas — 331

38. Quantum Field Theory — 333
39. Some Basic Aspects of Fractional Quantum Numbers — 359

Grand Occasions — 385

40. From Concept to Reality to Vision — 387
41. Nobel Biography — 396

42. Asymptotic Freedom: From Paradox to Paradigm	400
43. Advice to Students	430

Breaking into Verse — 433

44. Virtual Particles	435
45. Gluon Rap	436
46. Reply In Sonnet Form	437
47. From Beneath An e-Avalanche	438
48. Frog Sonnet	439
49. Archaeopteryx	440

Another Dimension	441
Nobel Blog: A Year in the Life	442
Acknowledgments	519

Constructing This World, and Others

I hate "reductionism." Not the thing, the word. Its form belies its content. Explaining things is *augmentation*, not *reduction*. I much prefer Newton's phrase, Analysis and Synthesis, to describe the process of breaking problems or phenomena into simpler component parts (Analysis), and then building up from those parts (Synthesis).

In 600 B.C. Pythagoras set out a most ambitious goal for Analysis, with his visionary pronouncement "All things are number." His program has had, to say the least, its ups and downs. For two millennia his vision inspired aspiring illuminati, but with the rise of classical physics it became passé. Then, with the rise of quantum theory in the twentieth century, Pythagoras made a triumphant comeback. The amazing story of how the most sophisticated and accurate rendition of reality came to embody a discredited ancient fantasy is recounted in "The World's Numerical Recipe."

The four-part series *Analysis and Synthesis* is a brief but wide-ranging survey of the deepest levels of our present Analysis of Nature, indicating its unfinished character and, in the end, its profound limitations.

In "Analysis and Synthesis 1" I hit on the happy idea of a succession of idealized worlds, each precisely defined in terms of its building blocks and their behavior, each closed and complete in its own way. The later versions include additional features, like successive versions of Word. So we have World 1.0, World 2.0, and so forth. As with Word, if you don't need the new features, the earlier, simpler World versions, which avoid bloat, can be superior. Six useful versions of World are proposed in "Analysis and Synthesis 1." Actually World 6.0, the full-featured latest version, is a real kluge. I've seen ads for World 7.0, but it's vaporware.

To understand the one world we observe, it probably won't be enough just to know its basic equations. Since those equations almost certainly will allow many solutions, we'll need to have some additional principles that tell us which solutions apply to reality. In "Analysis and Synthesis 2" and "Analysis and Synthesis 3" I discuss how modern cosmology allows Analysis to penetrate deeply into that problem. Given a handful of numerical inputs, it constructs a specific world-solution.

There's been a lot of discussion of "the limits of reductionism." Most of it's been vague, emotional, or both. Yet it's possible to be specific, since the progress of scientific Analysis has revealed several perfectly objective, profound effects that limit the power of scientific Analysis. In "Analysis and Synthesis 4" I identify three such phenomena: projection (i.e., the fact that we don't get to see everything), the probabilistic logic of quantum mechanics, and chaos (i.e., extreme sensitivity to initial conditions). These effects will make it impossible, in practice, to go from the solution (if we had it) of the basic equations (if we had them) to a complete description of experience. The technical word for this is contingency — you can't analyze everything, some things are just accidents. On the positive side, I think it's likely that *all* contingency in the world will be found to arise from just those three sources mentioned. So we can analyze their effects, to sort out what we can and can't ultimately Analyze. Thus reason discovers its own limitations.

It is liberating to recognize these limitations honestly. It frees us of any illusion that Analysis can converge on a Theory of Everything that will answer all questions, and gives appropriate (wide!) scope to more empirically based theoretical enterprises.

I'm fond of an image I invented for "Analysis and Synthesis 4":

> to score a bull's-eye by painting the target around your shot

Archery is better sport if you paint the bull's-eye first, and stand within the range of your bow.

The World's Numerical Recipe

Twentieth-century physics began around 600 B.C., when Pythagoras of Samos proclaimed an awesome vision.

By studying the notes sounded by plucked strings, Pythagoras discovered that the human perception of harmony is connected to numerical ratios. He examined strings made of the same material, having the same thickness, and under the same tension, but of different lengths. Under these conditions, he found that the notes sound harmonious precisely when the ratio of the lengths of string can be expressed in small whole numbers. For example, the length ratio 2:1 sounds a musical octave, 3:2 a musical fifth, and 4:3 a musical fourth.

The vision inspired by this discovery is summed up in the maxim "All Things are Number." This became the credo of the Pythagorean Brotherhood, a mixed-sex society that combined elements of an archaic religious cult and a modern scientific academy.

The Brotherhood was responsible for many fine discoveries, all of which it attributed to Pythagoras. Perhaps the most celebrated and profound is the Pythagorean Theorem. This theorem remains a staple of introductory geometry courses. It is also the point of departure for the Riemann–Einstein theories of curved space and gravity.

Unfortunately, that very theorem undermined the Brotherhood's credo. Using the Pythagorean Theorem, it is not hard to prove that the ratio of the hypotenuse of an isosceles right triangle to either of its two shorter sides cannot be expressed in whole numbers. A member of the Brotherhood who revealed this dreadful secret drowned shortly afterwards, in suspicious circumstances. Today, when we say $\sqrt{2}$ is irrational, our language still reflects those ancient anxieties.

Still, the Pythagorean vision broadly understood, and stripped of cultic, if not entirely of mystical, trappings remained a touchstone for pioneers of mathematical science for centuries. Those working within this tradition did not insist on whole numbers, but continued to postulate that the deep structure of the physical world could be captured in purely conceptual constructions. Considerations of symmetry and abstract geometry were allowed to supplement simple numerics.

In the work of the German astronomer Johannes Kepler (1570–1630), this program reached a remarkable apotheosis — only to unravel completely.

Students today still learn about Kepler's three laws of planetary motion. But before formulating these celebrated laws, that great speculative thinker had announced another law — we can call it Kepler's zeroth law — of which we hear much less, for the very good reason that it is entirely wrong. Yet it was his discovery of the zeroth law that fired Kepler's enthusiasm for planetary astronomy, and in particular for the Copernican system, and launched his extraordinary career. Kepler's zeroth law concerns the relative size of the orbits of different planets. To formulate it, we must imagine that the planets are carried about on concentric spheres around the Sun. His law states that the successive planetary spheres are of such proportions that they can be inscribed within and circumscribed about the five Platonic solids. These five remarkable solids — tetrahedron, cube, octahedron, dodecahedron, icosahedron — have faces that are congruent equilateral polygons. The Pythagoreans studied them, Plato employed them in the speculative cosmology of the *Timaeus*, and Euclid climaxed his *Elements* with the first known proof that only five such regular polyhedra exist.

Kepler was enraptured by his discovery. He imagined that the spheres emitted music as they rotated, and he even speculated on the tunes. (This is the source of the phrase "music of the spheres.") It was a beautiful realization of the Pythagorean ideal. Purely conceptual, yet sensually appealing, the zeroth law seemed a production worthy of a mathematically sophisticated Creator.

To his great credit as an honest man and — though the concept is anachronistic — as a scientist, Kepler did not wallow in mystic rapture, but actively strove to see whether his law accurately matched reality. He discovered that it does not. In wrestling with the precise observations of

Tycho Brahe, Kepler was forced to give up circular in favor of elliptical orbits. He couldn't salvage the ideas that first inspired him.

After this, the Pythagorean vision went into a long, deep eclipse. In Newton's classical synthesis of motion and gravitation, there is no sense in which structure is governed by numerical or conceptual constructs. All is dynamics. Given the positions, velocities, and masses of gravitating bodies at one time, Newton's laws inform us how they will move in the future. Those laws do not fix a unique size or structure for the Solar System. Indeed, recent discoveries of planetary systems around distant stars have revealed quite different patterns. The great developments of nineteenth-century physics, epitomized in Maxwell's equations of electrodynamics, brought many new phenomena within the scope of physics, but they did not alter this situation essentially. There is nothing in the equations of classical physics that can fix a definite scale of size, whether for planetary systems, atoms, or anything else. The world-system of classical physics is divided between initial conditions, which can be assigned arbitrarily, and dynamical equations. In those equations, neither whole numbers nor any other purely conceptual elements play a distinguished role.

Quantum mechanics changed everything.

Emblematic of the new physics, and decisive historically, was Niels Bohr's atomic model of 1913. Though it applies in a vastly different domain, Bohr's model of the hydrogen atom bears an uncanny resemblance to Kepler's system of planetary spheres. The binding force is electrical rather than gravitational, the players are electrons orbiting around protons rather than planets orbiting the Sun, and the size is a factor 10^{-22} smaller; but the leitmotif of Bohr's model is unmistakably "Things are Number."

Through Bohr's model, Kepler's idea that the orbits occurring in Nature are precisely those that embody a conceptual ideal emerged from its embers, reborn like a phoenix, after three hundred years' quiescence. If anything, Bohr's model conforms more closely to the Pythagorean ideal than Kepler's, since its preferred orbits are defined by whole numbers rather than geometric constructions. Einstein responded with great empathy and enthusiasm, referring to Bohr's work as "the highest form of musicality in the sphere of thought."

Later work by Heisenberg and Schrödinger, which defined modern quantum mechanics, superseded Bohr's model. This account of subatomic matter is less tangible than Bohr's, but ultimately much richer. In the Heisenberg–Schrödinger theory, electrons are no longer particles moving in space, elements of reality that at a given time are "just there, and not anywhere else." Rather they define oscillatory, space-filling wave patterns always "here, there, and everywhere." Electron waves are attracted to a positively charged nucleus, and can form localized standing wave patterns around it. The mathematics describing the vibratory patterns that define the states of atoms in quantum mechanics is identical to the mathematics describing the resonance of musical instruments. The stable states of atoms correspond to pure tones. I think it's fair to say that the musicality Einstein praised in Bohr's model is, if anything, heightened in its progeny (though Einstein himself, notoriously, withheld his approval from the new quantum mechanics).

The big difference between Nature's instruments and those of human construction is that Her designs depend not on craftsmanship refined by experience, but rather on the ruthlessly precise application of simple rules. Now if you browse through a textbook on atomic quantum mechanics, or look at atomic vibration patterns using modern visualization tools, "simple" might not be the word that leaps to mind. But it has a precise, objective meaning in this context. A theory is simpler, the fewer non-conceptual elements, which must be taken from observation, enter into its construction. In this sense Kepler's zeroth law provided a simpler (as it turns out, too simple) theory of the Solar System than Newton's, because in Newton's theory the relative sizes of planetary orbits must be taken from observation, whereas in Kepler's they are determined conceptually.

From this perspective, modern atomic theory is extraordinarily simple. The Schrödinger equation, which governs electrons in atoms, contains just two non-conceptual quantities. These are the mass of the electron and the so-called fine-structure constant, denoted α, that specifies the overall strength of the electromagnetic interaction. By solving this one equation, finding the vibrations it supports, we make a concept-world that reproduces a tremendous wealth of real-world data, notably the accurately measured spectral lines of atoms, which encode their inner structure. The marvelous theory of electrons and their interactions with light is called quantum electrodynamics, or QED.

In the initial modeling of atoms, the focus was on their accessible, outlying parts, the electron clouds. The nuclei of atoms, which contain most of their mass and all of their positive charge, were treated as so many tiny (but very heavy!) black boxes, buried in the core. There was no theory for the values of nuclear masses or their other properties; these were simply taken from experiment. That pragmatic approach was extremely fruitful, and to this day provides the working basis for practical applications of physics in chemistry, materials science, and biology. But it failed to provide a theory that was in our sense simple, and so it left the ultimate ambitions of a Pythagorean physics unfulfilled.

Starting in the early 1930s, with electrons under control, the frontier of fundamental physics moved inward, to the nuclei. This is not the occasion to recount the complex history of the heroic constructions and ingenious deductions that at last, after fifty years of strenuous international effort, fully exposed the secrets of this inaccessible domain. Fortunately the answer is easier to describe, and it advances and consummates our theme.

The theory that governs atomic nuclei is quantum chromodynamics, or QCD. As its name hints, QCD is firmly based on quantum mechanics. Its mathematical basis is a direct generalization of QED, incorporating a more intricate structure supporting enhanced symmetry. Metaphorically, QCD stands to QED as an icosahedron stands to a triangle. The basic players in QCD are quarks and gluons. For constructing an accurate model of ordinary matter just two kinds of quarks, called up and down or simply u and d, need to be considered. (There are four other kinds, at least, but they are highly unstable, and not important for ordinary matter.) Protons, neutrons, π mesons, and a vast zoo of very short-lived particles, called resonances, are constructed from these building blocks. The particles and resonances observed in the real word match the resonant wave-patterns of quarks and gluons in the concept-world of QCD, much as states of atoms match the resonant wave patterns of electrons. You can predict their masses and properties directly by solving the equations.

A peculiar feature of QCD, and a major reason why it was hard to discover, is that the quarks and gluons are never found in isolation, but always in complex associations. QCD actually predicts this "confinement" property, but that's not easy to prove.

Considering how much it accounts for, QCD is an amazingly simple theory, in our objective sense. Its equations contain just three non-conceptual ingredients: the masses of the u and d quarks, and the strong coupling constant α_s, analogous to the fine structure constant of QED, which specifies how powerfully quarks couple to gluons. The gluons are automatically massless.

Actually even three is an overestimate. The quark-gluon coupling varies with distance, so we can trade it in for a unit of distance. In other words, mutant QCDs with different values of α_s generate concept-worlds that behave exactly identically, but use different-sized meter sticks. Also, the masses of the u and d quarks turn out not to be very important, quantitatively. Most of the mass of strongly interacting particles is due to the pure energy of the moving quarks and gluons they contain, according to the converse of Einstein's equation, $m = E/c^2$. The masses of the u and d quarks are much smaller than the masses of the protons and other particles that contain them.

Putting all this together, we arrive at a most remarkable conclusion. To the extent that we are willing to use the proton itself as a meter-stick, and ignore the small corrections due to the u and d quark masses, QCD becomes a theory with *no non-conceptual elements whatsoever*.

Let me summarize. Starting with precisely four numerical ingredients, which must be taken from experiment, QED and QCD cook up a concept-world of mathematical objects whose behavior matches, with remarkable accuracy, the behavior of real-world matter. These objects are vibratory wave patterns. Stable elements of reality — protons, atomic nuclei, atoms — correspond, not just metaphorically but with mathematical precision, to pure tones. Kepler would be pleased.

This tale continues in several directions. Given two more ingredients, Newton's constant G_N and Fermi's constant G_F, which parametrize the strength of gravity and of the weak interaction respectively, we can expand our concept-world beyond ordinary matter, to describe virtually all of astrophysics. There is a brilliant series of ideas involving unified field theories and supersymmetry that might allow us to get by with just five ingredients. (Once you're down to so few, each further reduction marks an epoch.) These ideas will be tested decisively in coming years, especially as the Large Hadron Collider (LHC) at CERN, near Geneva, swings into operation around 2007.

On the other hand, if we attempt to do justice to the properties of many exotic, short-lived particles discovered at high-energy accelerators, things get much more complicated and unsatisfactory. We have to add pinches of many new ingredients to our recipe, until it may seem that rather than deriving a wealth of insight from a small investment of facts, we are doing just the opposite. That's the state of our knowledge of fundamental physics today — simultaneously triumphant, exciting, and a mess.

The last word I leave to Einstein:

I would like to state a theorem which at present can not be based upon anything more than upon a faith in the simplicity, i.e., intelligibility, of Nature: there are no arbitrary constants ... that is to say, Nature is so constituted that it is possible logically to lay down such strongly determined laws that within these laws only rationally completely determined constants occur (not constants, therefore, whose numerical value could be changed without destroying the theory).

Analysis and Synthesis 1: What Matters for Matter

Query 31 of Isaac Newton's *Opticks* was his last word in science. It begins:

> Have not the small Particles of Bodies certain Powers, Virtues, or Forces, by which they act … upon one another for producing a great Part of the Phenomena of Nature?

After a lengthy sketch of how this concept might lead to explanations of various phenomena in what we would now call chemistry and condensed matter physics, Query 31 concludes with a statement of methodological faith:

> As in Mathematicks, so in Natural Philosophy, the Investigation of difficult Things by the Method of Analysis ought ever to precede the Method of Composition. … By this way … we may proceed from Compounds to Ingredients, and from Motions to the Forces producing them; and in general, from Effects to their Causes … And the Synthesis consists in assuming the Causes discovered, and established as Principles, and by them explaining the Phenomena proceeding from them, and proving the Explanations.

"Reductionism" is modern jargon for the program that Newton advocated in Query 31. But this is an ugly and misleading term, and it has become almost a term of abuse in fashionable intellectual circles. So in this series of four columns. I'll avoid the R-word and stick with Newton's "Analysis and Synthesis." The phrase has the virtue of emphasizing that both procedures, breaking down and building back up, form essential elements in scientific understanding — and that such understanding is therefore not reduction, but rather enrichment.

Whatever you call it, as a tool for understanding the physical world, the method of analysis and synthesis has been astoundingly successful. In the description of matter, it has worked out more or less along the lines Newton suggested; in cosmology, along lines quite different from anything Newton imagined. (As also appears in Query 31, Newton believed that the world was fashioned by God, whose active intervention is required for its proper functioning.) In both cases, analysis has brought us to foundational models that come close to achieving perfection.

The history of physics emphasizes that synthesis is a challenging and deeply creative activity in its own right. It is more open-ended, however, and less easy to summarize, so I'll only be able to mention a few illustrative high points.

We shall also see that, in the course of its triumphal advance, ironically the method of analysis and synthesis has itself discovered profound, sharply defined limits to its explanatory power.

Electronic Levels of Description

It is instructive to work up to our most complete analysis of matter through some intermediate models that are extremely important and useful in their own right.

The first model builds a world in which the only active players are non-relativistic electrons obeying the laws of quantum mechanics. To flesh out this world, we also allow highly localized, static sources of charge, concentrated in positive multiples of $|e|$, where e is the electron charge. Those sources provide a schematic, "black-box" description of atomic nuclei. This is the world captured in Schrödinger's original wave equation, using Coulomb's law for the forces. With hindsight, we realize that this world is an approximation to ours, in which atomic nuclei are regarded as being infinitely heavy compared to electrons (instead of merely thousands of times heavier), and light as moving infinitely faster (instead of merely a hundred times faster).

This model is an extremely economical construction. An objective measure of its parsimony is how few parameters it involves. Superficially, there appear to be three: the charge unit e, Planck's constant $\hbar$, and the electron mass m_e. But we can swap those parameters for a system of units: length ($\hbar^2/m_e e^2$), time ($\hbar^3/m_e e^4$), and mass (m_e). When we express physical results using these units, the original parameters no longer appear at all. So our model contains no genuine parameters whatsoever. Its prediction for any dimensionless quantity is unambiguously fixed by its conceptual structure. Analysis can go no further.

Yet this Spartan framework supports an extremely rich and complex world construction. We can use it to compute what kinds of molecules should exist, and what their shapes are, by identifying local minima of the energy as a function of the positions of the source charges. (Strictly speaking, we must also specify some symmetry rules to take into account the quantum statistics of electrons.) This framework gives us a parameter-free foundation for the vast subject of structural chemistry.

A second model refines the first by incorporating special relativity, that is, by drawing out the consequences of the finite speed of light, $c < \infty$. We pass from the Schrödinger to the Dirac equation, and from Coulomb's law to the Maxwell equations, and thereby introduce real and virtual photons. By passing from Schrödinger–Coulomb to Dirac–Maxwell, we arrive at quantum electrodynamics (QED) with sources. A pure number now enters the game, namely the fine structure constant $\alpha = e^2/4\pi\hbar c$. The refined model is in that respect less perfect, but in compensation, it is both more accurate and much more comprehensive. It now includes the dynamical effects of electron spin, the Lamb shift, radiation phenomena, and much more.

These two wondrous models have a big shortcoming, however: The molecules can't move. Reactions, diffusion, and thermal phenomena are completely missing, as are vibrational and rotational spectra. To allow motion, we must be less schematic about the nuclei. In doing so, however, we open ourselves up to *many* more input parameters. At a minimum we need the masses of the different nuclei and isotopes, and for accuracy in details we need their spins, magnetic moments, and other properties. Literally hundreds of nuclear parameters are required.

A practical approach is simply to take all the needed quantities from experiment. Of course, with that step we acknowledge that there are many measurements whose results we do not attempt to predict. It is a major compromise in analysis — a strategic retreat. But all is not lost, to say the least, because synthesis can do wonders with this material. Indeed our third model, with its strategic retreat, provides a comprehensive foundation for quantum chemistry and condensed matter physics. Within that scope one can contemplate hundreds of thousands — if not an infinite number — of significant measurements, so a working model containing hundreds of parameters can remain extremely useful and predictive. It is precisely this world construction that Paul Dirac famously described as containing "all of chemistry and most of physics."

Quark-Gluon Levels of Description

Thanks to advances made over the past 30 years, we now can carry the analysis of matter significantly further, and reconquer most of the lost ground. Now we can base our description of nuclei on a theory whose economy and elegance rivals that of QED. Here, of course, I speak of quantum chromodynamics (QCD).

The fourth model combines the Dirac–Maxwell theory of electrons and photons (as in the second model) with a truncated version of QCD, which I call QCD Lite. QCD Lite builds atomic nuclei using as ingredients only massless particles: the color gluons and two kinds of quarks, up u and down d. There are good reasons to believe that QCD Lite is capable of reproducing the important nuclear parameters at least crudely, perhaps at the 10–20% level. (I'll explain in a moment why I've stated this claim so gingerly.) That is astonishing, because QCD Lite is a parameter-free theory! Indeed, its equations can be formulated using just $\hbar$, c, and a mass Λ_{QCD}. Those parameters can be exchanged for a system of units, similar to what we did in our first model, the Schrödinger–Coulomb model. (For more on QCD Lite and the origin of mass, see "Mass without Mass 1: Most of Matter.") The fifth and final model is a simple refinement of the fourth where we introduce nonzero masses m_u and m_d for the quarks. That modification produces a much more accurate representation of reality, at some cost in economy.

With this fifth model, the modern analysis of terrestrial matter is essentially complete. A fundamental theory that offers an extremely complete and accurate set of equations governing the structure and behavior of ordinary matter in ordinary conditions, with a very liberal definition of "ordinary," requires the parameters $\hbar$, e, m_e, c, Λ_{QCD}, m_u, and m_d. Three of these can be traded in for units, and two (m_u and m_d) play a relatively minor role. Thus we have a rough analysis of matter down to two parameters, and an accurate one with four. Two more are required for astrophysics, as I'll discuss in the next installment.

Here, a confession: Our faith in QCD does not stem from our ability to use it to calculate nuclear parameters — in fact, nobody knows a practical way to do such calculations. The best quantitative tests of the theory occur in an entirely different domain, ultra-high energy experiments where the underlying quarks and gluons and their couplings show up clearly. Our faith relies, for empirical support, on the successful outcome of those tests, together with encouraging but as yet fairly crude results from massive numerical calculations of proton structure and the hadron spectrum. These successes are highly leveraged, because the theory that describes them is extremely rigid. It will not bend without breaking. If we want to stay consistent with requirements of quantum mechanics and special relativity, the possibilities for couplings among quarks and gluons are very restricted. Only a few parameters can possibly appear in QCD — just the quark masses and Λ_{QCD}. Nothing else is permitted.

So we have rigidly defined equations, tested rigorously and quantitatively at high energy. We know that solving the equations will yield specific values for the nuclear parameters. Thus we are led to believe these values will be accurate even though, at present, we can carry out the necessary calculations only in the simplest cases. I think the expectation is perfectly reasonable, not so different from our faith that QED applies to complicated chemistry. Unfortunately, however, the "reduction" we offer to nuclear physicists, or for that matter to chemists, is not of much use to them in practice. It is a big standing challenge to design better algorithms.

Toolkits and Artworks

To include everything we know about matter, the tidy framework of the fifth model must be expanded considerably. A natural next step is to the full-blown standard model. Within that framework, we can accommodate an astonishing variety of phenomena that have been discovered in cosmic rays and at accelerators. On the downside, this step opens us up to many more parameters, about two dozen, mostly describing the masses and weak mixing angles of various elusive or highly unstable particles. It is reminiscent of the earlier step from the second model to the third model. But while that earlier step has now been assimilated into a beautiful and economical theory, as I just discussed, there is at present nothing comparable "Beyond the Standard Model." At the frontier of analysis, a considerable tension exists between elegance and accuracy.

Tension also is apparent between ease of use and completeness. A profound joke is that one can measure the progress of physics by the problems that can't be solved. In Newtonian gravity, the three-body problem is difficult, but the two-body problem can be done exactly; in general relativity, two bodies are difficult but one body can be done exactly; in quantum gravity, the vacuum is intractable.

Faced with such choices, wisdom opts for "all of the above." Different levels of description have different virtues and can be used for different purposes. The Schrödinger–Coulomb theory, for all its limitations, is an amazing work of art. If you doubt it, let me urge you to contemplate the wave-mechanical hydrogen atom using Dean Dauger's remarkable shareware "Atom in a Box"[1] — while remembering that the software animation represents deep aspects of reality directly. Then try to imagine a carbon atom or a water molecule. And the third model, for all its compromises, is an amazing toolkit. It is used to design new generations of lasers and microelectronic devices, among many other applications.

Real understanding is a much subtler and suppler affair than doing analysis at the finest possible resolution. A "Theory of Everything," were it constructed, wouldn't be.

Dirac is sometimes invoked as a mythological patron saint of purity and reductionism. The flesh-and-blood Dirac was trained as an electrical engineer, and in his scientific autobiography[2] he pays tribute to the value of that training:

> The engineering course influenced me very strongly ... I've learned that, in the description of Nature, one has to tolerate approximations, and that even work with approximations can be interesting and can sometimes be beautiful.

References

1. See the Atom in a Box link online at http://www.dauger.com. The program requires a Macintosh computer.
2. P. A. M. Dirac in *History of Twentieth Century Physics*, ed. C. Weiner, Academic Press, New York (1977).

Analysis and Synthesis 2: Universal Characteristics

Quantum electrodynamics (QED) runs the show outside atomic nuclei, and quantum chromodynamics (QCD) runs the show inside. This Q*D dynasty governs ordinary matter. The Q*Ds, which are purely based on abstract concepts, provide good impressionistic models of matter when supplemented with just two numerical parameters, and quite a lifelike rendering using four parameters. From these ingredients, we can synthesize a universe of mathematical possibilities that we believe, on good evidence, accurately mirrors the physical universe of materials and their chemistry.

The Polished Core of Astrophysics

In astrophysics, we study the behavior of very large amounts of matter over very long periods of time. Small but cumulative effects (gravity), or rare but transformative ones (weak interactions), which are negligible for most terrestrial and laboratory concerns, must be taken into account when we so widen our horizons. Yet by adding just two more parameters, we can extend our analysis of matter to cover most of astrophysics.

We have an excellent theory for gravity, namely Einstein's general theory of relativity. It contains just one new parameter, Newton's constant G_N. As I explained in an earlier column ("Scaling Mount Planck 3"), there is absolutely no *practical* problem in combining the successful theories of matter with general relativity. For example, the intricate global positioning system (GPS), which defines space-times operationally with precision and versatility, works just fine. The GPS is so accurate that it is sensitive to the effect of Earth's gravitational redshift, a direct reflection of the warping of time by matter. While recognizing Einstein's theory, so far the GPS seems perfectly oblivious to ongoing, much-heralded crises and revolutions in space-time concepts. This lack of awareness is consistent with expectations from the working theory of quantum gravity I sketched in "Scaling Mount Planck" — which is also the theory tacitly assumed throughout astrophysics.

The weak interaction powers the energy release of stars and drives their evolution. For most purposes, it is enough to include the basic interaction $d \to ue\bar{v}$ that converts a d quark into a u quark, an electron, and an antineutrino. The overall strength of that process is governed by one more new parameter, the Fermi constant G_F. (Strictly speaking, what appears is G_F times the squared cosine of the Cabibbo angle. This numerical factor is about 0.98, that is, very nearly unity.) This basic quark-conversion process underlies both radioactive β decays on Earth and the complex forms of nuclear cooking that occur in normal stars. Inclusion of the weak interaction also allows us to smooth out a slight imperfection in the logical structure of our account of matter based on QED and QCD. By destabilizing nuclei that would otherwise be stable, the weak interaction plays an important negative role. It defines the boundary of the periodic table for practical chemistry at the line of β stability, and removes spurious isotopes that would otherwise appear.

And the Ragged Edge

Other significant weak interaction processes involve strange (in the technical sense) particles or neutral currents producing neutrinoantineutrino pairs. Such processes become important in the later states of stellar evolution and especially during the cataclysmic explosions of supernovae and their aftermath. Analysis of this physics brings in three more numerical parameters: the Cabibbo angle, the Weinberg angle, and the strange quark mass. With these, we can set up the governing equations. Of course, as in previous cases, solving the equations poses problems of a different order.

In the aftermath of supernova explosions, and perhaps in other extreme astrophysical environments, a few particles get accelerated to extremely high energies — they become cosmic rays. When cosmic rays collide with interstellar material, or impact our atmosphere, the debris of the collisions contains muons, taus, and heavy quarks. To describe the properties of all those particles, we must introduce many new parameters, specifically their masses and weak mixing angles. Recently we've learned, also mainly through the study of cosmic rays (including those emanating from the Sun) that neutrinos oscillate, so we must include masses and mixings for them, too.

At the high-energy frontier, parameters begin to proliferate more rapidly than major new phenomena. At lower energies, our use of intermediate, truncated models sustained a magnificent price to earnings ratio of parameters to results. But at the high-energy frontier, I think, we've reached the point of diminishing returns. Our most accurate understanding of the laws of Nature, including all known details — the standard model, supplemented with nonzero neutrino masses and mixings, merged with general relativity — is clearly an inventory of raw materials, not a finished product.

That batch of raw materials is where the program of understanding matter by analysis and synthesis stands today. The analysis, though clearly unfinished, is already stunningly successful. It supports the synthesis of a remarkably economical conceptual system encompassing, in Dirac's phrase, "All of chemistry and most of physics." And the "most" now includes, as it did not in Dirac's time, nuclear physics and astrophysics.

Cosmology by Numbers

Remarkably, a parallel program of analysis and synthesis can be carried through for cosmology. We can construct conceptual models that use a very small number of parameters to describe major aspects of the universe as a whole.

The first model treats the universe as homogeneous and isotropic, or in plain English uniform. It isn't, of course. Indeed, superficially the distribution of matter in the universe appears to be anything but uniform. Matter is concentrated in stars, separated by vast nearly empty spaces, gathered into galaxies separated by still emptier spaces. We'll be well rewarded for temporarily ignoring such embarrassing details, however.

The parameters of the model specify a few average properties of matter, taken over large spatial volumes. Those are the densities of ordinary matter (baryons), dark matter, and dark energy. We know quite a lot about ordinary matter, as I've been discussing, and can detect it at great distances by several methods. It contributes about 3% of the total density. About dark (actually, transparent) matter we know much less. It has been seen only indirectly, through the influence of its gravity on the motion of visible matter. Dark matter is observed to exert very little pressure, and it contributes about 30% of the total density. Dark (actually, transparent) energy contributes about 67% of the total density. It has a large *negative* pressure. Dark energy is most mysterious and disturbing, as I'll elaborate next time.

Fortunately, our nearly total ignorance concerning the nature of most of the mass of the universe does not bar us from modeling its evolution. That's because the dominant interaction on large scales is gravity, and gravity does not care about details. According to general relativity, only total energy–momentum counts — or equivalently, for uniform matter, total density and pressure.

We can use the equations of general relativity to extrapolate the present expansion of the universe back to earlier times, assuming the observed relative densities for matter, dark matter, and dark energy. We treat the geometry of space as flat, and the distribution of matter as uniform. This extrapolation defines the standard Big Bang scenario. It successfully predicts several things that would otherwise be very difficult to understand, including the redshift of distant galaxies, the existence of the microwave background radiation, and the relative abundance of light nuclear isotopes. The procedure is also internally consistent, and even self-validating, in that the microwave background is observed to be uniform to high accuracy, namely a few parts in 10^5.

A second model, refining the first, allows for small departures from uniformity in the early universe and follows the dynamical evolution of some assumed spectrum of initial fluctuations. The seeds grow by gravitational instability, with overly dense regions attracting more matter, and thereby increasing their density enhancement with time. Starting from very small seeds, this growth process plausibly could eventually trigger the formation of galaxies, stars, and other structures observed today. *A priori*, one might consider all kinds of assumptions about the initial fluctuations, and over the years many hypotheses have been proposed. But recent observations, especially the gorgeous WMAP measurements of microwave background anisotropies, favor what in many ways is the simplest possible guess, the so-called Harrison–Zeldovich spectrum. In that theory the fluctuations are assumed to be strongly random — uncorrelated and Gaussian with a spatially scale-invariant spectrum, to be precise — and to affect both ordinary and dark matter equally (adiabatic). Given such strong assumptions, just one parameter, the overall amplitude of fluctuations, defines the statistical distribution completely. With the appropriate value for that amplitude, the second cosmological model fits the WMAP data and other measures of large-scale structure remarkably well.

Yearning, Opportunity, Discontent

As I have just sketched, cosmology has been reduced to some general hypotheses and just four new continuous parameters. It is an amazing development. Yet I think that most physicists will not, and should not, feel entirely satisfied with it. The parameters appearing in the cosmological models, unlike those in the comparable models of matter, do not describe the fundamental behavior of simple entities. Rather, they appear as summary descriptors of averaged properties of macroscopic (VERY macroscopic) agglomerations. They are neither key players in a varied repertoire of phenomena nor essential elements in a beautiful mathematical theory.

Due to these shortcomings we are left wondering why just these parameters appear necessary to make a working description of existing observations, and uncertain whether we'll need to include more as observations are refined. We'd like to carry the analysis to another level, where the four working parameters will give way to different ones that are closer to fundamentals.

A different limitation to our insight penetrates modern cosmology so deeply that most physicists and cosmologists, inured by long familiarity, aren't as discontented as they ought to be. Modern cosmology consigns everything about the world, apart from a handful of statistical regularities, to chance and contingency. In Dante's universe everything had its reason for being and its proper place. Kepler aspired to explain the specific form of the Solar System. Now it appears that the number of planets in the Solar System, their masses, and the shape of their orbits — and more generally every specific fact about every specific object or group of objects in the universe — are mere accidents, not susceptible to fundamental explanation. Some are born modest and some work to be modest, but cosmologists have had modesty thrust upon them.

There are genuine, exciting opportunities for carrying the analysis of cosmological parameters further. It seems most unlikely, however, that the pervasive indeterminacy and randomness of the core of our model of the universe will go away. That uncertainty confronts us with an opportunity of another sort: the opportunity to expand our vision of what constitutes fundamental explanation. I'll say more about these opportunities next time.

Analysis and Synthesis 3: Cosmic Groundwork

Most of cosmology can be captured in a few parameters, but those parameters are themselves mysterious, as I discussed in my previous column. Now I'll survey some prospects for understanding them better.

Digging Deeper

A big reason for excitement and optimism among physicists is that models of inflation suggest new links between fundamental physics and cosmology. Several assumptions in our cosmological models, specifically uniformity, spatial flatness, and the Harrison–Zeldovich spectrum, were originally suggested on grounds of simplicity, expediency, or aesthetics. They can be supplanted with a single dynamical hypothesis: that very early in its history, the universe underwent a period of superluminal expansion, or inflation.

Such a period could be driven by a matter field that was excited coherently out of its ground state permeating the universe. Possibilities of that kind are easy to imagine in models of fundamental physics. For example, scalar fields are used to implement symmetry breaking even in the standard model. Theoretically, such fields can easily find themselves unable to shed energy quickly enough to stay close to their ground state as the universe expands. Inflation will occur if the approach to the ground state is slow enough. Fluctuations will be generated because the relaxation process is not quite synchronized across the universe.

Inflation is a wonderfully attractive, logically compelling idea, but very basic challenges remain. Can we be specific about the cause of inflation, and ground it in explicit, well-founded models of fundamental physics? To be concrete, can we calculate the correct amplitude of fluctuations? Existing implementations have a problem here: getting the amplitude sufficiently small takes some nice adjustment.

More promising, perhaps, than the difficult business of extracting hard quantitative predictions from the broadly flexible idea of inflation is to follow up on the essentially new and surprising possibilities it suggest. The violent restructuring of space-time attending inflation should generate detectable gravitational waves, and the nontrivial dynamics of relaxation should generate some detectable deviation from a strictly scale-invariant spectrum of fluctuations. Future precision measurements of polarization in the microwave background radiation and of the large-scale distribution of matter will be sensitive to these effects. Stay tuned!

There are many ideas for how an asymmetry between matter and antimatter might be generated in the early universe. Then after much mutual annihilation of particles and antiparticles, the asymmetry could be left as the present baryon density. Several of those ideas seem able to accommodate the observed density. Unfortunately the answer generally depends on details of particle physics at energies that are unlikely to be accessible experimentally any time soon. So for a decision among the models we may be reduced to waiting for a functioning Theory of (Nearly) Everything.

Dark Matter

I'm much more optimistic about the dark matter problem. Here we have the unusual situation that two good ideas exists — which, according to William of Occam (the razor guy), is one too many.

The symmetry of the standard model can be enhanced, and some of its aesthetic shortcomings can be overcome, if we extend it to a larger theory. Two proposed extensions, logically independent of one another, are particularly specific and compelling. One incorporates a symmetry suggested by Roberto Peccei and Helen Quinn in 1977. Peccei–Quinn symmetry rounds out the logical structure of quantum chromodynamics by removing QCD's potential to support strong violation of time-reversal symmetry, which is not observed. This extension predicts the existence of a remarkable new kind of very light, feebly interacting particle: the axion.

The other proposal incorporates supersymmetry, an extension of special relativity, to include quantum space-time transformations. Supersymmetry serves several important qualitative and quantitative purposes in modern thinking about unification: it relieves difficulties with understanding why W bosons are as light as they are and why the couplings of the standard model take the values they do. In many implementations of supersymmetry, the lightest supersymmetric particle, or LSP, interacts rather feebly with ordinary matter (though much more strongly than do axions) and is stable on cosmological time scales.

The properties of the particles, axion or LSP, are just right for dark matter. Moreover, you can calculate how abundantly each would be produced in the Big Bang. For both particles, the predicted abundance is also quite promising. Vigorous, heroic experimental searches are under way to observe dark matter in either of those forms. We will also get crucial information about supersymmetry once the Large Hadron Collider starts running in 2007. I will be disappointed — and surprised — if, a decade from now, we don't have a much more personalized portrait of the dark matter.

Dark Energy

Now for a few words about the remaining parameter, the density of dark energy. Why is it so small? Why is it so big?

The standard model provides a great lesson: What our senses evolved to perceive as empty space is in fact a richly structured medium. It contains symmetry-breaking condensates associated with both electroweak superconductivity and spontaneous chiral symmetry breaking in QCD, an effervescence of virtual particles, and probably much more.

Because gravity is sensitive to all forms of energy, it really ought to see this stuff, even if our eyes and instruments don't. A straightforward estimation suggests that empty space should weigh several orders of magnitude of orders of magnitude (no misprint here!) more that it does. It "should" be much denser than a neutron star, for example. The expected energy of empty space acts like dark energy, with its weird negative pressure, but far more is expected than is found.

To me, the discrepancy concerning the density of empty space is the most mysterious fact in all of physical science, the fact with the greatest potential to rock the foundations. We're obviously missing some major insight here. Given that, it's hard to know what to make of the ridiculously small amount of dark energy that presently dominates the universe.

Possible Worlds

Discovery of definitive mathematical equations that govern the behavior of matter, and even mastery of techniques for solving them, if we ever achieved it, would by no means complete the program of understanding Nature through analysis and synthesis. We would still need to address the problem of selection. How, among all possible solutions, do those that actually describe reality get selected out?

Niels Bohr defined a profound truth as a truth whose opposite is also a (profound) truth. In that spirit, I'd like to define a deep question as a question that doesn't make clear sense until after you've answered it. The selection problem is a deep question. In posing it, we take it for granted that we can make a distinction between what is "possible" and what is "real." On the face of it, that sounds pretty unscientific: the goal of science is to understand the real world, and only the real world is possible! But in practice a clear and useful distinction does emerge.

A famous episode from the early history of science will serve to illuminate the issues. In Johannes Kepler's first attempt to understand the Solar System, he postulated that the relative sizes of planetary orbits are determined as a system of concentric spheres, centered at the Sun, inscribed and circumscribed around a sequence of the five regular (Platonic) solids. To Kepler, the "fact" that there were six planets provided an impressive numerical confirmation of his idea. In light of later developments Kepler's model rates a bemused smile, but purely as a logical matter, it might have been right. And if such a model had provided the ultimate account of the Solar System, no useful distinction would have developed between possible solutions of its governing equations and the solution actually realized. Indeed, there wouldn't be equations to solve, as such — just the solution, perfect in its own terms, and incapable of further analysis.

Kepler's geometric model was soon made obsolete by the development of classical celestial mechanics, beginning with Kepler's own discoveries and culminating in Isaac Newton's world-system. That framework gives a clean separation between the governing laws and their specific realizations. The equations of classical mechanics can be solved for any number of planets, with any initial assignment of positions and velocities. So in the physics of 1700 — as in the physics we use today — it was easy to imagine perfectly consistent planetary systems with different sizes and shapes than the Solar System. Galileo had already observed a very differently textured planetary system around Jupiter, and today astronomers are discovering new types surrounding distant stars.

Newton thought that God determined the initial conditions, by an act of will and creation. Nowadays most physicists consider the question of why the Solar System is precisely what it is as a bad question, or at least one that cannot be addressed from fundamentals, since it seems pretty clear that the answer depends largely on chance and accidents of history.

The Big Bang's Peculiar Foundation

The ultimate question of selection, of course, is how to select among possible candidate solutions to describe the universe as a whole.

Conventional Big Bang cosmology begins by assuming that, early on, matter throughout the universe was in thermal equilibrium at some high temperature, while at the same time space was (almost) perfectly uniform and had negligible spatial curvature. Those assumptions are consistent with everything we know and they lead to several successful predictions. They surely contain a large element of truth.

Yet from a fundamental perspective, these assumptions are most peculiar. According to general relativity, which of course underpins the whole discussion, space-time curvature is a dynamical entity whose shape yields gravity. It would appear natural, therefore, to assume that space-time too achieves some sort of equilibrium, through gravitational interaction. But gravity's action on matter produces universal attraction, and if allowed to run to completion, that attraction will agglomerate the matter into lumps. The long-term equilibrium result is the very opposite of the smooth starting conditions commonly — and quite successfully — assumed.

Putting it another way, if the present universe were eventually to start contracting and evolve toward a Big Crunch, matter as it was squeezed together would tend to heap into bigger and bigger black holes, and the last moments would look nothing like the time-reverse of our best reconstruction of the beginning. That beginning, a contrasting mix of maximum disorder for matter and perfect quiescence for gravity is, on the face of it, extremely difficult to reconcile with the idea that ultimately gravity is unified with the other interactions. If so, it ought to be treated equally with them in describing extreme conditions in the earliest universe. Where there's no distinction, there can be no difference!

In particle physics, we like to unify gravity with the other interactions, but in cosmology, we apparently have to assume they behaved very differently. Simple models of inflation begin to address this issue. During the inflationary phase, wrinkles in space-time get smoothed out, while energy is frozen into a scalar "inflaton" field. The inflaton field eventually melts, and its energy pours into more-or-less conventional forms of matter, which interact and thermalize. As long as the melting does not produce too high a temperature, few gravitons are produced, which is to say that space-time, having been stretched smooth, remains so. Some elements of this picture are quite speculative, of course. But it gives us a tangible, accessible approach to the problem of certifying the soundness of cosmology's foundation.

Even if this program of cosmic analysis and synthesis succeeds, questions will remain. In particular, we still lack a fundamental explanation of why the inflationary phase occurred. The general problem of how we select among solutions appears as one of the ultimate limits to the Analysis and Synthesis. Ultimate limitations will be my concluding subject in this series.

Analysis and Synthesis 4: Limits and Supplements

The earlier columns in this series recounted extraordinary triumphs of physicists using analysis and synthesis or, alternatively, (the dreaded epithet) reductionism to account for the behavior of matter and the structure of the universe as a whole. But I left off by admitting that we scored a bull's-eye only by painting the target around our shot. There's quite a lot that physicists might once have hoped to derive or explain based on fundamental principles, for which that hope now seems dubious or forlorn. In this column I explore some sources of these limitations and the role different sorts of explanations might play in filling the voids.

Selection by Energy

One important limitation, which I started to explore last time, concerned the lack of a principle that could lead to a unique choice among different seemingly possible solutions of the fundamental equations and could select out the universe we actually observe.

To get oriented, it's very instructive to consider the corresponding problem for atoms and matter. Like the classical equations governing planetary systems, the equations of quantum mechanics for electrons in a complex atom allow all kinds of solutions. In fact, the quantum equations allow even more freedom of choice in the initial conditions than the classical equations do. The wavefunctions for N particles live in a much larger space than the particles do: They inhabit a full-bodied $3N$-dimensional configuration space, as opposed to $2N$ copies of three-dimensional space. (For example, the quantum description of the state of two particles requires a wavefunction $\psi(x_1, x_2)$ that depends on six variables, whereas the classical description requires 12 numbers, namely their positions and velocities.)

Yet the atoms we observe are always described by the same solutions — otherwise we wouldn't be able to do stellar spectroscopy, or even chemistry. Why? A proper answer involves combining insights from quantum field theory, mathematics, and a smidgen of cosmology.

Quantum field theory tells us that electrons — unlike planets — are all rigorously the same. Then the mathematics of the Schrödinger equation, or its refinements, tells us that the low-energy spectrum is discrete, which is to say that if our atom has only a small amount of energy, then only a few solutions are available for its wavefunction.

But because energy is conserved, this explanation begs another question: What made the energy small in the first place? Well, the atoms we study are not closed systems; they can emit and absorb radiation. So the question becomes: Why do they emit more often than they absorb, and thereby settle down into a low-energy state? That's where cosmology comes in. The expanding universe is quite an effective heat sink. In excited atoms, energy radiated as photons eventually leaks into the vast interstellar spaces and redshifts away. By way of contrast, a planetary system has no comparably efficient way to lose energy — gravitational radiation is ridiculously feeble — and it can't relax.

So one selection principle that applies to many important cases is to choose solutions with low energy. In the same spirit, when the residual energy can't be neglected, one should choose thermal equilibrium solutions. This simplifying assumption is appropriate for systems and degrees of freedom that have relaxed, though not in general.

Stability: Answer or Question?

The selection procedure that dominates the literature of high-energy physics and string theory is energy based, along the lines we just discussed for atoms. It is traditional to identify the lowest-energy solution with the physical vacuum, and to model the content of the world using low-energy excitations above that state. For the solution to count as successful, the possible excitations should include at least the particles of the standard model.

One can't seriously quarrel with that selection procedure as a practical *necessary* criterion. The universe, slowly evolving and mostly empty, does appear to be a low-energy excitation around a stable state. But why?

For atoms, selection of low-energy solutions was justified by their tendency to dissipate energy into radiation that ultimately finds its way into interstellar space and does not return. That mechanism can't work for the universe as a whole, of course, but a different form of dissipation comes into play, depending on a different aspect of cosmology. The universe has been expanding for a long time — a *very* long time, many orders of magnitude longer than the natural time scales associated with elementary interactions. For present purposes, the mismatch of time scales has two profound consequences. First, as the universe expands, the matter in it cools. We can think of the cooling as a sort of radiation leak into the future. In any case, it takes the universe toward a local minimum of the energy. Second, the mismatch gives many sorts of instabilities time to play out. (Not all, however. Instabilities involving quantum tunneling can have absurdly long time scales, and as a logical possibility they might figure in our future.) Altogether, then, the vastness of cosmological time serves to justify focus on low-energy excitations around stable "vacuum" configurations that are at least local minima of the energy density.

This answer to the question of why the universe can be described as a low-energy excitation, though, frames a new question: What sets the cosmic time scale to be so different from the scale of fundamental interactions? And that question is a thinly veiled form of the question for which we have no good answer: Why is the cosmological term so small? In that same dark neck of the woods lurks the specter of the observed dark energy, which might well expose the provisional nature of our selection criteria. Indeed, a leading hypothesis for this energy, known under the rubric of "quintessence," is that it reflects a difference between empty space as it actually exists and the ideal stable vacuum.

In string theory, as currently understood, the selection problem poses severe foundational challenges, because in string theory the criterion of stability is by no means sufficient to select a unique solution. Many stable solutions apparently exist, including flat 10-dimensional space-time, anti-deSitter space that supports a huge negative cosmological term, and tens of thousands of other possibilities the vast majority of which bear no resemblance whatsoever to the world as we know it. The pragmatic response, which dominates the literature, has been to add phenomenological inputs (effective space-time dimension, gauge group, number of families, and so on) to the selection criteria. One can then try to derive relations among the remaining parameters. Success along those lines could be very significant, reassuring us that the theory was on the right track. So far, that hasn't happened. But even if it did, it would hardly fulfill the promise of a final analysis of Nature.

So, how does the solution that actually describes the world get selected out? It remains a deep question.

From Questionable Questions

The analysis and synthesis program, in the course of its pursuit, exposed other limitations. Consider these three questions, whose elucidations were once considered major goals for science:

Why is the Solar System as it is?

When will a given radioactive nucleus decay?

What will the weather be in Boston a year from today?

They are questions that suffered an unusual fate: instead of being answered, they were discredited. In each case, the scientific community at first believed that unique, clear-cut answers would be possible. And in each case, it was a major advance, with wide implications, to discover that there are good fundamental reasons why a clean answer is not possible.

I discussed the solar-system question in "Analysis and Synthesis 3" with reference to Johannes Kepler's struggles. That question is questionable because it suffers from a problem I call projection. If there were only one solar system, or if all such systems were the same, tracing that system's properties to a unique cause would be important. Those hypotheses once seemed reasonable, but now we see that they are illegitimate projections, which ascribe universal significance to what is actually a very limited slice of the world. There are serious premonitions that we physicists might need to relearn this lesson on a cosmic scale. Popular speculations that we live in a "multiverse" or inhabit "braneworlds" involve the idea that the world is extremely inhomogeneous on large scales — that the values of fundamental parameters, the structure of the standard model, or even the effective dimensionality of space-time varies from place to place. If so, attempts to predict such things from fundamentals are as misguided as Kepler's polyhedra.

The second question, of course, was rendered questionable by quantum mechanics. The precise time that any particular radioactive nucleus will decay is postulated to be inherently random — a postulate that has acquired, by now, tremendous empirical support. Erwin Schrödinger, with his macroscopic cat, thought he was reducing quantum mechanics to absurdity. But the emerging standard account of cosmology traces the origin of structure in the universe back to quantum fluctuations in an inflation field! If that account holds up, it will mean that attempting to predict the specific pattern of structures we observe from fundamentals is as misguided as trying to predict when a nucleus will decay or a Schrödinger cat expire.

The third question was rendered questionable by the discovery of chaos: the phenomenon that the solution of perfectly deterministic, innocuous-seeming equations can have solutions whose long-term behavior depends extremely sensitively on exquisite details of the initial conditions. Chaos raises another barrier that can separate ideal analysis from successful synthesis.

... To Answerable Answers

It is possible, I suppose, that these apparent limitations will prove illusory and that, in the end, the vision of a unique, deterministic Universe fully accessible to rational analysis, championed by Baruch Spinoza and Albert Einstein, will be restored. But to me it seems wise to accept what appears to be overwhelming evidence that projection, quantum uncertainty, and chaos are inherent in the nature of things, and to build on those insights. With such acceptance, new constructive principles appear, supplementing pure logical deduction from fine-grained analysis as irreducible explanations of observed phenomena.

By accepting the occurrence of projection, we license anthropic explanations. How do we understand why Earth is at the distance it is from the Sun, or why our sun is the kind of star it is? Surely important insights into these questions follow from our existence as intelligent observers. Some day, we may be able to check such arguments by testing their predictions for exobiology.

By accepting quantum uncertainty, we license, well ... quantum mechanics. Specifically, in the spirit of this column, we can test the hypothesized quantum origin of primordial fluctuations by checking whether those fluctuations satisfy statistical criteria for true randomness.

By accepting the implications of chaos, broadly defined, we license evolutionary explanations. Outstanding examples include an explanation of why the Moon always faces us due to the long-term action of tidal friction, and of the structure of gaps in the asteroid belt due to resonance with planetary periods. Also from considerations internal to the program of analysis and synthesis, we motivate the search for emergent properties of complex systems that Philip Anderson has advocated under the rubric "More is different." For these emergent properties can form the elements of robust descriptions — they transcend the otherwise incapacitating sensitivity to initial conditions.

In constructing explanations based on anthropics, randomness, and dynamical evolution, we must use intermediate models incorporating many things that can't be calculated. Such necessary concessions to reality compromise the formal purity of the ideal of understanding the world by analysis and synthesis, but in compensation, they allow its spirit much wider scope.

Musing on Mechanics

In preparing to teach mechanics to MIT freshman, instead of doing sensible things like getting comfortable with the textbook, deciding what to leave out, and planning a schedule, I read a lot of mechanics books, mostly much too advanced, and worried about the foundations of the subject. (I also worked an enormous number of problems, which really was sensible preparation.) I paid the price by having to scramble and improvise through the semester, but I did get to understand mechanics much better, or at least to develop some unorthodox opinions about it.

Eventually I wrote a series of three *Physics Today* Reference Frame articles expressing some of those opinions. The columns generated a large number of thoughtful responses, both pro (mostly) and con; probably more than all my 15 other Reference Frames put together. Classical mechanics is central to the work of many *Physics Today* readers, many teach it, and almost all have taken courses on it and keep it in their repertoire. So the columns hit close to home. Some people were relieved to find their secret dissatisfactions aired in public; others were testy to be disturbed from their dogmatic slumber.

In one respect I was surprised, myself, to discover how far my critique led me. Conservation of mass appeared as a fundamental assumption in Newtonian mechanics and became, for Lavoisier, the key to opening modern chemistry. Yet a great triumph of modern physics is to construct ordinary matter using building blocks with zero mass ("The Origin of Mass" and "Mass without Mass 1"). Of course that construction trashes conservation of mass, since $0 + 0 + 0 = 0$, while ordinary matter has non-vanishing mass. In "Whence the Force of $F = ma$? 2" I explain how in modern physics the Newton–Lavoisier principle appears as an *approximate, emergent* consequence of specific deep facts in QCD and QED.

Three unsystematic addenda:

After reading Usher's *History of Mechanical Invention* I realize that my stab at the history of metric time in "Whence the Force of $F = ma$? 3" does scant justice (actually, injustice) to a rich and fascinating subject. Mechanical clocks were the high tech of Medieval times, and their development changed how people lived and how they viewed the world.

In the sentence that concludes the whole series the word "statuized," meaning made into a statue, is used. The editors queried it, but I was able to produce this from the *OED*:

> **statuize**
> *nonce-wd.*
> [f. STATUE *n.* + -IZE.]
> *trans.* To make a statue of, to commemorate by means of a statue. 1719
>
> **1719** OZELL tr. *Misson's Trav. Eng.* 309 James II. did also statuize himself [orig. *s'est aussi fait Statuër*] in Copper, in one of the Courts of Whitehall.

and they had to admit it was English.

I'm not done with this subject. A major theme of the columns is that energy and momentum appear to be more fundamental concepts than force. Roughly speaking, force is the space derivative of energy and the time derivative of momentum. You can take one more step up the ladder: energy and momentum themselves are both derivatives of *action*: energy is its time derivative, momentum its space derivative. Coming soon: *Where the Action Is*.

Whence the Force of $F = ma$?
1: Culture Shock

When I was a student, the subject that gave me the most trouble was classical mechanics. That always struck me as peculiar, because I had no trouble learning more advanced subjects, which were supposed to be harder. Now I think I've figured it out. It was a case of culture shock. Coming from mathematics, I was expecting an algorithm. Instead I encountered something quite different — a sort of culture, in fact. Let me explain.

Problems with $F = ma$

Newton's second law of motion, $F = ma$, is the soul of classical mechanics. Like other souls, it is insubstantial. The right-hand side is the product of two terms with profound meanings. Acceleration is a purely kinematical concept, defined in terms of space and time. Mass quite directly reflects basic measurable properties of bodies (weights, recoil velocities). The left-hand side, on the other hand, has no independent meaning. Yet clearly Newton's second law is full of meaning, by the highest standard: it proves itself useful in demanding situations. Splendid, unlikely looking bridges, like the Erasmus Bridge (known as the Swan of Rotterdam), do bear their loads; spacecraft do reach Saturn.

The paradox deepens when we consider force from the perspective of modern physics. In fact, the concept of force is conspicuously absent from our most advanced formulations of the basic laws. It doesn't appear in Schrödinger's equation, or in any reasonable formulation of quantum field theory, or in the foundations of general relativity. Astute observers commented on this trend to eliminate force even before the emergence of relativity and quantum mechanics.

In his 1895 *Dynamics*, the prominent physicist Peter G. Tait, who was a close friend and collaborator of Lord Kelvin and James Clerk Maxwell, wrote

> In all methods and systems which involve the idea of force there is a leaven of artificiality ... there is no necessity for the introduction of the word "force" nor of the sense-suggested ideas on which it was originally based.[1]

Particularly striking, since it is so characteristic and so over-the-top, is what Bertrand Russell had to say in his 1925 popularization of relativity for serious intellectuals, *The ABC of Relativity*:

> If people were to learn to conceive the world in the new way, without the old notion of "force," it would alter not only their physical imagination, but probably also their morals and politics ... . In the Newtonian theory of the solar system, the sun seems like a monarch whose behests the planets have to obey. In the Einsteinian world there is more individualism and less government than in the Newtonian.[2]

The 14th chapter of Russell's book is entitled "The Abolition of Force."

If $F = ma$ is formally empty, microscopically obscure, and maybe even morally suspect, what's the source of its undeniable power?

The Culture of Force

To track that source down, let's consider how the formula gets used.

A popular class of problems specifies a force and asks about the motion, or vice versa. These problems look like physics, but they are exercises in differential equations and geometry, thinly disguised. To make contact with physical reality, we have to make assertions about the forces that actually occur in the world. All kinds of assumptions get snuck in, often tacitly.

The zeroth law of motion, so basic to classical mechanics that Newton did not spell it out explicitly, is that mass is conserved. The mass of a body is supposed to be independent of its velocity and of any forces imposed on it; also total mass is neither created nor destroyed, but only redistributed, when bodies interact. Nowadays, of course, we know that none of that is quite true.

Newton's third law states that for every action there's an equal and opposite reaction. Also, we generally assume that forces do not depend on velocity. Neither of those assumptions is quite true either; for example, they fail for magnetic forces between charged particles.

When most textbooks come to discuss angular momentum, they introduce a fourth law, that forces between bodies are directed along the line that connects them. It is introduced in order to "prove" the conservation of angular momentum. But this fourth law isn't true at all for molecular forces.

Other assumptions get introduced when we bring in forces of constraint, and friction.

I won't belabor the point further. To anyone who reflects on it, it soon becomes clear that $F = ma$ by itself does not provide an algorithm for constructing the mechanics of the world. The equation is more like a common language, in which different useful insights about the mechanics of the world can be expressed. To put it another way, there is a whole culture involved in the interpretation of the symbols. When we learn mechanics, we have to see lots of worked examples to grasp properly what force really means. It is not just a matter of building up skill by practice; rather, we are imbibing a tacit culture of working assumptions. Failure to appreciate this is what got me in trouble.

The historical development of mechanics reflected a similar learning process. Isaac Newton scored his greatest and most complete success in planetary astronomy, when he discovered that a single force of quite a simple form dominates the story. His attempts to describe the mechanics of extended bodies and fluids in the second book of *The Principia*[3] were path breaking but not definitive, and he hardly touched the more practical side of mechanics. Later physicists and mathematicians including notably Jean d'Alembert (constraint and contact forces), Charles Coulomb (friction), and Leonhard Euler (rigid, elastic, and fluid bodies) made fundamental contributions to what we now comprehend in the culture of force.

Physical and Psychological Origins

Many of the insights embedded in the culture of force, as we've seen, aren't completely correct. Moreover, what we now think are more correct versions of the laws of physics won't fit into its language easily, if at all. The situation begs for two probing questions: How can this culture continue to flourish? Why did it emerge in the first place?

For the behavior of matter, we now have extremely complete and accurate laws that in principle cover the range of phenomena addressed in classical mechanics and, of course, much more. Quantum electrodynamics (QED) and quantum chromodynamics (QCD) provide the basic laws for building up material bodies and the nongravitational forces between them, while general relativity gives us a magnificent account of gravity. Looking down from this exalted vantage point, we can get a clear perspective on the territory and boundaries of the culture of force.

Compared to earlier ideas, the modern theory of matter, which really only emerged during the 20th century, is much more specific and prescriptive. To put it plainly, you have much less freedom in interpreting the symbols. The equations of QED and QCD form a closed logical system: They inform you what bodies can be produced at the same time as they prescribe their behavior; they govern your measuring devices — and you, too! — thereby defining what questions are well posed physically; and they provide such questions with answers — or at least with algorithms to arrive at answers. (I'm well aware that QED + QCD is not a complete theory of Nature, and that, in practice, we can't solve the equations very well.) Paradoxically, there is much less interpretation, less culture involved in the foundations of modern physics than in earlier, less complete syntheses. The equations really do speak for themselves: They are algorithmic.

By comparison to modern foundational physics, the culture of force is vaguely defined, limited in scope, and approximate. Nevertheless it survives the competition, and continues to flourish, for one overwhelmingly good reason: It is much easier to work with. We really do not want to be picking our way through a vast Hilbert space, regularizing and renormalizing ultraviolet divergences as we go, then analytically continuing Euclidean Green's functions defined by a limiting procedure, ... working to discover nuclei that clothe themselves with electrons to make atoms that bind together to make solids, ... all to describe the collision of two billiard balls. That would be lunacy similar in spirit to, but worse than, trying to do computer graphics from scratch, in machine code, without the benefit of an operating system. The analogy seems apt: Force is a flexible construct in a high-level language, which, by shielding us from irrelevant details, allows us to do elaborate applications relatively painlessly.

Why is it possible to encapsulate the complicated deep structure of matter? The answer is that matter ordinarily relaxes to a stable internal state, with high energetic or entropic barriers to excitation of all but a few degrees of freedom. We can focus our attention on those few effective degrees of freedom; the rest just supply the stage for those actors.

While force itself does not appear in the foundational equations of modern physics, energy and momentum certainly do, and force is very closely related to them: Roughly speaking, it's the space derivative of the former and the time derivative of the latter (and $F = ma$ just states the consistency of those definitions!). So the concept of force is not quite so far removed from modern foundations as Tait and Russell insinuate: It may be gratuitous, but it is not bizarre. Without changing the content of classical mechanics, we can cast it in Lagrangian terms, wherein force no longer appears as a primary concept. But that's really a technicality; the deeper questions remains: What aspects of fundamentals does the *culture of force* reflect? What approximations lead to it?

Some kind of approximate, truncated description of the dynamics of matter is both desirable and feasible because it is easier to use and focuses on the relevant. To explain the rough validity and origin of specific concepts and idealizations that constitute the culture of force, however, we must consider their detailed content. A proper answer, like the culture of force itself, must be both complicated and open-ended. The molecular explanation of friction is still very much a research topic, for example. I'll discuss some of the simpler aspects, addressing the issues raised above, in my next column, before drawing some larger conclusions.

Here I conclude with some remarks on the psychological question, why force was — and usually still is — introduced in the foundations of mechanics, when from a logical point of view energy would serve at least equally well, and arguably better. The fact that changes in momentum — which correspond, by definition, to forces — are visible, whereas changes in energy often are not, is certainly a major factor. Another is that, as active participants in statics — for example, when we hold up a weight — we definitely feel we are doing something, even though no mechanical work is performed. Force is an abstraction of this sensory experience of exertion. D'Alembert's substitute, the virtual work done in response to small displacements, is harder to relate to. (Though ironically it is this sort of virtual work, continually made real, that explains our exertions. When we hold a weight steady, individual muscle fibers contract in response to feedback signals they get from spindles; the spindles sense small displacements, which must get compensated before they grow.[4]) Similar reasons might explain why Newton used force. A big part of the explanation for its continued use is no doubt (intellectual) inertia.

References

1. P. G. Tait, *Dynamics*, Adam & Charles Black, London (1895).
2. B. Russell, *The ABC of Relativity*, 5th rev. ed., Routledge, London (1997).
3. I. Newton, *The Principia*, I. B. Cohen, A. Whitman, trans., U. of Calif. Press, Berkeley (1999).
4. S. Vogel, *Prime Mover: A Natural History of Muscle*, W. W. Norton, New York (2001), p. 79.

Whence the Force of $F = ma$?
2: Rationalizations

In my previous column, I discussed how assumptions about F and m give substance to the spirit of $F = ma$. I called this set of assumptions the culture of force. I mentioned that several elements of the culture, though often presented as "laws," appear strange from the perspective of modern physics. Here I discuss how, and under what circumstances, some of those assumptions emerge as consequences of modern fundamentals — or don't!

Critique of the Zeroth Law

Ironically, it is the most primitive element of the culture of force — the zeroth law, conservation of mass — that bears the subtlest relationship to modern fundamentals.

Is the conservation of mass as used in classical mechanics a consequence of the conservation of energy in special relativity? Superficially, the case might appear straightforward. In special relativity we learn that the mass of a body is its energy at rest divided by the speed of light squared ($m = E/c^2$); for slowly moving bodies, it is approximately that. Since energy is a conserved quantity, this equation appears to supply an adequate candidate, E/c^2, to fill the role of mass in the culture of force.

That reasoning won't withstand scrutiny, however. The gap in its logic becomes evident when we consider how we routinely treat reactions or decays involving elementary particles.

To determine the possible motions, we must specify the mass of each particle coming in and of each particle going out. Mass is an intrinsic property of elementary particles — that is, all protons have one mass, all electrons have another, and so on. (For experts: "Mass" labels irreducible representations of the Poincaré group.) There is no separate principle of mass conservation. Rather, the energies and momenta of such particles are given in terms of their masses and velocities, by well-known formulas, and we constrain the motion by imposing conservation of energy and momentum. In general, it is simply not true that the sum of the masses of what goes in is the same as the sum of the masses of what goes out.

Of course when everything is slowly moving, then mass does reduce to approximately E/c^2. It might therefore appear as if the problem, that mass as such is not conserved, can be swept under the rug, for only inconspicuous (small and slowly moving) bulges betray it. The trouble is that as we develop mechanics, we want to focus on those bulges. That is, we want to use conservation of energy again, subtracting off the mass–energy exactly (or rather, in practice, ignoring it) and keeping only the kinetic part $E - mc^2 \cong \frac{1}{2}mv^2$. But you can't honestly squeeze two conservation laws (for mass and nonrelativistic energy) out of one (for relativistic energy). Ascribing conservation of mass to its approximate equality with E/c^2 begs an essential question: Why, in a wide variety of circumstances, is mass–energy accurately walled off, and not convertible into other forms of energy?

To illustrate the problem with an important example, consider the reaction ^{2}H + ^{3}H → ^{4}He + n, which is central for attempts to achieve controlled fusion. The total mass of the deuterium plus tritium exceeds that of the alpha plus neutron by 17.6 MeV. Suppose that the deuterium and tritium are initially at rest. Then the alpha emerges at .04 c; the neutron at .17 c.

In that (D, T) reaction, mass is not accurately conserved, and (nonrelativistic) kinetic energy has been produced from scratch, even though no particle is moving at a speed very close to the speed of light. Relativistic energy is conserved, of course, but there is no useful way to divide it up into two pieces that are separately conserved. In thought experiments, by adjusting the masses, we could make this problem appear in situations where the motion is arbitrarily slow. Another way to keep the motion slow is to allow the liberated mass–energy to be shared among many bodies.

Recovering the Zeroth Law

Thus, by licensing the conversion of mass into energy, special relativity nullifies the zeroth law, in principle. Why is Nature so circumspect about exploiting this freedom? How did Antoine Lavoisier, in the historic experiments that helped launch modern chemistry, manage to reinforce a central principle (conservation of mass) that isn't really true?

Proper justification of the zeroth law requires appeal to specific, profound facts about matter.

To explain why most of the energy of ordinary matter is accurately locked up as mass, we must first appeal to some basic properties of nuclei, where almost all the mass resides. The crucial properties of nuclei are persistence and dynamical isolation. The persistence of individual nuclei is a consequence of baryon number and electric charge conservation, and of the properties of nuclear forces that result in a spectrum of quasi-stable isotopes. The physical separation of nuclei and their mutual electrostatic repulsion — Coulomb barriers — guarantee their approximate dynamical isolation. That approximate dynamical isolation is rendered completely effective by the substantial energy gaps between the ground state of a nucleus and its excited states. Since the internal energy of a nucleus cannot change by a little bit, then in response to small perturbations, it doesn't change at all.

Because the overwhelming bulk of the mass–energy of ordinary matter is concentrated in nuclei, the isolation and integrity of nuclei — their persistence and effective lack of internal structure — go most of the way toward justifying the zeroth law. But note that to get this far, we needed to appeal to quantum theory and special aspects of nuclear phenomenology! For it is quantum theory that makes the concept of energy gaps available, and it is only particular aspects of nuclear forces that insure substantial gaps above the ground state. If it were possible for nuclei to be very much larger and less structured — like blobs of liquid or gas — the gaps would be small, and the mass–energy would not be locked up so completely.

Radioactivity is an exception to nuclear integrity, and more generally the assumption of dynamical isolation goes out the window in extreme conditions, such as we study in nuclear and particle physics. In those circumstances, conservation of mass simply fails. In the common decay $\pi^0 \to \gamma\gamma$, for example, a massive π^0 particle evolves into photons of zero mass.

The mass of an electron is a universal constant, as is its charge. Electrons do not support internal excitations, and the number of electrons is conserved (if we ignore weak interactions and pair creation). These facts are ultimately rooted in quantum field theory. Together, they guarantee the integrity of electron mass–energy.

In assembling ordinary matter from nuclei and electrons, electrostatics plays the dominant role. We learn in quantum theory that the active, outer-shell electrons move with velocities of order $\alpha c = e^2/4\pi\hbar \approx .007\,c$. This indicates that the energies in play in chemistry are of order $m_e(\alpha c)^2/m_e c^2 = \alpha^2 \approx 5\times 10^{-5}$ times the electron mass–energy, which in turn is a small fraction of the nuclear mass–energy. So chemical reactions change the mass–energy only at the level of parts per billion, and Lavoisier rules!

Note that inner-shell electrons of heavy elements, with velocities of order $Z\alpha$, can be relativistic. But the inner core of a heavy atom — nucleus plus inner electron shells — ordinarily retains its integrity, because it is spatially isolated and has a large energy gap. So the mass–energy of the core is conserved, though it is *not* accurately equal to the sum of the mass–energy of its component electrons and nucleus.

Putting it all together, we justify Isaac Newton's zeroth law for ordinary matter by means of the integrity of nuclei, electrons, and heavy atom cores, together with the slowness of the motion of these building blocks. The principles of quantum theory, leading to large energy gaps, underlie the integrity; the smallness of α, the fine-structure constant, underlies the slow motion.

Newton defined mass as "quantity of matter," and assumed it to be conserved. The connotation of his phrase, which underlies his assumption, is that the building blocks of matter are rearranged, but neither created nor destroyed, in physical processes; and that the mass of a body is the sum of the masses of its building blocks. We've now seen, from the perspective of modern foundations, why ordinarily these assumptions form an excellent approximation, if we take the building blocks to be nuclei, heavy atom cores, and electrons.

It would be wrong to leave the story there, however. For with our next steps in analyzing matter, we depart from this familiar ground: first tumbling off a cliff, then soaring into glorious flight. If we try to use more basic building blocks (protons and neutrons) instead of nuclei, then we discover that the masses don't add accurately. If we go further, to the level of quarks and gluons, we can largely derive the mass of nuclei from pure energy, as I've discussed in earlier columns.

Mass and Gravity

On the face of it, this complex and approximate justification of the mass concept used in classical mechanics poses a paradox: How does that rickety construct manage to support stunningly precise and successful predictions in celestial mechanics? The answer is that it is bypassed. The forces of celestial mechanics are gravitational, and so proportional to mass, and m cancels from the two sides of $F = ma$. This cancellation in the equation for motion in response to gravity becomes a foundational principle in general relativity, where the path is identified as a geodesic in curved space-time, with no mention of mass.

In contrast to a particle's *response* to gravity, the gravitational *influence* that the particle exerts is only approximately proportional to its mass; the rigorous version of Einstein's field equation relates space-time curvature not to mass but to energy-momentum density. As far as gravity is concerned, there is no separate measure of quantity of matter apart from energy; that the energy of ordinary matter is dominated by mass–energy is immaterial.

The Third and Fourth Laws

The third and fourth laws are approximate versions of conservation of momentum and conservation of angular momentum, respectively. (Recall that the fourth law stated that all forces are two-body central forces.) In the modern foundations of physics these great conservation laws reflect the symmetry of physical laws under translation and rotation symmetry. Since the conservation laws are more accurate and profound than the assumptions about forces commonly used to "derive" them, those assumptions have truly become anachronisms. I believe that they should, with due honors, be retired.

Newton argued for his third law by observing that a system with unbalanced internal forces would begin to accelerate spontaneously, "which is never observed." But this argument really motivates the conservation of momentum directly. Similarly, one can "derive" conservation of angular momentum from the observation that bodies don't spin up spontaneously. Of course, as a matter of pedagogy, one would point out that action–reaction systems and two-body central forces provide especially simple ways to satisfy the conservation laws.

Tacit Simplicities

Some tacit assumptions about the simplicity of F are so deeply embedded that we easily take them for granted. But they have profound roots.

In calculating the force, we take into account only nearby bodies. Why can we get away with that? Locality in quantum field theory, which embodies basic requirements of special relativity and quantum mechanics, gives us expressions for energy and momentum density — and thereby for force — that depend only on the position of nearby bodies. Even so-called long-range electric and gravitational forces (actually $1/r^2$ — still falling rapidly with distance) reflect the special properties of locally coupled gauge fields and their associated covariant derivatives.

Similarly, the absence of significant multibody forces is connected to the fact that sensible (renormalizable) quantum field theories can't support them.

In this column I've stressed, and maybe strained, the relationship between the culture of force and modern fundamentals. In the final column of this series, I'll discuss its importance both as a continuing, expanding endeavor and as a philosophical model.

Whence the Force of $F = ma$?
3: Cultural Diversity

The concept of force, as we have seen, defines a culture. In the previous columns of this series I've indicated how $F = ma$ acquires meaning through interpretation of — that is, additional assumptions about — F. This body of interpretation is a sort of folklore. It contains both approximations that we can derive, under appropriate conditions, from modern foundations, and also rough generalizations (such as "laws" of friction and of elastic behavior) abstracted from experience.

In the course of that discussion it became clear that there is also a smaller, but nontrivial, culture around m. Indeed, the conservation of m for ordinary matter provides an excellent, instructive example of an emergent law. It captures in a simple statement an important consequence of broad regularities whose basis in modern fundamentals is robust but complicated. In modern physics, the idea that mass is conserved is drastically false. A great triumph of modern quantum chromodynamics (QCD) is to build protons and neutrons, which contribute more than 99% of the mass of ordinary matter, from gluons that have exactly zero mass, and from u and d quarks that have very small masses. To explain from a modern perspective why conservation of mass is often a valid approximation, we need to invoke specific, deep properties of QCD and quantum electrodynamics (QED), including the dynamical emergence of large energy gaps in QCD and the smallness of the fine structure constant in QED.

Isaac Newton and Antoine Lavoisier knew nothing of all this, of course. They took conservation of mass as a fundamental principle. And they were right to do so, because by adopting that principle they were able to make brilliant progress in the analysis of motion and of chemical change. Despite its radical falsity, their principle was, and still is, an adequate basis for many quantitative applications. To discard it is unthinkable. It is an invaluable cultural artifact and a basic insight into the way the world works despite — indeed, in part, because of — its emergent character.

The Culture of a

What about a? There's a culture attached to acceleration, as well. To obtain a, we are instructed to consider the change of the position of a body in space as a function of time, and to take the second derivative. This prescription, from a modern perspective, has severe problems.

In quantum mechanics, bodies don't have definite positions. In quantum field theory, they pop in and out of existence. In quantum gravity, space is fluctuating and time is hard to define. So evidently serious assumptions and approximations are involved even in making sense of a's definition.

Nevertheless, we know very well where we're going to end up. We're going to have an emergent, approximate concept of what a body is. Physical space is going to be modeled mathematically as the Euclidean three-dimensional space $\mathbf{R}^3$ that supports Euclidean geometry. This tremendously successful model of space has been in continuous use for millennia, with applications in surveying and civil engineering that even predate Euclid's formalization.

Time is going to be modeled as the one-dimensional continuum $\mathbf{R}^1$ of real numbers. This model of time, at a topological level, goes into our primitive intuitions that divide the world into past and future. I believe that the metric structure of time — that is, the idea that time can be not only ordered but divided into intervals with definite numerical magnitude — is a much more recent innovation. That idea emerged clearly only with Galileo's use of pendulum clocks (and his pulse!).

The mathematical structures involved are so familiar and fully developed that they can be, and are, used routinely in computer programs. This is not to say they are trivial. They most definitely aren't. The classical Greeks agonized over the concept of a continuum. Zeno's famous paradoxes reflect these struggles. Indeed, Greek mathematics never won through to comfortable algebraic treatment of real numbers.

Continuum quantities were always represented as geometric intervals, even though that representation involved rather awkward constructions to implement simple algebraic operations.

The founders of modern analysis (René Descartes, Newton, Gottfried Wilhelm Leibniz, Leonhard Euler, and others) were on the whole much more freewheeling, trusting their intuition while manipulating infinitesimals that lacked any rigorous definition. (In his *Principia*, Newton did operate geometrically, in the style of the Greeks. That is what makes the *Principia* so difficult for us to read today. The *Principia* also contains a sophisticated discussion of derivatives as limits. From that discussion I infer that Newton and possibly other early analysts had a pretty good idea about what it would take to make at least the simpler parts of their work rigorous, but they didn't want to slow down to do it.) Reasonable rigor, at the level commonly taught in mathematics courses today — the much-bemoaned epsilons and deltas — entered into the subject in the 19th century.

"Unreasonable" rigor entered in the early 20th century, when the fundamental notions from which real numbers and geometry are constructed were traced to the level of set theory and ultimately symbolic logic. In their *Principia Mathematica* Bertrand Russell and Alfred Whitehead develop 375 pages of dense mathematics before proving $1+1=2$. To be fair, their treatment could be slimmed down considerably if attaining that particular result were the ultimate goal. But in any case, an adequate definition of real numbers from symbolic logic involves some hard, complicated work. Having the integers in hand, you then have to define rational numbers and their ordering. Then you must complete them by filling in the holes so that any bounded increasing sequence has a limit. Then finally — this is the hardest part — you must demonstrate that the resulting system supports algebra and is consistent.

Perhaps all that complexity is a hint that the real-number model of space and time is an emergent concept that some day will be derived from physically motivated primitives that are logically simpler. Also, scrutiny of the construction of real numbers suggests natural variants, notably John Conway's surreal numbers, which include infinitesimals (smaller than any rational number!) as legitimate quantities.[1] Might such quantities, whose formal properties seem no less natural and elegant than those of ordinary real numbers, help us to describe Nature? Time will tell.

Even the unreasonable rigor of symbolic logic does not reach ideal strictness. Kurt Gödel demonstrated that this ideal is unattainable: No reasonably complex, consistent axiomatic system can be used to demonstrate its own consistency.

But all the esoteric shortcomings in defining and justifying the culture of *a* clearly arise on an entirely different level from the comparatively mundane, immediate difficulties we have in doing justice to the culture of *F*. We can translate the culture of *a*, without serious loss, into C or FORTRAN. That completeness and precision give us an inspiring benchmark.

The Computational Imperative

Before they tried to do it, most computer scientists anticipated that to teach a computer to play chess like a grand master would be much more challenging than to teach one to do mundane tasks like drive a car safely. Notoriously, experience has proved otherwise. A big reason for that surprise is that chess is algorithmic, whereas driving a car is not. In chess the rules are completely explicit; we know very concretely and unambiguously what the degrees of freedom are and how they behave. Car driving is quite different: Essential concepts like "other driver's expectations" and "pedestrian," when you start to analyze them, quickly burgeon into cultures. I wouldn't trust a computer driver in Boston's streets because it wouldn't know how to interpret the mixture of intimidation and deference that human drivers convey by gestures, maneuvers, and eye contact.

The problem with teaching a computer classical mechanics is, of course, of more than academic interest: We'd like robots to get around and manipulate things; computer gamesters want realistic graphics; engineers and astronomers would welcome smart silicon collaborators — up to a point, I suppose.

The great logician and philosopher Rudolf Carnap made brave, pioneering attempts to create axiomatic systems for elementary mechanics, among many other things.[2] Patrick Hayes issued an influential paper, "Naive Physics Manifesto," challenging artificial-intelligence researchers to codify intuitions about materials and forces in an explicit way.[3] Physics-based computer graphics is a lively, rapidly advancing endeavor, as are several varieties of computer-assisted design. My MIT colleagues Gerald Sussman and Jack Wisdom have developed an intensely computational approach to mechanics,[4] supported every step of the way with explicit programs. The time may be ripe for a powerful synthesis, incorporating empirical properties of specific materials, successful known designs of useful mechanisms, and general laws of mechanical behavior into a fully realized computational culture of $F = ma$. Functioning robots might not need to know a lot of mechanics explicitly, any more than most human soccer players do; but *designing* a functioning robotic soccer player may be a job that can best be accomplished by a very smart and knowledgeable man-machine team.

Blur and Focus

An overarching theme of this series has been that the law $F = ma$, which is sometimes presented as the epitome of an algorithm describing Nature, is actually not an algorithm that can be applied mechanically (pun intended). It is more like a language in which we can easily express important facts about the world. That's not to imply it is without content. The content is supplied, first of all, by some powerful general statements in that language — such as the zeroth law, the momentum conservation laws, the gravitational force law, the necessary association of forces with nearby sources — and then by the way in which phenomenological observations, including many (though not all) of the laws of material science can be expressed in it easily.

Another theme has been that $F = ma$ is not in any sense an ultimate truth. We can understand, from modern foundational physics, how it arises as an approximation under many circumstances. Again, that does not prevent it from being extraordinarily useful; indeed, one of its primary virtues is to shield us from the unnecessary complexity of irrelevant precision!

Viewed this way, the law of physics $F = ma$ comes to appear a little softer than is commonly considered. It really does bear a family resemblance to other kinds of laws, like the laws of jurisprudence or of morality, wherein the meaning of the terms takes shape through their use. In those domains, claims of ultimate truth are wisely viewed with great suspicion; yet nonetheless we should actively aspire to the highest achievable level of coherence and explicitness. Our physics culture of force, properly understood, has this profoundly modest but practically ambitious character. And once it is no longer statuized, put on a pedestal, and seen in splendid isolation, it comes to appear as an inspiring model for intellectual endeavor more generally.

References

1. D. Knuth, *Surreal Numbers*, Addison-Wesley, Reading, Mass. (1974).
2. R. Carnap, *Introduction to Symbolic Logic and Its Application*, Dover, New York (1958).
3. P. Hayes, in *Expert Systems in the Microelectronic Age*, ed. D. Michie, Edinburgh U. Press, Edinburgh, UK (1979).
4. G. Sussman, J. Wisdom, *Structure and Interpretation of Classical Mechanics*, MIT Press, Cambridge, Mass. (2001).

Making Light of Mass

Every year the MIT physics department puts out a beautifully produced glossy magazine, the *MIT Physics Annual*. In addition to department news, it contains a few articles by department members describing their research. It's meant to help generate support for the department, and my impression is that it's been very effective.

In recent years Carol Breen has been the driving force behind the *Annual*. Among other things Carol also runs our Papallardo lunches, which are wonderful events that I attend regularly. To vindicate the theorem that there's no such thing as a free lunch, Carol asked me to write a big article for 2003 edition. "The Origin of Mass" is that article.

Carol insisted on understanding everything. Now she's a clever person, so if it was reasonably possible to understand something, she would, but of course if it wasn't, she wouldn't. In either case, she'd tell me, so that I knew what worked, and what I had to work on some more. What could be more useful, and in the end gratifying? She also paid close attention to visual design, so that the end product really looks good.

I began "The Origin of Mass" by explaining why it makes sense to ask what is the origin of mass — and why that's not at all obvious. Then comes an elementary presentation of what QCD is, and how we know it's right, that Carol understood. I think it's the best of its kind. Thank you, Carol.

The explanation of the origin of (most) mass from pure energy falls right out.

"Mass without Mass 1: Most of Matter" is an earlier, shorter, and less self-contained discussion of the same central result, written as a Reference Frame for *Physics Today*. As befits the different audience, there is less exposition and more reference to other physics ideas. In particular, I discussed the relationship between the successful modern understanding of the origin of mass and the pioneering but unsuccessful attempts of the recent past (Lorentz' model of the electron and its progeny).

Although QCD explains the origin of most of the mass of ordinary matter, there are several other forms of mass that it does not explain. "Mass without Mass 2: The Medium is the Mass-age" takes up these issues. They are developed at much greater length in "Masses and Molasses" and "In Search of Symmetry Lost."

The Origin of Mass

Everyday work at the frontiers of modern physics usually involves complex concepts and extreme conditions. We speak of quantum fields, entanglement, or supersymmetry, and analyze the ridiculously small or conceptualize the incomprehensibly large. Just as Willie Sutton famously explained that he robbed banks because "that's where the money is," so we do these things because "that's where the Unknown is." It is an amazing and delightful fact, however, that occasionally this sophisticated work gives answers to child-like questions about familiar things. Here I'd like to describe how my own work on subnuclear forces, the world of quarks and gluons, casts brilliant new light on one such child-like question: What is the origin of mass?

Has Mass an Origin?

That a question makes grammatical sense does not guarantee that it is answerable, or even coherent.

The concept of mass is one of the first things we discuss in my freshman mechanics class. Classical mechanics is, literally, unthinkable without it. Newton's second law of motion says that the acceleration of a body is given by dividing the force acting upon it by its mass. So a body without mass wouldn't know how to move, because you'd be dividing by zero. Also, in Newton's law of gravity, the mass of an object governs the strength of the force it exerts. One cannot build an object that gravitates out of material that does not, so you can't get rid of mass without getting rid of gravity. Finally, the most basic feature of mass in classical mechanics is that it is conserved. For example, when you bring together two bodies, the total mass is just the sum of the individual masses. This assumption is so deeply ingrained that it was not even explicitly formulated as a law. (Though I teach it as Newton's Zeroth Law.) Altogether, in the Newtonian framework it is difficult to imagine what would constitute an "origin of mass," or even what this phrase could possibly mean. In that framework mass just is what it is — a primary concept.

Later developments in physics make the concept of mass seem less irreducible. Einstein's famous equation $E = mc^2$ of special relativity theory, written in that way, betrays the prejudice that we should express energy in terms of mass. But we can write the same equation in the alternative form $m = E/c^2$. When expressed in this form, it suggests the possibility of explaining mass in terms of energy. Einstein was aware of this possibility from the beginning. Indeed, his original 1905 paper is entitled, "Does the Inertia of a Body Depend on Its Energy Content?" and it derives $m = E/c^2$, not $E = mc^2$. Einstein was thinking about fundamental physics, not bombs.

At modern particle accelerators, $m = E/c^2$ comes to life. For example, in the Large Electron-Positron collider (LEP), at the CERN laboratory near Geneva, beams of electrons and antielectrons (positrons) were accelerated to enormous energies. Powerful, specially designed magnets controlled the paths of the particles, and caused them to circulate in opposite directions around a big storage ring. The paths of these beams intersected at a few interaction regions, where collisions could occur. (After more than a decade of fruitful operation, in which MIT scientists played a leading role, the LEP machine was dismantled in 2000. It is making way for the Large Hadron Collider (LHC), which will use the same tunnel. LHC will collide protons instead of electrons, and will operate at much higher energy. Hence the past tense.)

When a high-energy electron collides with a high-energy positron we often observe that many particles emerge from the event. (See Figures 2a and 2b.) The total mass of these particles can be thousands of times the mass of the original electron and positron. Thus mass has been created, physically, from energy.

What Matters for Matter

Having convinced ourselves that the question of the origin of mass might make sense, let us now come to grips with it, in the very concrete case of ordinary matter.

Ordinary matter is made from atoms. The mass of an atom is overwhelmingly concentrated in its nucleus. The surrounding electrons are of course crucial for discussing how atoms interact with each other — and thus for chemistry, biology, and electronics. But they provide less than a part in a thousand of the mass! Nuclei, which provide the lion's share of mass, are assembled from protons and neutrons. All this is a familiar, well-established story, dating back seventy years or more.

Newer and perhaps less familiar, but by now no less well-established, is the next step: protons and neutrons are made from quarks and gluons. So most of the mass of matter can be traced, ultimately, back to quarks and gluons.

QCD: What It Is

The theory of quarks and gluons is called quantum chromodynamics, or QCD. QCD is a generalization of quantum electrodynamics (QED). For a nice description of quantum electrodynamics, written by an MIT grad who made good, I highly recommend *QED: The Strange Theory of Electrons and Light*, by Richard Feynman.

The basic concept of QED is the response of photons to electric charge. Figure 1a shows a space-time picture of this core process. Figure 1b shows how it can be used to describe the effect of one electric charge on another, through exchange of a "virtual" photon. (A virtual photon is simply one that gets emitted and absorbed without ever having a significant life of its own. So it is not a particle you can observe directly, but it can have effects on things you do observe.) In other words, Figure 1b describes electric and magnetic forces!

Pictures like these, called Feynman diagrams, may look like childish scribbles, but their naive appearance is misleading. Feynman diagrams are associated with definite mathematical rules that specify how likely it is for the process they depict to occur. The rules for complicated processes, perhaps involving many real and virtual charged particles and many real and virtual photons, are built up in a completely specific and definite way from the core process. It is like making constructions with TinkerToys®. The particles are different kind of sticks you can use, and the core process provides the hubs that join them. Given these elements, the rules for construction are completely determined. In this way all the content of

Maxwell's equations for radio waves and light, Schrödinger's equation for atoms and chemistry, and Dirac's more refined version including spin — all this, and more, is faithfully encoded in the squiggle (Figure 1a).

At this most primitive level QCD is a lot like QED, but bigger. The diagrams look similar, and the rules for evaluating them are similar, but there are more kinds of sticks and hubs. More precisely, while there is just one kind of charge in QED — namely, electric charge — QCD has three different kinds of charge. They are called colors, for no good reason. We could label them red, white, and blue; or alternatively, if we want to make drawing easier, and to avoid the colors of the French flag, we can use red, green, and blue.

Every quark has one unit of one of the color charges. In addition, quarks come in different "flavors." The only ones that play a role in ordinary matter are two flavors called u and d, for up and down. (Of course, quark "flavors" have nothing to do with how anything tastes. And, these names for u and d don't imply that there's any real connection between flavors and directions. Don't blame me; when I get the chance, I give particles dignified scientific-sounding names like axion and anyon.) There are u quarks with a unit of red charge, d quarks with a unit of green charge, and so forth, for six different possibilities altogether.

And instead of one photon that responds to electric charge, QCD has eight color gluons that can either respond to different color charges or change one into another.

So there is quite a large variety of sticks, and there are also many different kinds of hubs that connect them. It seems as if things could get complicated and messy. And so they would, were it not for the overwhelming symmetry of the theory. If you interchange red with blue everywhere, for example, you must still get the same rules. The more complete symmetry allows you to mix the colors continuously, forming blends, and the rules must come out the same for blends as for pure colors. I won't be able to do justice to the mathematics here, of course. But the final result is noteworthy, and easy to convey: there is one and only one way to assign rules to all the possible hubs so that the theory comes out fully symmetric. Intricate it may be, but messy it is not!

With these understandings, QCD is faithfully encoded in squiggles like Figure 1c, and the force between quarks emerges from squiggles like Figure 1d. We have definite rules to predict how quarks and gluons behave and interact. The calculations involved in describing specific processes, like the organization of quarks and gluons into protons, can be very

FIGURES 1A, 1B, 1C, AND 1D. QED and QCD in pictures. The physical content of quantum electrodynamics (QED) is summarized in the algorithm that associates a probability amplitude with each of its Feynman graphs, depicting a possible process in space-time. The Feynman graphs are constructed by linking together hubs, more conventionally called interaction vertices, of the form shown in 1a. The solid line depicts the world-line of an electrically charged particle, and the squiggly line straddles the world-line of a photon. By connecting hubs together we can describe physical processes such as the interaction between electrons, as shown in 1b.

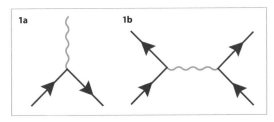
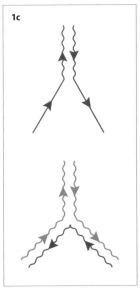
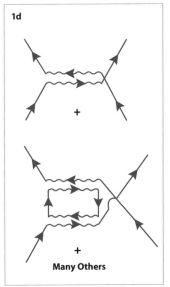

Quantum chromodynamics (QCD) can be summarized similarly, but with a more elaborate set of ingredients and hubs.

There are three kinds of charges, called colors. Quarks resemble electrons in their mechanical properties (technically, they are spin-1/2 fermions), but their interactions are quite different, because they carry a unit of color charge. Quarks come in several flavors — u, d, s, c, b, and t — so we have **u u u d d** and so forth. Only u and d, which have very small masses, are important in ordinary matter. The others are heavy and unstable. Gluons resemble photons in their mechanical properties (technically, they are massless spin-1 bosons), but their interactions are quite different. There are eight different types of color gluons, which respond to and change the color charges of quarks they interact with. A typical hub for a quark-gluon interaction is shown in 1c, along with a hub for gluon-gluon interaction. The latter has no analog in QED, because the photon carries no electric charge. Asymptotic freedom, and all the drastic differences between how particles with and without color charges are observed to behave, ultimately arise from these new gluon-gluon interactions.

In principle we can try to use Feynman diagrams to calculate the quark-quark interaction, as shown in 1d. But unlike QED, in QCD contributions from graphs containing many hubs are not small, and this method is impractical.

THE ORIGIN OF MASS 61

difficult to carry through, but there is no ambiguity about the outcome. The theory is either right or wrong — there's nowhere to hide.

How We Know It's Right

Experiment is the ultimate arbiter of scientific truth. There are many experiments that test the basic principles of QCD. Most of them require rather sophisticated analysis, basically because we don't get to see the underlying simple stuff, the individual quarks and gluons, directly. But there is one kind of experiment that comes very close to doing this, and that is what I'd like to explain to you now.

I'll be discussing what was observed at LEP. But before entering into details, I'd like to review a fundamental point about quantum mechanics, which is necessary background for making any sense at all of what happens. According to the principles of quantum mechanics, the result of an individual collision is unpredictable. We can, and do, control the energies and spins of the electrons and positrons precisely, so that precisely the same kind of collision occurs repeatedly; nevertheless, different results emerge. By making many repetitions, we can determine the probabilities for different outcomes. These probabilities encode basic information about the underlying fundamental interactions; according to quantum mechanics, they contain *all* the meaningful information.

When we examine the results of collisions at LEP, we find there are two broad classes of outcomes. Each happens about half the time.

In one class, the final state consists of a particle and its antiparticle moving rapidly in opposite directions. These could be an electron and an antielectron (e^-e^+), a muon and an antimuon ($\mu^-\mu^+$), or a tau and an antitau ($\tau^-\tau^+$). The little superscripts denote signs of their electric charges, which are all of the same absolute magnitude. These particles, collectively called leptons, are all closely similar in their properties.

Leptons do not carry color charges, so their main interactions are with photons, and thus their behavior should be governed by the rules of QED. This is reflected, first of all, in the simplicity of their final states. Once produced, any of these particles could — in the language of Feynman diagrams — attach a photon using a QED hub, or alternatively, in physical terms, radiate a photon. The basic coupling of photons to a unit charge is fairly weak, however. Therefore each attachment is predicted to decrease the probability of the process being described, and so the most usual case is no attachment.

In fact, the final state $e^-e^+\gamma$, including a photon, does occur, with about 1% of the rate of simply e^-e^+ (and similarly for the other leptons). By studying the details of these 3-particle events, such as the probability for the photon to be emitted in different directions (the "antenna pattern") and with different energy, we can check all aspects of our hypothesis for the underlying hub. This provides a wonderfully direct and incisive way to check the soundness of the basic conceptual building block from which we construct QED. We can then go on to address the extremely rare cases (.01%) where two photons get radiated, and so forth.

For future reference, let's call this first class of outcomes "QED events."

The other broad class of outcomes contains an entirely different class of particles, and is in many ways far more complicated. In these events the final state typically contains ten or more particles, selected from a menu of pions, rho mesons, protons and antiprotons, and many more. These are all particles that in other circumstances interact strongly with one another, and they are all constructed from quarks and gluons. Here, they make a smorgasbord of Greek and Latin alphabet soup. It's such a mess that physicists have pretty much given up on trying to describe all the possibilities and their probabilities in detail.

Fortunately, however, some simple patterns emerge if we change our focus from the individual particles to the overall flow of energy and momentum.

Most of the time — in about 90% of the cases — the particles emerge all moving in either one of two possible directions, opposite to one another. We say there are back-to-back jets. (Here, for once, the scientific jargon is both vivid and appropriate.) About 9% of the time, we find flows in three directions; about .9% of the time, four directions; and by then we're left with a very small remainder of complicated events that are hard to analyze this way.

I'll call the second broad class of outcomes "QCD events." Representative 2-jet and 3-jet QCD events, as they are actually observed, are displayed in Figure 2.

Now if you squint a little, you will find that the QED events and the QCD events begin to look quite similar. Indeed, the pattern of energy flow is qualitatively the same in both cases, that is, heavily concentrated in a few narrow jets. There are two main differences. One, relatively trivial, is that multiple jets are more common in QCD than in QED. The other is much more profound. It is that, of course, in the QED events the jets are just single particles, while in the QCD events the jets are sprays of several particles.

2a

FIGURES 2A AND 2B. Real jets.

These are pictures of the results of electron-positron collisions at LEP, taken by the L3 collaboration led by Professors Ting, Becker, and Fisher. The alignment of energetic particles in jets is visible to the naked eye.

2b

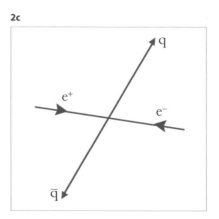

2c

FIGURES 2C AND 2D. Conceptual jets.

These diagrams represent our conceptual model of the deep structure beneath jet production as it is observed. Electrons and positrons annihilate into "pure energy" (a virtual photon, actually), which materializes into a quark-antiquark pair. The quark and antiquark usually dress themselves with soft radiation, as described in the text, and we observe a two-jet event. About 10% of the time, however, a hard gluon is radiated. The quark, antiquark, and gluon all dress themselves with soft radiation, and we see three jets. Figures 2c and 2d have been drawn to parallel the geometry of the observations shown in Figures 2a and 2b. (N.B. To keep things simple, I have not tried to maintain the full color scheme from Figure 1.)

2d

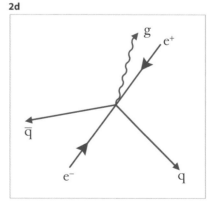

64   FANTASTIC REALITIES

In 1973, while I was a graduate student working with David Gross at Princeton, I discovered the explanation of these phenomena. We took the attitude that the deep similarities between the observed basic behaviors of leptons (based on QED) and the strongly interacting particles might indicate that the strongly interacting particles are also ultimately described by a simple, rule-based theory, with sticks and hubs. In other words, we squinted.

To bring our simplified picture of the QCD events into harmony with the observations, we relied on a theoretical discovery I'll describe momentarily, which we christened ASYMPTOTIC FREEDOM. (Please notice that our term is not "cute.") Actually, our discovery of asymptotic freedom preceded these specific experiments, so we were able to predict the results of these experiments before they were performed. As a historical matter, we discovered QCD and asymptotic freedom by trying to come to terms with the MIT-SLAC "scaling" experiments done at the Stanford Linear Collider in the late 1960s, for which Jerome Friedman, Henry Kendall, and Richard Taylor won the Nobel Prize in 1990. Since our analysis of the scaling experiments using QCD was (necessarily) more complicated and indirect, I've chosen to focus here on the later, but simpler to understand, experiments involving jets.

The basic concept of asymptotic freedom is that the probability for a fast-moving quark or gluon to radiate away some of its energy in the form of other quarks and gluons depends on whether this radiation is "hard" or "soft." Hard radiation is radiation that involves a substantial deflection of the particle doing the radiating, while soft radiation is radiation that does not cause such a deflection. Thus hard radiation changes the flow of energy and momentum, while soft radiation merely distributes it among additional particles, all moving together. Asymptotic freedom says that hard radiation is rare, but soft radiation is common.

This distinction explains why on the one hand there are jets, and on the other hand why the jets are not single particles. A QCD event begins as the materialization of quark and antiquark, similar to how a QED event begins as the materialization of lepton-antilepton. They usually give us two jets, aligned along the original directions of the quark and antiquark, because only hard radiation can change the overall flow of energy and momentum significantly, and asymptotic freedom tells us hard radiation is rare. When a hard radiation does occur, we have an extra jet! But we don't see the original quarks or antiquarks, individually, because they are always accompanied by their soft radiation, which is common.

By studying the antenna patterns of the multi-jet QCD events we can check all aspects of our hypotheses for the underlying hubs. Just as for QED, such antenna patterns provide a wonderfully direct and incisive way to check the soundness of the basic conceptual building blocks from which we construct QCD.

Through analysis of this and many other applications, physicists have acquired complete confidence in the fundamental correctness of QCD. By now experimenters use it routinely to design experiments searching for new phenomena, and they refer to what they're doing as "calculating backgrounds" rather than "testing QCD"!

Many challenges remain, however, to make full use of the theory. The difficulty is always with the soft radiation. Such radiation is emitted very easily, and that makes it difficult to keep track of. You get a vast number of Feynman graphs, each with many attachments, and they get more and more difficult to enumerate, let alone calculate. That's very unfortunate, because when we try to assemble a proton from quarks and gluons none of them can be moving very fast for very long (they're supposed to be inside the proton, after all), so all their interactions involve soft radiation.

To meet this challenge, a radically different strategy is required. Instead of calculating the paths of quarks and gluons through space and time, using Feynman graphs, we let each segment of space-time keep track of how many quarks and gluons it contains. We then treat these segments as an assembly of interacting subsystems.

Actually in this context "we" means a collection of hardworking computers. Skillfully orchestrated, and working full time at teraflop speeds, they manage to produce quite a good account of the masses of protons and other strongly interacting particles, as you can see from Figure 3. The equations of QCD, which we discovered and proved from very different considerations, survive this extremely intense usage quite well. There's a big worldwide effort, at the frontiers of computer technology and human ingenuity, to do calculations like this more accurately, and to calculate more things.

The Ingredients of QCD, Lite and Full-Bodied

With the answer in hand, let's examine what we've got. For our purposes it's instructive to compare two versions of QCD, an idealized version I call QCD Lite, and the realistic Full-Bodied version.

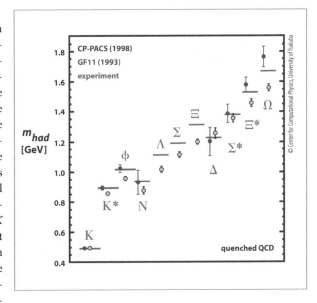

FIGURE 3. Its from bits. This plot, taken from the CP-PACS collaboration, shows a comparison between the predictions of QCD and the masses of particles. The green level lines indicate observed values of particle masses, while the circles within intervals indicate computational results and their statistical uncertainties. The K meson is the left-most entry, while the proton and neutron are N. The calculations employ cutting-edge computer technology with massive parallelism, and even then some approximations must be introduced to make the computations feasible. These results are a remarkable embodiment of the vision that elements of reality can be reproduced by purely conceptual constructions — "Its from Bits" — because the underlying theory, based on profoundly symmetrical equations, contains very few adjustable parameters.

QCD Lite is cooked up from massless gluons, massless u and d quarks, and nothing else. (Now you can fully appreciate the wit of the name.) If we use this idealization as the basis for our calculation, we get the proton mass low by about 10%.

Full-Bodied QCD differs from QCD Lite in two ways. First, it contains four additional flavors of quarks. These do not appear directly in the proton, but they do have some effect as virtual particles. Second, it allows for non-zero masses of the u and d quarks. The realistic value of these masses, though, turns out to be small, just a few percent of the proton mass. Each of these corrections changes the predicted mass of the proton by about 5%, as we pass from QCD Lite to Full-Bodied QCD. So we find that 90% of the proton (and neutron) mass, and therefore 90% of the mass of ordinary matter, emerges from an idealized theory whose ingredients are entirely massless.

The Origin of (most) Mass

Now I've shown you the theory that describes quarks and gluons, and therefore has to account for most of the mass of matter. I've described some of the experiments that confirm the theory. And I've displayed successful calculations of hadron masses, including the masses of protons and neutrons, using this theory.

In a sense, these calculations settle the question. They tell us the origin of (most) mass. But simply having a computer spit out the answer, after gigantic and totally opaque calculations, does not satisfy our hunger for understanding. It is particularly unsatisfactory in the present case, because the answer appears to be miraculous. The computers construct for us massive particles using building blocks — quarks and gluons — that are themselves massless. The equations of QCD Lite output Mass without Mass, which sounds suspiciously like Something for Nothing. How did it happen?

The key, again, is asymptotic freedom. Previously, I discussed this phenomenon in terms of hard and soft radiation. Hard radiation is rare, soft radiation is common. There's another way of looking at it, mathematically equivalent, that is useful here. From the classical equations of QCD, one would expect a force field between quarks that falls off as the square of the distance, as in ordinary electromagnetism (Coulomb's law). Its enhanced coupling to soft radiation, however, means that when quantum mechanics is taken into account a "bare" color charge, inserted into empty space, will start to surround itself with a cloud of virtual color gluons. These color gluons fields themselves carry color charge, so they are sources of additional soft radiation. The result is a self-catalyzing enhancement that leads to runaway growth. A small color charge, in isolation, builds up a big color thundercloud.

All this structure costs energy, and theoretically the energy for a quark in isolation is infinite. That's why we never see individual quarks. Having only a finite amount of energy to work with, Nature always finds a way to short-circuit the ultimate thundercloud.

One way is to bring in an antiquark. If the antiquark could be placed right on top of the quark, their color charges would exactly cancel, and the thundercloud would never get started. There's also another more subtle way to cancel the color charge by bringing together three quarks, one of each color.

In practice these exact cancellations can't quite happen, however, because there's a competing effect. Quarks obey the rules of quantum mechanics. It is wrong to think of them simply as tiny particles, rather they are quantum-mechanical wavicles. They are subject, in particular, to Heisenberg's uncertainty principle, which implies that if you try to pin down their position too precisely, their momentum will be wildly uncertain. To support the possibility of large momentum, they must acquire large energy. In other words, it takes work to pin quarks down. Wavicles want to spread out.

So there's a competition between two effects. To cancel the color charge completely, we'd like to put the quark and antiquark at precisely the same place; but they resist localization, so it's costly to do that.

This competition can result in a number of compromise solutions, where the quark and antiquark (or three quarks) are brought close together, but are not perfectly coincident. Their distribution is described by quantum mechanical wave functions. Many different stable wave patterns are possible, and each corresponds to a different kind of particle that you can observe. There are patterns for protons, neutrons, and for each entry in the whole Greek and Latin smorgasbord. Each pattern has some characteristic energy, because the color fields are not entirely cancelled particles and because the wavicles are somewhat localized. And that, through $m = E/c^2$, is the origin of mass.

A similar mechanism, though much simpler, works in atoms. Negatively charged electrons feel an attractive electric force from the positively charged nucleus, and from that point of view they'd like to snuggle right on top of it. Electrons are wavicles, though, and that inhibits them. The result, again, is a series of possible compromise solutions. These are what we observe as the energy levels of the atom. When I give the talk on which this article is based, at this point I use Dean Dauger's marvelous "Atom in a Box" program to show the lovely, almost sensuous patterns of undulating waves that describe the possible states of that simplest of atoms, hydrogen. I hope you will explore "Atom in a Box" for yourself. You can link to it at http://www.dauger.com. In its absence, I will substitute a classic metaphor.

The wave patterns that describe protons, neutrons, and their relatives resemble the vibration patterns of musical instruments. In fact the mathematical equations that govern these superficially very different realms are quite similar.

Musical analogies go back to the prehistory of science. Pythagoras, partly inspired by his discovery that harmonious notes are sounded by strings whose lengths are in simple numerical ratios, proposed that "All things are Number." Kepler spoke of the music of the spheres, and his longing to find their hidden harmonies sustained him through years of tedious calculations and failed guesses before he identified the true patterns of planetary motions.

Einstein, when he learned of Bohr's atomic model, called it "the highest form of musicality in the sphere of thought." Yet Bohr's model, wonderful as it is, appears to us now as a very watered-down version of the true wave-mechanical atom; and the wave-mechanical proton is more intricate and symmetric by far!

I hope that some artist/nerd will rise to the challenge, and construct a "Proton in a Box" for us to play with and admire.

The World as Concept, Algorithm, and Number

I will conclude with a few words concerning the broader significance of these developments for our picture of the world.

A major goal of theoretical physics is to describe the world with the smallest number of concepts. For that reason alone, it is an important result that we can largely eliminate mass as a property necessary for describing matter. But there is more. The equations that describe the behavior of elementary particles become fundamentally simpler and more symmetric when the mass of the particles is zero. So eliminating mass enables us to bring more symmetry into the mathematical description of Nature.

The understanding of the origin of mass that I've sketched for you here is the most perfect realization we have of Pythagoras' inspiring vision that the world can be built up from concepts, algorithms, and numbers. Mass, a seemingly irreducible property of matter, and a byword for its resistance to change and sluggishness, turns out to reflect a harmonious interplay of symmetry, uncertainty, and energy. Using these concepts, and the algorithms they suggest, pure computation outputs the numerical values of the masses of particles we observe.

Still, as I've already mentioned, our understanding of the origin of mass is by no means complete. We have achieved a beautiful and profound understanding of the origin of *most* of the mass of ordinary matter, but not of *all* of it. The value of the electron mass, in particular, remains deeply mysterious even in our most advanced speculations about unification and string theory. And ordinary matter, we have recently learned, supplies only a small fraction of mass in the Universe as a whole. More beautiful and profound revelations await discovery. We continue to search for concepts and theories that will allow us to understand the origin of mass in all its forms, by unveiling more of Nature's hidden symmetries.

Mass without Mass 1: Most of Matter

With his unique talent for the paradoxical profundity, John Wheeler coined the phrase "mass without mass" to advertise the goal of removing any mention of mass from the basic equations of physics.[1] Can we really hope to do this? How far have we come? Why should we try? In this piece, I answer the first question and part of the second; in my next column, I'll round out the story and look ahead.

As commonly used, the words "massive" and "weighty" suggest things that are too obvious and significant to ignore, as in a massive fraud or a weighty opinion. Thus our very language conditions us to think of the mass of a physical object as one of its primary characteristics. So does our everyday experience, and even our early education in physics. Indeed, the concept of mass lies at the heart of Newtonian physics. It appears explicitly both in the foundational equation $F = ma$ and in the law of universal gravitation $F = GMm/r^2$.

Later developments in physics made the concept of mass seem less irreducible, and less basic. This undermining process started in earnest with the theories of relativity. The famous equation $E = mc^2$ of special relativity theory, written that way, betrays the prejudice that we should express energy in terms of mass. But it doesn't take an Einstein to derive from that equation $m = E/c^2$, which suggests the possibility of explaining mass in terms of energy. And the conceptual hub of the general theory of relativity, the equivalence principle, is the observation that the response of a body to gravitation is independent of its mass. Consistent with this observation, Newton's two laws can be combined into $a = GM/r^2$, wherein *m* does not appear. The central equation of general relativity theory,

$$R_{\mu\nu} - \frac{1}{2}g_{\mu\nu}R = T_{\mu\nu}$$

(in appropriate units), equates the curvature of space-time to the energy–momentum of matter. Einstein referred to the left-hand side as a palace of gold, and to the right-hand side as a hovel of wood, thus expressing his ambition to make improvements on the right-hand side, to root it in concepts of depth and beauty comparable to Riemannian geometry. Of course, it is only on the right-hand, wooden side that masses of particles occur, raw and unadorned. Can we replace them with finer material?

Quantum field theory simplifies our task by vastly reducing the inventory of different parts we need to replace. In quantum field theory, the primary elements of reality are not individual particles, but underlying fields. Thus, for example, all electrons are but excitations of an underlying field, naturally called the electron field, which fills all space and time. This formulation explains why all electrons everywhere and for all time have exactly the same properties, including, of course, the same mass. If one constructs all matter from excitations of a few fields, as we do in the modern Standard Model, the challenge of mass takes a new and profoundly simpler form. At worst, we will have to specify a few numerical parameters — one for each fundamental field — to account for mass in general.

In practice, we do much better. The bulk of the mass of ordinary matter (better than 99%) comes from the masses of protons and neutrons. In quantum chromodynamics (QCD), the protons and neutrons appear as secondary, composite structures built up from quarks and gluons. We can maintain an excellent approximation to reality while working with a truncated version of QCD, which contains only the color gluons plus up and down quark fields. The heavier quarks play an extremely minor role in the structure of the proton and neutron.

Our theory of the color gluons is derived from a powerful symmetry principle — non-Abelian, or Yang–Mills, gauge symmetry — similar in many respects to the general covariance of general relativity. Gauge symmetry forbids mass terms for the gluon fields. Thus color gluons, like gravitons and photons, and for similar reasons, have no mass. Furthermore, there is much phenomenological evidence that the mass terms associated with up and down quarks are quite small. Let us set them to zero. Now our resulting truncated, approximate version of QCD contains no mass terms at all. (In a genuine sense, it has no free parameters whatsoever.[2]) Yet, if we use it to calculate the mass of protons and neutrons — the mass of ordinary matter — we find it is accurate[3] to within 10%!

How is it possible that massive protons and neutrons can be built up out of strictly massless quarks and gluons? The key is $m = E/c^2$. There is energy stored in the motion of the quarks, and energy in the color gluon fields that connect them. This bundling of energy makes the proton's mass.

The emergent picture of the proton mass realizes, in a different context, a modified form of the dream of Hendrik A. Lorentz[4] (pursued by many others including Henri Poincaré, P. A. M. Dirac, Wheeler, and Richard Feynman) to account for the electron's mass entirely in terms of its electromagnetic fields. A classical point electron is surrounded by an electron field varying as $1/r^2$. The energy in this field is infinite, due to a divergent contribution around $r \to 0$. Lorentz hoped that in a correct model of electrons, they would emerge as extended objects, and that the energy in the Coulomb field would come out finite and in fact account for all or most of the inertia of electrons.

Later progress in the quantum theory of electrons rendered this program moot, by showing that the charge of an electron — and therefore, of course, the singularity of its associated electric field — is intrinsically smeared out by quantum fluctuations in the electron's position. As a result, the electric field energy of an electron makes only a small correction to its total mass. Thus Lorentz's dream, in its original form, is not realized. But beautiful ideas are rarely entirely wrong, and something close to Lorentz's idea is embodied in modern QCD. Quarks carry color charge, and generate color electric fields analogous to the ordinary electric fields around electrons. The potentially diverging energy of color electric fields close to the quark is removed by quantum mechanics, just as for ordinary electric fields around electrons. But although ordinary electric fields fall off rapidly far from their sources, color electric fields do not. This property explains why quarks are never observed in isolation.

Triples of quarks, however, can cunningly contrive to generate fields that cancel at very large distances. To build protons and neutrons, they must do this. Even so, at finite distances the fields do not cancel exactly, and so a finite field energy remains. According to QCD, it is precisely this color field energy that mostly makes us weigh. It thus provides, quite literally, "mass without mass."

Thus QCD takes us a long stride toward the Einstein–Wheeler ideal of "mass without mass." For ordinary matter, *quantitatively*, it brings us amazingly close. If your friend puts on a few pounds yet complains, "But I never eat anything heavy," modern physics sanctions you to give her (or him) the benefit of the doubt.

References

1. J. A. Wheeler, *Geometrodynamics*, Academic, New York (1962), p. 25.
2. F. Wilczek, *Nature* **397**, 303 (1999).
3. C. Bernard et al., *Nucl. Phys. Proc. Suppl.* **73**, 198 (1999), and references therein.
4. H. A. Lorentz, *Proc. Acad. Sci. Amsterdam* **6** (1904), reprinted in A. Einstein et al., *The Principle of Relativity*, Dover, New York (1952), p. 24. See also, especially, R. P. Feynman, *Lectures in Physics*, vol. 2, Addison-Wesley, Reading, Mass. (1964), chap. 28.

Mass without Mass 2: The Medium is the Mass-age

In "Mass without Mass 1" I discussed how most of the mass of ordinary matter arises from the energy associated with quark motion and color gluon fields. That mass can be computed with good accuracy using the theory I call QCD Lite — a simplified form of quantum chromodynamics (QCD), in which the masses of the up and down quarks are set to zero, and the other quarks are omitted. In QCD Lite, protons and neutrons are assembled entirely from massless building blocks. For the bulk of ordinary matter, we thereby realize John Wheeler's goal of deducing mass as a secondary property, instead of having to introduce mass as a primary property. In Wheeler's phrase, we've achieved "mass without mass."

The elimination of mass as a primary property of matter is a delightful and important consequence of QCD Lite. However, in QCD this feature is a luxury, not a necessity. If the up and down quarks had larger masses, QCD Lite would be a poor approximation, and the masses of protons and neutrons would arise mostly from the masses of the quarks that made them, but QCD itself would remain a perfectly consistent theory. Indeed, many of its central results — for examples, its predictions for the probabilities of jet production in high-energy collisions — would not be significantly different.

By contrast, in the other part of the Standard Model of particle physics, the electroweak sector, mass without mass is indispensable. The modern theory of electroweak interactions couldn't work without it. For the core principle of this theory is chiral gauge symmetry, and chiral gauge symmetry abhors mass. This tight knot of jargon entangles some profound and far-reaching ideas, which I'll now try to untangle.

First, *chiral*, as used in this context, refers to an intrinsic distinction between right and left. In 1956–57, the physics world was rocked by the discovery that the weak interaction makes such a distinction. All previous experience in atomic and nuclear physics was consistent with fundamental symmetry between left and right. Theorists had elevated this observation into a principle: parity. The weak interactions flout parity. The electrons emitted in beta decay, for example, are almost always left-handed: if you point your left thumb in the direction of such an electron's motion, usually you'll find that it's spinning in the same direction as your fingers curl. The same is true for electrons, or for muons, emitted in other weak decays, whereas positrons and antimuons are almost always right-handed. Detailed study revealed that the weak interactions violate parity between left and right systematically, and in a very big way. Theorists attempted to make the best of this situation by proposing a new principle: maximal parity noninvariance (for the weak interaction only).

The ideal, straightforward statement of maximal parity noninvariance would be single-handedness, or chirality. The statement would read: Only left-handed particles and only right-handed antiparticles participate in the weak interaction. This formulation, however, is inconsistent with the principle of relativity. Indeed, what appears as a left-handed electron to a stationary observer will look right-handed to an observer moving in the electron's direction, but faster — the direction of motion appears reversed, but the direction of spin remains the same. If electrons were massless, this inconsistency would not arise, for then they would move at the speed of light, and no observer could overtake them. The nonzero mass of electrons, and other particles, greatly complicates the task of converting maximal parity noninvariance from a rough rule of thumb into a precise principle. Chirality's natural habitat is a massless world.

Second, *gauge symmetry*. Gauge symmetry was first discovered as a property of quantum electrodynamics (QED). We learn in freshman physics that electromagnetic waves are purely transverse: that the fields in such waves are excited only in directions perpendicular to the direction of wave propagation. When we come to quantize the electromagnetic field, it turns out to be quite difficult to ensure this behavior. Quantum fluctuations will explore all possible field configurations, including longitudinal waves. To reproduce the laws of freshman physics, one must make sure that these longitudinal waves do not represent any physical effects. So adding or subtracting longitudinal waves must leave the equations of

QED unchanged. This property implies a very large symmetry, which is what we call gauge symmetry.

Subtle but powerful theoretical arguments tell us that gauge symmetry is a necessary property of consistent quantum field theories containing vector (spin-1) particles. This is required because the longitudinal waves have very unpleasant properties (negative probabilities!), and must be avoided. Consistent with this idea, not only QED but also QCD (which is mediated by vector particles, the color gluons) is based on equations with gauge symmetry. Gauge symmetry requires that the vector particles be massless. The same argument used for chirality also applies here: If an observer could catch up with an electromagnetic wave so that it appeared stationary, there would be no distinction between longitudinal and transverse excitations. To prevent this ambiguity, electromagnetic waves had better move at the limiting velocity — not coincidentally, the speed of light!

The weak interaction is mediated by vector particles, the W and Z bosons. Unfortunately, they're massive. Oops.

From two independent perspectives, therefore, we see that the theory of the weak interaction would much prefer a world built from massless particles. Ours is not such a world. There is, however, a happy resolution.

Amazingly enough the phenomenon of superconductivity, which seems a world away from these problems, points the way. The limitation of magnetic fields to a thin layer — the Meissner effect — is the essence of superconductivity. The microscopic theory of superconductivity ascribes this effect to a pervasive condensate of Cooper pairs in the superconducting material. When electromagnetic fields attempt to invade the superconductor, they disturb the condensate it houses. This perturbation costs energy, and the condensate does its best to expel the fields. Thus photons no longer come cheap, and the fields associated with them are short-ranged. Within a superconductor, in a word, photons have become massive.

The modern theory of the weak interaction postulates a condensate that plays, for W and Z bosons in empty space, the same role played by Cooper pairs in a superconductor. This condensate can serve double duty, also gumming up the propagation of leptons and quarks. According to this conception, the basic equations of electroweak theory apply to a (nonexistent) external world of massless particles, within which we happily inhabit a weird superconductor.

At present, there is no independent or direct evidence for the required condensate. Its existence is simply a hypothesis that allows the known facts to snap into place, while introducing no contradictions. No known form of matter has the right properties to produce the required condensate. Something new must be added — the so-called Higgs field. What we call empty space, or vacuum, is filled with a condensate spawned by that field.

Fortunately, this fantasy has observable consequences. By agitating the Higgs field, we should be able to produce its quanta — Higgs particles. The spin of the Higgs particle (namely 0 — it's a chip off the vacuum), and its couplings to matter are unambiguously predicted. Indeed, if the Higgs field is doing the job we hired it for — that is, generating masses of quarks, leptons, and W and Z bosons — the coupling strength of the Higgs particle to each of the other particles ought to be proportional to their own mass. These properties are quite distinctive, so we'll certainly be able to recognize the Higgs particle, if and when it's found.

Ironically the one property of the Higgs particle that's not predicted precisely is its mass. There are excellent reasons, however, to think that the mass is such as to render the particle accessible to near-future accelerators — the Large Hadron Collider (LHC), if not the Large Electron-Positron (LEP) collider or the Tevatron. It will be very satisfying — and a great triumph for theoretical physics — if experimentalists find a particle with the predicted properties. Failure to find it could be still more instructive (but I don't expect that to occur).

Let me summarize. We've come a long way toward dethroning mass as a primary, irreducible property of matter. Most of the mass of ordinary matter, for sure, is the pure energy of moving quarks and gluons. The remainder, a quantitatively small but qualitatively crucial remainder — it includes the mass of electrons — is all ascribed to the confounding influence of a pervasive medium, the Higgs field condensate. There is already much indirect evidence for this concept, and crucial direct tests are in the offing.

So much for mass, in the Standard Model. Physics as a whole offers additional mass parameters that still await comparable assimilation. Perhaps the most profound is the Planck mass, which characterizes quantum gravity. Another is the recently discovered nonzero neutrino mass. They are, respectively, much larger (factor $\sim 10^{17}$) and much smaller (factor $\sim 10^{-13}$) than the Higgs mass. A major objective for a unified theory of fundamental physics must be to give a convincing account of these extraordinary ratios. At present, one can venture important connections, but they're too loose for comfort, and tenuous to boot.

Astronomy may jazz the chord. Indeed, notoriously, the source of most of the gravitational mass in the universe is yet to be identified. "Mass without matter"? As we realize past dreams, we awaken to new realities.

QCD Exposed

The title of "QCD Made Simple" is a play on words that might be lost to future generations, so I'd better explain it here. It recalls a popular series of books with the generic title *Made Simple, like *Astronomy Made Simple* (I read that one), *Electronics Made Simple*, and so forth. Long before *For Dummies, was Made Simple. And of course, in "QCD Made Simple" I've tried to explain the basics of the theory in a limpid (that is, clear and simple) way.

The joke is that a big point of "QCD Made Simple" is that while the behavior of QCD is usually very complicated, it can be *made simple* by studying it in special, extreme physical situations: high virtuality (explained in the article), high temperature, or high density. Each of these opens a window into very interesting worlds of phenomena, in high-energy accelerator physics, cosmology, and neutron star astrophysics, respectively. At high virtuality or high temperature, the fundamental quark and gluon degrees of freedom, which ordinarily are confined within complicated bound states, get exposed, thanks to asymptotic freedom. At high density the dynamical mechanisms behind confinement and chiral symmetry breaking, which ordinarily are hidden in opaque equations, get exposed, thanks to color superconductivity.

By the way the title of this whole section, "QCD Exposed," is also a play on words. But I don't suppose I had to tell you that.

"10^{12} Degrees in the Shade" was written for a magazine, *The Sciences*, that turned out to be too good for this world. They asked for substantial articles that put ongoing scientific research in its historical context, both past and future, for a broad audience. ("Lots of lawyers read it," I was told.) Also there could be no pictures, figures, or diagrams: instead, articles were ornamented with abstract art works!

Asked to write about heavy ion collisions, I began with a discussion of the historical role of high temperature studies in physics, including Planck's investigation of blackbody radiation and its amazing cosmological realization. That was easy and well-trodden terrain, but then came the challenging part: explaining chiral symmetry breaking to lawyers. I gave it my best shot.

"Back to Basics at Ultrahigh Temperatures," a *Physics Today* Reference Frame, is mostly a breezier take on the same issues. But the first five paragraphs, together with the final four, add a methodological moral and a deep cosmological riddle. I returned to this riddle, and proposed a possible answer, five years later, in the last part of "Analysis and Synthesis 4."

QCD Made Simple

Quantum chromodynamics, familiarly called QCD, is the modern theory of the strong interaction.[1] Historically its roots are in nuclear physics and the description of ordinary matter — understanding what protons and neutrons are and how they interact. Nowadays QCD is used to describe most of what goes on at high-energy accelerators.

Twenty or even fifteen years ago, this activity was commonly called "testing QCD." Such is the success of the theory, that we now speak instead of "calculating QCD backgrounds" for the investigation of more speculative phenomena. For example, discovery of the heavy W and Z bosons that mediate the weak interaction, or of the top quark, would have been a much more difficult and uncertain affair if one did not have a precise, reliable understanding of the more common processes governed by QCD. With regard to things still to be found, search strategies for the Higgs particle and for manifestations of supersymmetry depend on detailed understanding of production mechanisms and backgrounds calculated by means of QCD.

Quantum chromodynamics is a precise and beautiful theory. One reflection of this elegance is that the essence of QCD can be portrayed, without severe distortion, in Figure 1. But first, for comparison, let me remind you that the essence of quantum electrodynamics (QED), which is a generation older than QCD, can be portrayed by the single picture at the top of the box, which represents the interaction vertex at which a photon responds to the presence or motion of electric charge.[2] This is not just a metaphor. Quite definite and precise algorithms for calculating

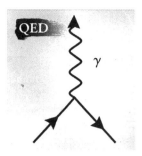

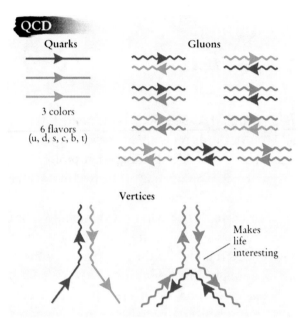

FIGURE 1. QED and QCD in pictures.

The physical content of quantum electrodynamics is summarized in the algorithm that associates a probability amplitude with each of its Feynman graphs, depicting a possible process in space-time. The Feynman graphs are constructed by linking together interaction vertices of the type at left, which represents a point charged particle (lepton or quark) radiating a photon. To get the amplitude, one multiplies together a kinematic "propagator" factor for each line and an interaction factor for each vertex. Reversing a line's direction is equivalent to replacing a particle by its antiparticle.

Quantum chromodynamics can be similarly summarized, but with a more elaborate set of ingredients and vertices, as shown below. Quarks (antiquarks) carry one positive (negative) unit of color charge. Linear superpositions of the 9 possible combinations of gluon colors shown below form an SU(3) octet of 8 physical gluon types.

A qualitatively new feature of QCD is that there are vertices describing direct interactions of color gluons with one another. Photons, by contrast, couple only to electric charge, of which they carry none themselves.

physical processes are attached to the Feynman graphs of QED, constructed connecting just such interaction vertices.

In the same pictorial language, QCD appears as an expanded version of QED. Whereas in QED there is just one kind of charge, QCD has three different kinds of charge, labeled by "color." Avoiding chauvinism, we might choose red, green, and blue. But, of course, the color charges of QCD have nothing to do with physical colors. Rather, they have properties analogous to electric charge. In particular, the color charges are conserved in all physical processes, and there are photon-like massless particles, called color gluons, that respond in appropriate ways to the presence or motion of color charge, very similar to the way photons respond to electric charge.

Quarks and Gluons

One class of particles that carry color charge are the quarks. We know of six different kinds, or "flavors," of quarks — denoted u, d, s, c, b, and t, for: up, down, strange, charmed, bottom, and top. Of these, only u and d quarks play a significant role in the structure of ordinary matter. The other, much heavier quarks are all unstable. A quark of any one of the six flavors can also carry a unit of any of the three color charges. Although the different quarks flavors all have different masses, the theory is perfectly symmetrical with respect to the three colors. This color symmetry is described by the Lie group SU(3).

Quarks, like electrons, are spin-1/2 point particles. But instead of electric charge, they carry color charge. To be more precise, quarks carry *fractional* electric charge ($+2e/3$ for the u, c, and t quarks, and $-e/3$ for the d, s, and b quarks) in addition to their color charge.

For all their similarities, however, there are a few crucial differences between QCD and QED. First of all, the response of gluons to color charge, as measured by the QCD coupling constant, is much more vigorous than the response of photons to electric charge. Second, as shown in the box, in addition to just responding to color charge, gluons can also change one color charge into another. All possible changes of this kind are allowed, and yet color charge is conserved. So the gluons themselves must be able to carry unbalanced color charges. For example, if absorption of a gluon changes a blue quark into a red quark, then the gluon itself must have carried one unit of red charge and minus one unit of blue charge.

All this would seem to require $3 \times 3 = 9$ different color gluons. But one particular combination of gluons — the color-SU(3) singlet — which responds equally to all charges, is different from the rest. We must remove it if we are to have a perfectly color-symmetric theory. Then we are left with only 8 physical gluon states (forming a color-SU(3) octet). Fortunately, this conclusion is vindicated by experiment!

The third difference between QCD and QED, which is the most profound, follows from the second. Because gluons respond to the presence and motion of color charge *and* they carry unbalanced color charge, it follows that gluons, quite unlike photons, respond directly to one another. Photons, of course, are electrically neutral. Therefore the laser sword fights you've seen in *Star Wars* wouldn't work. But it's a movie about the future, so maybe they're using color gluon lasers.

We can display QCD even more compactly, in terms of its fundamental equations (Figure 2). You should not necessarily be too impressed by that. After all, Richard Feynman showed that you could write down the Equation of the Universe in a single line: $U = 0$, where U, the total *unworldliness*,[3] is a definite function. It's the sum of contributions from *all* the laws of physics:

$$U = U_{\text{Newton}} + U_{\text{Gauss}} + \ldots,$$

where, for instance, $U_{\text{Newton}} = (\mathbf{F} - m\mathbf{a})^2$ and $U_{\text{Gauss}} = (\nabla \cdot \mathbf{E} - \rho)^2$.

So we can capture all the laws of physics we know, and all the laws yet to be discovered, in this one unified equation. But it's a complete cheat, of course, because there is no useful algorithm for unpacking U, other than to go back to its component parts. The equations of QCD, displayed

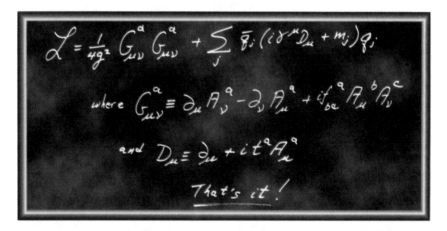

FIGURE 2. The QCD lagrangian $\mathcal{L}$ displayed here is, in principle, a complete description of the strong interaction. But, in practice, it leads to equations that are notoriously hard to solve. Here m_j and q_j are the mass and quantum field of the quark of jth flavor, and A is the gluon field, with space-time indices μ and ν and color indices a, b, c. The numerical coefficients f and t guarantee SU(3) color symmetry. Aside from the quark masses, the one coupling constant g is the only free parameter of the theory.

in Figure 2, are very different from Feynman's satirical unification. Their complete content is out front, and the algorithms that unpack them flow from the unambiguous mathematics of symmetry.

A remarkable feature of QCD, which we see in Figure 2, is how few adjustable parameters the theory needs. There is just one overall coupling constant g and six quark-mass parameters m_j for the six quark flavors. As we shall see, the coupling strength is a relative concept; and there are many circumstances in which the mass parameters are not significant. For example, the heavier quarks play only a tiny role in the structure of ordinary matter. Thus QCD approximates the theoretical ideal: From a few purely conceptual elements, it constructs a wealth of physical consequences that describe Nature faithfully.[4]

Describing Reality

At first sight it appears outrageous to suggest that the equations of Figure 2 or, equivalently, the pictures in the box, can describe the real world of the strongly interacting particles. None of the particles that we've actually seen appear in the box, and none of the particles that appear in the box has ever been observed. In particular, we've never seen particles carrying fractional electric charge, which we nonetheless ascribe to the quarks. And certainly we haven't seen anything like gluons — massless particles mediating long-range strong forces. So if QCD is to describe the world, it must explain why quarks and gluons cannot exist as isolated particles. That is the so-called confinement problem.

Besides confinement, there is another qualitative difference between observed reality and the fantasy world of quarks and gluons. This difference is quite a bit more subtle to describe, but equally fundamental. I will not be able here to do full justice to the phenomenological arguments, but I can state the problem in its final, sanitized theoretical form. The phenomenology indicates that if QCD is to describe the world, then the u and d quarks must have very small masses. But if these quarks do have very small masses, then the equations of QCD possess some additional symmetries, called chiral symmetries (after *chiros*, the Greek word for *hand*). These symmetries allow separate transformations among the right-handed quarks (spinning, in relation to their motion, like ordinary right-handed screws) and the left-handed quarks.

But there is no such symmetry among the observed strongly interacting particles; they do not come in opposite-parity pairs. So if QCD is to describe the real world, the chiral symmetry must be spontaneously broken, much as rotational symmetry is spontaneously broken in a ferromagnet.

Clearly, it's a big challenge to relate the beautifully simple concepts that underlie QCD to the world of observed phenomena. There have been three basic approaches to meeting this challenge:

▷ The first approach is to take the bull by the horns and just solve the equations. That's not easy. It had better not be too easy, because the solution must exhibit properties (confinement, chiral-symmetry breaking) that are very different from what the equations seem naively to suggest, and it must describe a rich, complex phenomenology. Fortunately, powerful modern computers have made it possible to calculate a few of the key predictions of QCD. Benchmark results are shown in Figure 3, where the calculated masses[5] of an impressive range of hadrons are compared with their measured values. The agreement is encouraging.

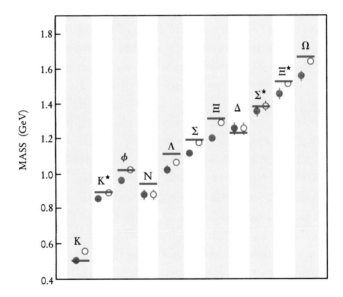

FIGURE 3. Mass spectrum of mesons and baryons, as predicted by QCD and intensive number-crunching.[5] Only two adjustable parameters went into this calculation: the coupling strength and the strange-quark mass (by way of the K or ϕ meson, both of which incorporate strange quarks). The up- and down-quark masses can be approximated by zero. Solid dots are for calculations tied to the measured K mass; open dots are tied to the ϕ mass. The agreement with the measured masses (red lines) is at the 10% level. Such calculations are improving as computer power and techniques get better.

Such calculations clearly demonstrate that confinement and chiral-symmetry breaking are consequences of solving the equations of QCD. The calculations show us no massless gluons, nor any fractionally charged particles, nor the enlarged multiplets that would indicate unbroken chiral symmetry. Just the observed particles, with the right properties — neither more nor less.

While these and other massive numerical calculations give impressive and useful results, they are not the end of all desire. There are many physically interesting questions about QCD for which the known numerical techniques become impractical. Also, it is not entirely satisfying to have our computers acting as oracles, delivering answers without explanations.

▷ The second approach is to give up on solving QCD itself, and to focus instead on models that are simpler to deal with, but still bear some significant resemblance to the real thing. Theorists have studied, for example, QCD-like models in fewer dimensions, or models incorporating supersymmetry or different gauge groups, and several other simplified variants. Many edifying insights have been obtained in this way. By their nature, however, such modelistic insights are not suited to hard-nosed confrontation with physical reality.

▷ The third approach, which is the subject of the rest of this article, is to consider physical circumstances in which the equations somehow become simpler.

Extreme Virtuality

The most fundamental simplification of QCD is illustrated in Figure 4. There we see, on the left, the jet-like appearance of collision events in which strongly interacting particles (hadrons) are produced in electron-positron annihilations at high energy. One finds many particles in the final state, but most of them are clearly organized into a few collimated "jets" of particles that share a common direction.[6] In about 90% of these hardron-producing events, there are just two jets, emerging in opposite directions. Occasionally — in about 9% of the hadronic final states — one sees three jets.

Compare those multiparticle hadronic events to collisions in which leptons, say muons, are produced. In that case, about 99% of the time one observes simply a muon and an antimuon, emerging in opposite directions. But occasionally — in about 1% of the muonic final states — a photon is emitted as well.

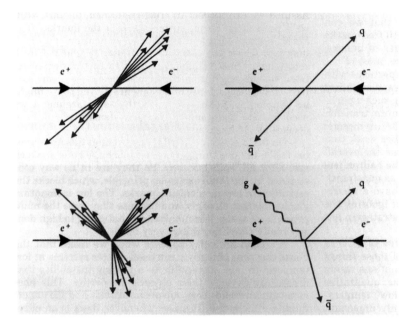

FIGURE 4. In high-energy e^+e^- annihilations into strongly interacting particles, the many-particle final state is observed (left) to consist of two or occasionally three (or, very rarely, four or more) "jets" of particles leaving the collision in roughly the same directions. QCD predicts their production rates and angular and energy distribution by assuming that (right) a single primary quark or gluon underlies each jet. The jets are explained by asymptotic freedom, which tells us that the probability is small for emitting a quark or gluon that drastically alters the flow of energy and momentum.

If history had happened in a different order, the observation of jet-like hadronic final states would surely have led physicists to propose that they manifest underlying phenomena like those displayed on the right-hand side of Figure 4. Their resemblance to leptonic scattering and QED would be too striking to ignore.

Eventually, by studying the details of how energy was apportioned among the jets, and the relative probabilities of different angles between them, physicists would have deduced directly from experiment the existence of light spin-1/2 and massless spin-1 objects, and how these covert objects couple to one another. By studying the rare 4-jet events, they could even have learned about the coupling of the spin-1 particles to each other. So all the basic couplings we know in QCD might have been inferred, more or less directly, from experiment. But there would still be one big puzzle: Why are there jets, rather than simply particles?

The answer is profound, and rich in consequences. It is that the strength with which gluons couple depends radically on their energy and momentum. "Hard" gluons, which carry a lot of energy and momentum, couple weakly; whereas the less energetic "soft" gluons couple strongly. Thus, only rarely will a fast-moving colored quark or gluon emit "radiation" (a gluon) that significantly redirects the flow of energy and momentum. That explains the collimated flows one sees in jets. On the other hand, there can be a great deal of soft radiation, which explains the abundant particle content of the jets. So, in a rigorous and very tangible sense, we really do get to see the quarks and gluons — but as flows of energy, not individual particles.

We refer to the phenomenon of weak coupling for hard gluons but strong coupling for soft gluons as "asymptotic freedom."[7] Despite its whimsical name, the concept is embodied in precise equations. It lets us make quantitative predictions of how often hard-radiation events occur in many strong-interaction processes of many different kinds, and at different energies. As we see in Figure 5, there is by now a wealth of direct evidence for the central prediction that the coupling strength of gluons *decreases* with increasing energy and momentum.[8] Note that several of the individual points in the figure summarize hundreds of independent measurements, all of which must be — and are — fitted with only one adjustable parameter (the quark-gluon coupling measured at the Z-boson mass).

The actual history was different. The need for asymptotic freedom in describing the strong interaction was deduced from much more indirect clues, and QCD was originally proposed as the theory of the strong interaction because it is essentially the unique quantum field theory having the property of asymptotic freedom.[9] From these ideas, the existence of jets, and their main properties, were predicted before their experimental discovery.[6]

High Temperature QCD

The behavior of QCD at high temperature is of obvious interest. It provides the answer to a childlike question: What happens if you keep making things hotter and hotter? It also describes the behavior of matter at crucial stages just after the Big Bang. And it is a subject that can be investigated experimentally with high-energy collisions between heavy nuclei. Brookhaven National Laboratory's Relativistic Heavy Ion Collider,

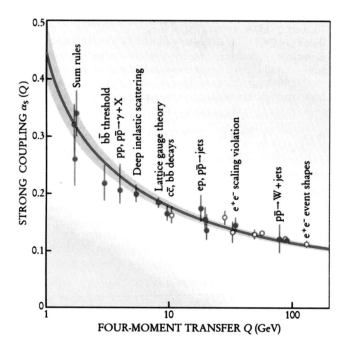

FIGURE 5. The running coupling "constant" α_s for the strong interaction is predicted by QCD to decrease with increasing energy and momentum. That's asymptotic freedom. The red curve is the predicted dependence of α_s on Q, the magnitude of the four-momentum transfer at a QCD vertex. An empirical input is the measured coupling of a quark pair to a virtual gluon at the Z boson mass; the orange swath reflects its uncertainty. The theory yields excellent agreement with a great variety of experiments,[14] shown by the data points and labels. The open points are results based on the general shapes of many-particle final states in momentum space.

where experiments are just getting under way, will be especially devoted to this kind of physics. (See Figure 6.)

To avoid confusion, I should state that, when I discuss high-temperature QCD in this article, I'm assuming that the net baryon density (quarks *minus* antiquarks) is very small. Conversely, when I discuss high-density QCD, I mean a high net density of quarks at low temperature, but well above the ordinary quark density of cold nuclear matter. Temperature and net baryon density are generally taken as the two independent variables of the phase diagram for hadronic matter.

Asymptotic freedom implies that QCD physics gets *simpler* at very high temperature. That would seem unlikely if you tried to build up the high-temperature phase by accounting for the production and

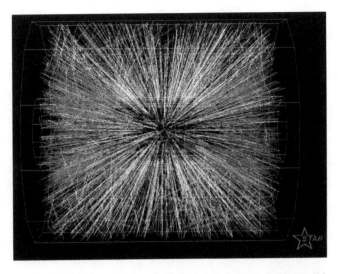

FIGURE 6. One of the first collisions between high-energy gold nuclei at Brookhaven's new Relativistic Heavy Ion Collider was recorded by the Star detector facility in June. In this reconstructed side view of the detector, the two 28 GeV-per-nucleon beams of gold nuclei enter from left and right, and collide at the center. About a thousand charged tracks were recorded emanating from this one collision. Significantly higher multiplicities are expected as RHIC works up to its design beam energy of 100 GeV-per-nucleon.

interaction of all the different mesons and baryon resonances that are energetically accessible at high temperature. Hoping to bypass this forbidding mess, we invoke a procedure that is often useful in theoretical physics. I call it the Jesuit Stratagem, inspired by what I'm told is a credal tenet of the Order: "It is more blessed to ask forgiveness than permission." The stratagem tells you to make clear-cut simplifying assumptions, work out their consequences, and check to see that you don't run into contradictions.

In this spirit we tentatively assume that we can describe high-temperature QCD starting with free quarks and gluons. In an ideal (noninteracting) gas of quarks, antiquarks, and gluons at high temperature, most of the energy and pressure will be contributed by particles with large energy and momentum. How do interactions affect these particles? Well, significantly deflecting such a particle requires an interaction with large momentum transfer. But such interactions are rare because, as asymptotic freedom tells us, they are governed by rather weak coupling. So interactions do not really invalidate the overall picture. To put it another way, if we treat the hadron jets generated by quarks, antiquarks, or gluons as quasiparticles "dressed" in hadronic garb, then we have a nearly ideal gas of quasiparticles. So it seems that ignoring the interactions was a valid starting point. The Jesuit Stratagem has succeeded.

Remarkably, the thermodynamic behavior of QCD as a function of temperature is another one of those things that can be calculated directly from the equations, using powerful computers.[10] Figure 7 shows the qualitative expectations dramatically vindicated. At "low" temperatures ($\lesssim 150$ MeV or 1.5×10^{12} K), the only important particles are the spinless pi mesons: π^+, π^-, and π^0. They represent 3 degrees of freedom. But from a quark–gluon description we come to expect many more degrees of freedom, because there are 3 flavors of light spin-1/2 quarks, each of which comes in 3 colors. If you then include 2 spin orientations, antiquarks, and 8 gluons, each with 2 polarization states, you end up with 52 degrees of freedom in place of the 3 for pions. So we predict a vast increase in

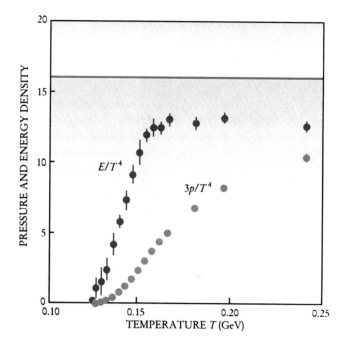

FIGURE 7. Steep rise of pressure p (blue points) and energy density E (red points) with increasing temperature T above 130 MeV indicates the opening of many quark–gluon degrees of freedom in this lattice-gauge QCD calculation of the thermodynamics of very hot nuclear matter.[15] For simplicity, the calculation assumes only two quark flavors. Normalized to T^4, both variables become dimensionless (in natural units) and asymptotically approach the green line.

the energy density, at a given temperature, as you go from a hadron gas to a quark–gluon plasma. And that is what the calculations displayed in Figure 7 show.

What about real experiments? Unfortunately our only access to the quark–gluon plasma is through the production of tiny, short-lived nuclear fireballs, of which we detect only the debris. Interpreting the data requires complicated modeling. In the quest for evidence of the quark–gluon plasma, there are two levels to which one might aspire. At the first level, one might hope to observe phenomena that are very difficult to interpret from a hadronic perspective but have a simple qualitative explanation based on quarks and gluons. Several such effects have been observed by the CERN heavy-ion program in recent years.[11] But there is a second, more rigorous level that remains a challenge for the future. Using fundamental aspects of QCD theory, similar to those I discussed in connection with jets, one can make quantitative predictions for the emission of various kinds of "hard" radiation from a quark–gluon plasma. We will not have done justice to the concept of a weakly interacting plasma of quarks and gluons until some of these predictions are confirmed by experiment.

High Density QCD

The behavior of QCD at large net baryon density (and low temperature) is also of obvious interest. It answers yet another childlike question: What will happen when you keep squeezing things harder and harder? It is also interesting for the description of neutron star interiors. But perhaps the most interesting and surprising thing about QCD at high density is that, by thinking about it, one discovers a fruitful new perspective on the traditional problems of confinement and chiral-symmetry breaking.

Why might we hope that QCD simplifies in the limit of large density? Again we use the Jesuit Stratagem. Assume we can neglect interactions. Then, to start with, we'll have large Fermi surfaces for all the quarks. (The Fermi surface bounds the smallest momentum-space volume into which you can pack all those fermions, even at zero temperature.) This means that the active degrees of freedom — the excitations of quarks near the Fermi surface — have large energy and momentum. And so we might be tempted to make essentially the same argument we used for the high-temperature, low-density regime and declare victory once again.

On further reflection, however, we find this argument too facile. For one thing, it doesn't touch the gluons, which are, after all, spin-1 *bosons*.

So they are in no way constrained by the Pauli exclusion principle, which blocks the excitation of low-momentum quarks. The low-momentum gluons interact strongly, and because they were the main problem all along, it is not obvious that going to high density really simplifies things very much.

A second difficulty appears when we recall that the Fermi surfaces of many condensed-matter systems at low temperature are susceptible to a pairing instability that drastically changes their physical properties. This phenomenon underlies both metallic superconductivity and the superfluidity of helium-3. It arises whenever there is an effective attraction between particles on opposite sides of the Fermi surface. As elucidated by John Bardeen, Leon Cooper, and Robert Schrieffer, even an arbitrarily weak attraction can, in principle, cause a drastic restructuring of the ground state.

A nominally small perturbation can have such a big effect because we're in the realm of degenerate perturbation theory. Low-energy excitation of pairs of particles on opposite sides of the Fermi surface, with total momentum zero, can be scattered into one another. By orchestrating a coherent mixture of such excitations, all pulling in the same direction, the system gains an energy advantage.

In condensed-matter physics, the occurrence of superconductivity is a difficult and subtle affair. That's because the fundamental interaction between electrons is Coulomb repulsion. In the classic metallic superconductors, an effective attraction arises from subtle retardation effects involving phonons. In the cuprate superconductors, the cause is still obscure.

In QCD, by contrast, the occurrence of what we might call "color superconductivity" is relatively straightforward.[12] That's because the fundamental interaction between two quarks, unlike that between two electrons, is already attractive! One can see this by a group-theoretical argument: Quarks form triplet representations of color SU(3). A pair of quarks, in the antisymmetric color state, form an antitriplet. So when two quarks are brought together, the effective color charge is reduced by a factor of two compared to when they were separated. The color flux emerging from them is reduced, lessening the energy in the color field. That implies an attractive force. So we should consider carefully what color superconductivity can do for us.

Two of the central phenomena of ordinary superconductivity are the Meissner effect and the energy gap. The Meissner effect is the inability of magnetic fields to penetrate far into the body of a superconductor. Supercurrents arise to cancel them out. Electric fields are, of course, also screened by the motion of charges. Thus electromagnetic fields in general become short-range. Effectively it appears as if the photon has acquired a mass. Indeed that is just what emerges from the equations. We can therefore anticipate that in a color superconductor, gluons will acquire mass. That's very good news, because it removes our problem with the low energy–momentum gluons.

The energy gap means that it costs a finite amount of energy to excite electrons from their superconducting ground state. That's quite unlike what we had for the free Fermi surface. So the original pairing instability, having run its course, is no longer present.

Now with both the sensitivity to small perturbations (pairing instability) and the bad actors (soft gluons) under control, the remaining effects of interactions really are small and under good theoretical control. Once again, the Jesuit Stratagem has served us well.

Color-Flavor Locking

The simplest and most elegant form of color superconductivity is predicted for a slightly idealized version of real-world QCD in which we imagine there are exactly three flavors of massless quarks: u, d, and s. The strange quark is in fact much lighter than c, b, or t. And neglecting small quark masses is an excellent approximation at extremely high density.

Here we discover the remarkable phenomenon of color–flavor locking.[13] Ordinarily the perfect symmetry among different quark colors is quite distinct and separate from the imperfect symmetry among different quark flavors. But in the imagined color–flavor locked state they become correlated. Both color symmetry and flavor symmetry, as separate entities, are spontaneously broken, and only a certain mixture of them survives unscathed.

Color–flavor locking in high-density QCD drastically affects the properties of quarks and gluons. As we have already seen, the gluons become massive. Due to the commingling of color and flavor, the electric charges of particles, which originally depended only on their flavor, are modified. Specifically, some of the gluons become electrically charged, and the quark charges are shifted. The electric charges of these particles all become integral multiples of the electron's charge!

Thus the most striking features of confinement — the absence of long-range color forces, and integer electric charge for all physical excitations — emerge as simple, rigorous consequences of color superconductivity. Also, because both left- and right-handed flavor symmetries are locked to color, they are also effectively locked to each other. Thus chiral symmetry, which required independent transformations among the left- and the right-handed components of the quarks, is spontaneously broken.

Altogether, there is a striking resemblance between the *calculated* properties of the low-energy excitations in the high-density limit of QCD and the *expected* properties — based on phenomenological experience and models — of hadronic matter at moderate density. This suggests the conjecture that there is no phase transition separating them.

Unfortunately both numerical and direct experimental tests of this conjecture seem out of reach at the moment. So we cannot be sure that the mechanisms of confinement and chiral-symmetry breaking in the calculable, high-density limit are the same as those that operate at moderate or low density. Still, I think it astonishing that these properties, which have long been regarded as mysterious and intractable, have been simply — yet rigorously — demonstrated to occur in a physically interesting limit of QCD.

I have tried to convince you of two things: first, that the fundamentals of QCD are simple and elegant, and second, that these fundamentals come into their own, and directly describe the physical behavior of matter, under various extreme conditions.

References

1. For a wide-ranging survey, see, H. Kastrup and P. Zerwas, eds., *QCD 20 Years Later*, World Scientific, Singapore (1993).
2. R. Feynman, *QED: The Strange Theory of Light and Matter*, Princeton University Press, Princeton (1985).
3. R. Feynman, *The Feynman Lectures on Physics*, Addison-Wesley, New York (1964), vol. II, pp. 25-10.
4. F. Wilczek, *Nature* **397**, 303 (1999).
5. S. Aoki *et al.*, http://xxx.lanl.gov/abs/hep-lat/9904012.
6. G. Hanson *et al.*, *Phys. Rev. Lett.* **35**, 1609 (1975).
7. D. Gross and F. Wilczek, *Phys. Rev. Lett.* **30**, 1343 (1973). H. Politzer, *Phys. Rev. Lett.* **30**, 1346 (1973).

8. For a recent review, see I. Hinchcliffe and A. Manohar, http://xxx.lanl.gov/abs/hep-ph/0004186.
9. D. Gross and F. Wilczek, *Phys. Rev.* **D9**, 980 (1974).
10. For a recent review, see F. Karsch, http://xxx.lanl.gov/abs/hep-lat/9909006.
11. U. Heinz and M. Jacob, CERN preprint nucl-th/0002042 (2000).
12. For a recent review, including extensive references, see T. Schäfer, http://xxx.lanl.gov/abs/hep-ph/9909574.
13. M. Alford, K. Rajagopal and F. Wilczek, *Nucl. Phys.* **B537**, 443 (1999).
14. M. Schmelling, http://xxx.lanl.gov/abs/hep-ex/9701002.
15. S. Gottlieb *et al.*, *Phys. Rev.* **D47**, 3619 (1993).

10^{12} Degrees in the Shade

Strange things happen at high temperatures. You need not look far for examples. When a pot of cold water is heated on a stove, the temperature rises steadily until the boiling point is reached. Then, counterintuitively, the temperature gets stuck. To unstick the temperature and raise it beyond the boiling point requires strenuous effort, in the form of a large amount of heat that must be supplied to the pot. Or consider one of the little bar magnets you might find in a toy store. Such a magnet is called permanent because under ordinary conditions it keeps its magnetism indefinitely. But at high temperatures something surprising takes place. Extensive studies by the "other" Curie — Pierre, Marie's husband — showed that when magnetic materials are heated past a certain critical temperature, the so-called Curie temperature, they lose their magnetism.

The strange behaviors of the pot of water and the bar magnet are rooted in the same underlying principle: during a transition between a low temperature and a high one, matter can undergo a basic reorganization. The temperature of the water gets stuck because, on the microscopic scale, to change liquid water into steam you must supply energy to rip apart clusters of water molecules. The energy needed to rearrange the molecules does not go into producing faster motion, and so it does not raise the temperature of the water. The temperature thus remains constant until the boiling is complete. A bar magnet, on a submicroscopic scale, is really a collection of iron atoms, each acting as a little elementary magnet. At low enough temperatures the magnets all are aligned with one another. Above the Curie temperature, however, the thermal energy available to the atomic magnets is enough to let them rotate freely, ignoring the efforts of their neighbors to bring them in line.

That basic principle of the reorganization of matter is richly intertwined with the history of physics, and it continues to enthrall physicists today. Indeed, it may even apply on the grandest scale imaginable: the universe itself. One of the most exciting frontiers of current research arises out of the possibility that the universe may have undergone a kind of reorganization shortly after the Big Bang exploded the universe into being. To test the theory, physicists at the Brookhaven National Laboratory in New York and at CERN, the European center for particle physics in Geneva, are preparing challenging experiments that will seek to mimic the high temperatures and extreme conditions of the universe in the first fraction of a second following the Big Bang. The temperatures in those experiments will reach a few times 10^{12} degrees Kelvin — a few thousand billion degrees Kelvin (10^{12} is a one followed by twelve zeros). To put such a temperature in perspective, the temperature at the surface of the sun is a few thousand degrees Kelvin, a billion times less.

According to modern cosmological ideas, the first few moments after the Big Bang were crucial for determining the structure of the present universe. To date, however, there have been few chances to test theoretical ideas about the properties of matter during that gestational period. Thus the new experiments afford physicists and cosmologists the prospect of checking their own understanding and, perhaps, of finding some surprises. In fact, surprises have already emerged from theoretical investigations. One is the strong possibility that new stable forms of matter (so-called strange matter) will arise; another is that the extreme conditions will give rise even to new forms of vacuum — a physical state that is commonly, but inaccurately, thought of as empty space. As compelling as those prospects are, however, the real attraction of the high-temperature frontier may be the simple challenge of exploring it "because it's there." Perhaps you will agree there is a certain intrinsic grandeur to the question of what takes place as things get hotter and hotter.

Before plunging into wild and poorly charted territory, it seems prudent to consult previous explorers. Twice before in this century the opening of new frontiers of high temperature has profoundly affected both physics and cosmology.

The first great divide affecting the basic properties of matter as the temperature rises occurs at about 10,000 degrees Kelvin. Below that, roughly speaking, is the usual domain of chemistry; above it is the realm of plasma physics. Below 10,000 degrees most atoms and many kinds of molecule retain their basic integrity. The thermal energies available to particles at those temperatures are not enough to tear electrons away from neutral atoms or to break strong chemical bonds.

But although the change is not abrupt, as temperatures approach and exceed 10,000 degrees, the typical energies available to electrons are enough to overcome the forces that normally bind them into orbits around positively charged atomic nuclei. Once the thermal velocity of electrons exceeds their "escape velocity" from atoms, the electrons break loose from their orbits and roam free. The presence of freely moving electrons is the defining characteristic of a plasma.

Plasmas at roughly 10,000 degrees are actually a familiar sight: both the interior of a fluorescent light and the surface of the sun are good examples. And it is no accident that both of those plasmas are excellent emitters of light. The Scottish physicist James Clerk Maxwell and other nineteenth-century physicists showed that light is a disturbance in the electromagnetic field. Disturbances in the field are generally caused by the motion of charged particles. When charged particles of opposite sign are tightly bound into atoms that are electrically neutral overall, their effects tend to cancel one another. Hence the coupling of ordinary atoms to the electromagnetic field is relatively weak. But when, on the contrary, the temperatures are so high that atoms are torn apart into charged nuclei and free electrons, their ability to radiate light is much enhanced. Thus plasma glows.

Max Planck, Albert Einstein and other physicists early in this century made it clear that light can be regarded as being made up of particles, now known as photons. From that perspective, to say that hot plasma glows is to say that the plasma contains, besides freely moving electrons and atomic nuclei stripped bare, a large number of photons. Thus supplying heat to a gas of atoms not only causes them to break apart into more elementary pieces: it also induces them to create something new: a gas of photons. The photon gas is every bit as real and tangible as an ordinary "material" gas. Indeed, as the temperature is further raised, the importance of the photon gas, by almost any measure, increases relative to the material gas that created it. The number of atomic nuclei and electrons is fixed, but the number of photons increases as the cube of the temperature.

What especially intrigued Planck about the photon gas (which, of course, is not what he called it) was its profoundly universal character.

The properties of the gas, which Planck and his contemporaries called blackbody radiation, are almost independent of the matter used to create it. Just as light is created in the collisions of charged particles, so it can also be absorbed. Eventually a dynamic equilibrium is established, characterized by a certain distribution of intensities of light of different colors, or, equivalently, densities of photons with different energies.

When he began his quest, Planck could show that the blackbody distribution was independent of such details as the density of atoms. (More atoms would mean more emission, but also more absorption, leading to the same net result.) He realized that one could even abstract the matter away altogether — or employ, as the old books quaintly put it, "a tiny speck of dust" as a catalyst — to arrive at the idea of a pure radiation gas. In its properties Planck, who was a deeply religious man, explicitly sought an absolute, an entity perfectly reflecting the fundamental laws of Nature, uncontaminated by the accidents of matter. But his many attempts to determine the distribution theoretically, based on the physics known in his day, failed.

The importance of the blackbody distribution of energy was not lost on experimenters, and Planck's colleagues, the German physicists Ferdinand Kurlbaum and Heinrich Rubens, obtained exquisitely accurate experimental data on its energy spectrum. With the new data in hand, Planck sought to fit them to various mathematical formulas; in 1901 he stumbled upon what is now known to be the correct formula describing the blackbody radiation. Working backward, he supplied a suggestive though unsound derivation of his empirical formula, in the course of which he introduced a new physical constant, h — now known as Planck's constant. That work triggered a cascade of revolutionary developments in physics, including Einstein's invention of the photon, and culminated in modern quantum mechanics, the theory of atomic physics, in which h plays a central role.

Planck could hardly have imagined that his quest would lead not only to a new mechanics but also to a new cosmology, yet that is precisely what happened. In the 1930s the American astronomer Edwin Powell Hubble discovered that distant galaxies are receding from the earth with velocities proportional to their distance. If you make a straightforward extrapolation of that motion backward in time, you find that the greater distance of the farthest galaxies is exactly compensated for by their greater velocities; hence, all of them appear to originate in the same place at the same time. On the basis of that extrapolation, the Belgian astrophysicist and Jesuit seminarian Georges-Henri Lemaître and others began to speculate that the universe originated in a much hotter, denser state — and that the expansion evident in the present Hubble motions is just what it looks like, the aftermath of an initial universal explosion. Lemaître called his model universe the primeval atom, but the term coined somewhat later by the English astrophysicist Fred Hoyle to mock that kind of theory has become standard: the Big Bang.

As Hoyle's term of derision implies, most physicists in the 1940s and 1950s regarded such ideas as speculations at or perhaps even beyond the border between science and mythology. Then, in 1964, Arno A. Penzias and Robert W. Wilson, working at the Bell Telephone Laboratories, somewhat accidentally observed a startling relic of those early times. In a large microwave receiver, they detected unexpected noise that seemed to come from all directions in space. A remarkable interpretation of the noise was immediately supplied by Robert H. Dicke and Philip James Edwin Peebles, both of Princeton University, who, along with their Princeton colleagues Peter G. Roll and David T. Wilkinson, had been planning a search for it. Dicke and Peebles proposed that Penzias and Wilson had discovered Planck's absolute — a perfect blackbody radiation that fills the universe.

In the light of the previous discussion one can appreciate the logic of the interpretation. If the material content of the universe was much hotter and denser early on, it must have existed in a plasma state at high temperatures, with an associated photon gas. As the material expanded and cooled to below 10,000 degrees Kelvin, the electrons and the nuclei must have combined into neutral atoms, which interact feebly with photons. The effect was that the universe rather suddenly became transparent. The photons did not suddenly disappear, however. The photon gas, though it too became cooler as the universe expanded, remained in existence. And it can be observed with suitable instruments, such as Penzias and Wilson's original microwave antenna. The existence of the blackbody radiation is one of the major supporting pillars of Big Bang cosmology.

Physicists learned an important lesson from that episode: to understand crucial events in the early history of the universe, they would do well to explore the high-temperature frontier.

Indeed, they soon realized that the passage of the universe through an earlier frontier, at still higher temperatures, had left a permanent imprint on its basic structure. That transition, the next big qualitative change in matter, occurs at about 10^{10} degrees Kelvin. It might be described as the boundary between nuclear chemistry and nuclear plasma physics. For just as atoms dissociate into their constituent electrons and nuclei at 10,000 degrees Kelvin, nuclei in turn dissociate into their constituent protons and neutrons at temperatures roughly a million times hotter. At temperatures higher than that the universe harbored none of the familiar nuclei observed today (other than lone protons, which are the nuclei of hydrogen). Those nuclei had to form later, as the hot stuff expanded and cooled.

Because physicists had gone to great lengths to understand the basic processes that occur at the relevant high energies, it was straightforward, at least in principle, to calculate how much of each kind of nucleus gets produced as the universe cools to below 10^{10} degrees Kelvin and the protons and neutrons begin to stick together. The nuclear abundances, calculated under the assumption that the material content of the universe at one time reached temperatures above 10^{10} degrees Kelvin, agree well with what is observed. The success of that calculation is the second pillar of Big Bang cosmology.

Those past successes give added impetus to the push to study ever higher temperatures. Can one make sensible predictions about what might take place at higher temperatures? Are there additional frontiers, beyond which qualitatively new things take place?

In fact, one can form reasonable expectations about what might take place at what I call the third frontier, at temperatures far beyond any that have been observed directly. Those theoretical expectations build on the results of much previous work devoted to pushing back the high-energy frontier. In particle accelerators at such laboratories as Brookhaven and CERN, the traditional goal has been to study the results of energetic collisions between simple projectiles — protons colliding with protons, for instance. To study conditions at high temperatures, however, one really must study how collections of particles behave, because the concept of temperature applies only to ensembles and not to individual particles. The complexity of that study dictates that the elementary processes up to a given energy scale be understood first; once that is done, the behavior of ensembles of large numbers of particles with that energy, repeatedly interacting with one another, can be explored.

The two laboratories will pursue the same basic strategy to push back the high-temperature frontier. Two beams made up of large atomic nuclei, each containing roughly 200 protons and neutrons, will be accelerated to velocities close to that of light in opposite directions around a ring. The paths of the particles will be controlled by cunningly designed magnetic fields, and at a few spots the beams will be focused and made to cross. When they cross, a few of the nuclei will collide head-on. Each collision will create a minute fireball, and inside the fireballs nuclear matter will be heated, for an unimaginably fleeting instant lasting about 10^{-22} second, to temperatures of a few times 10^{12} degrees Kelvin.

The new work will build on the old. One of the great triumphs of physics in the past fifty years has been the discovery and verification of theories that supply a detailed and rather complete picture of interactions up to energies corresponding to temperatures of 10^{15} degrees Kelvin. The most significant interactions in fireballs at 10^{12} degrees Kelvin are described by a theory known as quantum chromodynamics, or QCD. Physicists' understanding of QCD leads them to suspect that the coming high-temperature frontier will mark the passage from ordinary nuclear physics, with recognizable protons and neutrons, to something quite different — a state of matter in which the weird inner workings of QCD will be exposed.

QCD had its beginnings in the 1930s with the study of the interactions that hold protons and neutrons together in atomic nuclei and that govern the transformations of nuclei that take place inside stars and in nuclear reactors. Those forces are the most powerful ones known in Nature, and that explains their name: the strong forces.

As often happens in science, deeper investigation has altered the framework of discussion. Experiments done since 1950 have shown that protons and neutrons are not really elementary but are made up of simpler and more basic entities known as quarks and gluons. It appears from experiment that the forces among quarks and gluons, in contrast to the forces among protons and neutrons, are governed by simple, mathematically beautiful laws. QCD is precisely that body of laws.

There are several flavors, or species, of quark, which differ in mass and charge. The lightest quarks are the "up" quark, u, the "down" quark, d, and the "strange" quark, s; there are three heavier kinds as well. Only the u and d quarks are important constituents of ordinary matter. Each quark also has an antiquark: a quark that has the same mass but opposite electrical properties.

Each species of quark and antiquark in turn comes in three varieties. American physicists patriotically have labeled them red, white and blue, but they have nothing whatever to do with ordinary color. What the three varieties really resemble, physically and mathematically, are three kinds of charge, analogous to electric charge. And just as photons respond to ordinary electric charges, the color gluons of QCD respond to color charges.

When analyzed at the level of quarks and gluons, the strong force is fundamentally simple — only a little more complicated than the corresponding electromagnetic force. Ultimately those simple forces affecting quarks and gluons hold atomic nuclei together and give rise to all the other manifestations of the strong force.

A key theoretical property of QCD is known as asymptotic freedom: the interactions between quarks and gluons become weaker in a precisely calculable way at high energies or temperatures. Conversely, at low temperatures or energies, that is, under ordinary terrestrial conditions, the strong interaction is so strong that objects carrying unbalanced color charges cannot exist in isolation. The energy associated with the unbalanced color field is always sufficient to produce a neutralizing particle of the opposite charge. In particular, at low temperatures neither quarks nor gluons can exist in isolation. This peculiarity of QCD — that the basic entities of the theory cannot be isolated — is called confinement. It is, as you might imagine, one main reason the theory took so long to find.

With those basic ideas in mind, I can return to the discussion of the transformations of matter at high temperature. At 10^{10} degrees Kelvin the atomic nuclei have dissociated into individual protons and neutrons. But according to QCD, each proton is made of two u quarks and one d quark, one of each of the three colors; similarly, each neutron is made of one u quark and two d quarks. As the temperature approaches 10^{12} degrees Kelvin, the third frontier of high-temperature physics, another group of particles made up of quarks is produced in great numbers, just as photons are produced in great numbers in a plasma: the pi mesons, or pions.

The proliferation of pions is, however, merely the prelude to a change yet more profound. Asymptotic freedom predicts that the interaction among quarks and gluons gets weaker at high energy, and the detailed theory enables physicists to make a good estimate of the temperature at which thermal motion is sufficient to rip quarks and antiquarks apart. At the same time it becomes possible for single gluons to propagate freely, and they too are produced in great abundance. Thus, at temperatures above approximately 10^{12} degrees Kelvin, matter should take the form of a radically new kind of plasma, a quark–gluon plasma. In such a state of matter the basic entities of QCD, hitherto confined, roam free.

A plasma of free quarks, antiquarks and gluons differs in many ways from the gas of protons, neutrons and (mainly) pions from which it arises. One of the simplest and most basic differences is sheer density. When it is appropriate to describe things in terms of pions, there are basically only three kinds of particle to worry about, namely, the positively charged, the negatively charged and the electrically neutral pions. When the temperature rises just a little and it becomes appropriate to describe things in terms of free quarks and gluons, there are u and d quarks and their antiquarks, each of which comes in three colors, plus gluons that come in eight colors. And so suddenly there is a large proliferation of particles.

In equilibrium the distribution of energies for each kind of particle is roughly the same as the blackbody distribution Planck discovered for photons. But since those energies must now be distributed over so many more particles, it will suddenly cost a lot of energy to produce the small change in temperature over which the transition from the pion gas to the quark–gluon plasma takes place. In fact, the energy density must increase by more than tenfold. The situation in QCD is conceptually similar to what takes place when the pot of cold water is heated on the stove, but it takes place at a much higher temperature and energy scale. Instead of loose clusters of water molecules, it is pions that get ripped apart.

A second new effect in the quark–gluon plasma is that in the plasma it becomes much easier to create s (strange) quarks, together with their antiquarks. The mass of a free strange quark or its antiquark is about the same as that of a pion. Because of Einstein's equivalence between mass and energy, and because the energy density of the quark–qluon plasma is equivalent to the mass of a pion, strange quarks can simply materialize out of the energy background.

Finally, there is a more subtle difference between the low- and high-temperature regimes of QCD — probably the most profound and interesting difference of all. The concept is a bit much to swallow in one gulp, so I will introduce it by way of analogy.

Think once more about the little bar magnet from the toy store. From a microscopic perspective, magnetism arises because the total energy of all the atomic magnets is lower when the magnets are aligned in the same direction (whatever that may be) than it is when they point in different directions (whatever they may be). Because any material tends to assume its lowest-energy state, all the atomic magnets tend to point in the same direction; energetically, it does not matter what that common direction is. Small external influences, or random events that take place when the magnet is formed, determine the direction in which its magnetic field points.

The random nature of the direction of the magnetic field is easily seen when a magnet is heated past its Curie temperature and then cooled down again. Notice that when the magnet is heated and the individual atomic magnets point in many directions, the lump of metal is perfectly symmetrical in the sense that no direction is preferred. But when the magnet is cooled it generally settles down with its poles pointing in a direction different from the one they started with. Thus the original symmetry of the situation is broken by the emergence of a preferred direction; in physics one says that spontaneous symmetry breaking has taken place. Perhaps nothing could demonstrate more plainly the somewhat accidental character of the most obvious feature of a magnet: the direction of its magnetic field.

In QCD there is also a form of spontaneous symmetry breaking that is both conceptually and mathematically similar to the case of the magnet. I have already mentioned that the *u* and *d* quarks are the lightest quarks. For a moment let me adopt the convenient fiction that their masses are exactly zero. At zero mass, if the quarks and antiquarks did not interact with one another, it would cost no energy to fill space with them. The real situation is more dramatic: there is an attractive interaction among the quarks and antiquarks, and so one can achieve lower energy in a volume of space by filling it with quarks and antiquarks than one can by leaving it "empty."

Thus according to QCD, what people ordinarily think of as a vacuum — the state of lowest energy, or what remains when you take away everything that is takable — is actually a highly populated, complicated state. The "no-particle" state, devoid of the condensate of quarks and antiquarks, has a much higher energy than the true vacuum. If the no-particle state were produced it would immediately decay into a true vacuum containing a nonzero density of quarks and antiquarks, and large amounts of energy would be released in the process. That is not to say that the true vacuum has an indefinitely large number of quark–antiquark pairs: in fact, if too many pairs are crammed close together, they start to repel one another. Thus there is a definite best choice of the density of pairs that gives the lowest overall energy; that is the density of the vacuum.

There are four possible ways to pair off a quark and an antiquark: u, anti-u; u, anti-d; d, anti-u; and d, anti-d. The vacuum must be filled with a certain overall density of such pairs, but how much of each? The answer is, It does not matter. There is a perfect symmetry between the different kinds of quark and antiquark, in that the energy of the vacuum with one mix of relative densities is the same as the energy of any other vacuum. But here is the rub: to get the lowest possible energy you must choose some definite fixed ratio of densities, the same throughout all space.

Now you can begin to appreciate the sense in which the vacuum, according to quantum chromodynamics, is much like a magnet. Just as the magnet must "choose" one direction, spontaneously breaking the intrinsic symmetry of the atomic magnets, the pairs of quarks and antiquarks must "choose" some definite mix, a process that breaks the intrinsic symmetry among all the possible mixes of pairings. The process whereby the QCD vacuum acquires less symmetry than the intrinsic symmetry of the physical laws is called spontaneous chiral symmetry breaking.

One big difference between the QCD vacuum and a magnet, of course, is that one can get outside the magnet and see that it is a complicated object that spontaneously breaks a symmetry of the world — the equivalence of all directions. If there were creatures that lived inside a magnet and thought of it as their world, however, they would not perceive that anything was unusual or wrong about their surroundings. They would be accustomed to the idea that not all directions are equivalent: it would be the most obvious thing in their world. For such creatures to realize that the true symmetry of physical laws might be larger than what they perceive would require an enormous act of imagination. Perhaps then they would be inspired to try to create a state of higher symmetry than that of their vacuum, by exploring their own high-temperature frontier — beyond the Curie temperature.

That, in essence, is the prospect that lies before physicists at Brookhaven and CERN. The temperatures in prospect are beyond the "Curie temperature" of the QCD vacuum. There will be more than enough thermal energy to tear apart the quark–antiquark pairs abundantly present in the ordinary vacuum. A more perfect vacuum, exhibiting more nearly the full symmetry of physical law, will be established.

The theoretical ideas I have just discussed suggest that remarkable things will take place just beyond the third frontier of temperature.

Quarks, antiquarks and gluons will occur in great abundance and move freely. A generous sample of the ordinarily rare s quarks and antiquarks will be produced. The quark–antiquarks pair condensate that ordinarily fills the vacuum will vaporize, and the vacuum will regain its symmetry. But theoretical ideas, however beautiful and well motivated, become science only when one uses them in concrete ways to elucidate events in the natural world.

Is there any hope of gaining direct evidence that these extraordinary concepts describe reality? It is a challenge, because the extreme conditions of temperature needed to test them are generated only fleetingly and in a small volume. Fortunately, it does seem likely that experimenters will be able to find, through intelligent and diligent searching, signs of the remarkable goings-on in the early moments of the fireball. Each of the three qualitatively new features of the quark–gluon plasma mentioned above suggests some concrete goal for experimenters.

First, the great increase in the number of kinds of particle excited in the quark–gluon plasma should cause the temperature to stick and thus drastically affect the evolution of the fireball. That evolution, however, is quite complicated and hard to model precisely; only if the effect is large and qualitative is it likely to be discernible at all. Fortunately the evolution of the temperature can be traced from the energy of particles at the surface of the fireball as it cools. The higher the temperature of the fireball, the higher the energy of its emanations. If the temperature of the fireball sticks at a certain value, the emitted particles will bear the stamp of that temperature.

Second, the superheated plasma will give rise to many more s quarks and antiquarks than are produced at even slightly lower temperatures. The s quarks and antiquarks will eventually decay, but the processes responsible for their decay are relatively slow, and many of them will escape the fireball. They will escape inside particles such as K mesons and lambda particles, which experimenters are eminently equipped to observe.

An intriguing possibility, analyzed recently by Robert L. Jaffe of the Massachusetts Institute of Technology, is that new, long-lived quasi-atomic nuclei may exist, which contain many s quarks in the place of the u and d quarks in ordinary atomic nuclei. Jaffe calls the objects strangelets. He estimates that strangelets containing roughly equal numbers of each species — u, d and s quarks — are particularly favorable energetically. The quark–gluon plasma produced by a collision between two heavy ions, with its rich population of *s* quarks, provides for the first time an environment in which samples of that new form of matter, if it exists, might be produced.

Finally, the vaporization of the quark–antiquark condensate and the resultant loss of symmetry of the vacuum may have spectacular aftereffects. Recall that the poles of a magnet can change direction when the magnet is heated past its Curie temperature and then cooled. Similarly one might expect that the QCD vacuum, heated above the temperature at which its condensate of quark–antiquark pairs vaporizes, will generally recondense with a different mix of pairs.

At this point I must correct the convenient fiction I introduced earlier, namely, that the masses of the *u* and *d* quarks are zero. If those masses were actually zero, the analogy I made between the possible mixes of pairs in the QCD vacuum and the possible directions of the poles of a magnet would have been mathematically precise. But the effect of the real nonzero masses is also straightforward to visualize via the magnet analogy. It is as if there exists, in addition to the atomic magnets whose directions are arbitrary, a small additional magnetic field, external to the magnet itself. The additional external field defines a definite preferred direction, and to minimize its energy the bar magnet will eventually align with it. If the external field is weak, however, a heated lump of iron might cool and realign its poles at some angle to the external field, in the "wrong" direction. Such a magnet, suspended from a flexible string in a weak external field, could oscillate many times before settling down to point along the direction of the field.

The situation for the QCD vacuum may be closely analogous to the one for the magnet, according to recent work by Krishna Rajagopal of Harvard University and me. As the Russian physicist Aleksei A. Anselm of the Saint Petersburg Institute of Nuclear Research in Russia, among others, has pointed out, oscillations of the QCD vacuum as it seeks its true minimum value could lead to quite dramatic effects. Those oscillations, remember, are changes in the relative numbers of different kinds of quark–antiquark pairs in the condensate. As the wrong pairs in the misaligned condensate are converted into correct pairs, some pairs escape. They would be detected by experimenters as special, coherent collections of pions. The misaligned vacuum, in other words, creates a pion laser.

I have now described in some detail the high-temperature frontier on the immediate horizon, the frontier between the nuclear plasma and the quark–gluon plasma. It is likely exploring that frontier will be an important part of physics in the twenty-first century. But there are two even more remote frontiers, the first of which perhaps will be ripe for exploration in the twenty-second century — and the second. ... Who knows?

Those frontiers are associated with other episodes of spontaneous symmetry breaking, with the vaporization of other condensates. The fourth frontier, whose outlines can be perceived at least dimly, occurs at approximately 10^{15} degrees Kelvin. Above that temperature the force responsible for nuclear decay and for the energy of the sun, known as the weak interaction, is profoundly changed in character: the condensate responsible for making the weak interactions weak vaporizes. Under such conditions many processes that are rare or forbidden in the ordinary low-temperature vacuum become possible. Some recent theoretical work suggests that those processes, which would have unfolded in the moments immediately following the Big Bang, were responsible for generating the imbalance between matter and antimatter that characterizes the present universe. That is an imbalance without which human beings would not exist.

The fifth high-temperature frontier may occur around 10^{30} degrees Kelvin. Above that temperature theoretical calculations indicate that the condensate responsible for the differences among the strong, weak and electromagnetic interactions vaporizes. It is possible that events associated with the formation of that condensate in the moments immediately following the Big Bang induced a rapid inflation of the universe, as first proposed by the astrophysicist Alan H. Guth of MIT. If it were possible to reproduce such conditions — which would unfortunately require technology that is not in immediate prospect — it might be possible to grow new universes.

Such ideas go well beyond other frontiers: the frontier of knowledge and perhaps even the frontier of reasonable scientific speculation. But big questions about the origin of the universe will not be solved by refusing to consider them. Our successes so far make physicists optimistic that further progress on those big questions is possible.

Back to Basics at Ultrahigh Temperatures

Temperature is the epitome of an emergent property. It makes no sense to speak of the temperature of an elementary object, but in describing a system containing many elementary objects, one of the most important properties to mention is the system's temperature. Crude reductionism might suggest that elementary particle physicists would not find it useful to consider temperature. Instead, they could concentrate on more "fundamental" matters.

Things do not turn out that way. No one should turn up his or her nose at the childlike but intrinsically fascinating question: What happens to stuff as you make it hotter and hotter? Thinking about temperature in elementary particle physics has proved to be enormously fruitful. And, I argue here, it continues to provoke questions that are — by any reasonable definition — quite fundamental.

There are several reasons for this fruitfulness. Perhaps the most profound is related to comments made by Paul Dirac: "I understand what an equation means if I have a way of figuring out the characteristics of its solution without actually solving it" and Richard Feynman: "The physicist needs a facility in looking at problems from several points of view ... [A] physical understanding is a completely unmathematical, imprecise, and inexact thing, but absolutely necessary for a physicist," as well as to Einstein's marvelous use of thought experiments. These heroes all teach us that, in coming to grips with the meaning of abstract theories and matters far removed from everyday experience, it is important to try to visualize physical behavior the theories predict under different conditions. Our experience with problems in condensed matter and statistical mechanics suggests that some of the right questions to pose involve changes with temperature: Are there phase transitions? Is there latent heat? What are the useful degrees of freedom? When we can address such questions intelligently, we will be on our way to achieving understanding in the sense of Dirac and Feynman.

It's not just thought experiments that are at stake, however. Big Bang cosmology suggests that arbitrarily high temperatures were attained in the very early universe. So to understand the early universe, we must understand the behavior of matter at extremely high temperatures.

Finally, there is the prospect of doing, and analyzing, real experiments at the high-temperature frontier.

The most clear-cut challenge at the high-temperature frontier is to understand quantum chromodynamics, or QCD. We know precisely what the equations of the theory are. Furthermore, we have ample evidence that the theory is correct. Yet there is much we do not understand. This lack of understanding has hampered the application of QCD to nuclear physics, the physics of neutron stars, and many things of interest to high-energy physicists, such as the calculation of jet properties and of weak interaction matrix elements. These are all subjects that QCD in principle fully illuminates, but in practice each hides secrets, not only quantitative but also qualitative. (Specifically, in turn: Why do nuclear forces saturate? Is the interior of a neutron star made of quark matter? How do quark and gluon jets differ? And what explains the infamous $\Delta I = 1/2$ rule in non-leptonic weak decays? This is the striking old mystery, that one particular class of weak decay amplitudes is systematically 20–40 times larger than naive quark model estimates — usually fairly reliable — would suggest.)

The degrees of freedom basic to the theory — namely, gluons and (essentially) massless quarks — are obscured by two layers of structure. The first layer of structure is confinement. The force between non-neutral objects carrying equal and opposite color charges does not diminish as one separates them. Thus, it would take an infinite amount of energy to separate them completely. Instead, the stable, low-energy configurations of quarks and gluons are color neutral overall. The most important are baryons (three quarks), mesons (a quark plus an antiquark), and perhaps glueballs (two gluons). The quarks and gluons themselves, on the contrary, carry color charge, and they are never found in isolation.

The second layer of structure, chiral symmetry breaking, is more subtle but no less central. It comes about as follows. Massless spin-1/2 particles such as (slightly idealized) u and d quarks come in two forms, left- or right-handed, depending on the alignment of the spin with the direction of motion. As long as they are massless, they must travel at the speed of light, and there are no spontaneous transitions between the left- and the right-handed forms. Now it turns out that in QCD there is a very strong attraction between left-handed quarks and right-handed antiquarks (and between right-handed quarks and left-handed antiquarks). This attraction is so strong, in fact, that one can actually gain energy by creating large numbers of quarks and antiquarks and allowing them to pair off.

Thus, the vacuum of QCD — the state of lowest energy — is not the void, or "no-particle," state, but instead contains an ocean (usually called a condensate) of quark–antiquark pairs. That might sound, at first hearing, like a rather fantastic story, The idea that empty space should be full of complicated material is wilder than many crackpot theories, and more imaginative than most science fiction. Yet, the story of chiral symmetry breaking is firmly rooted in experimental facts, and has now been verified directly by numerical simulation. Because of confinement and chiral symmetry breaking, the usual world of QCD consists of complicated objects swimming in an ocean of other complicated objects.

A bit of thought affords the vision of a much simpler world, at high temperatures. At high enough temperatures, one has such a large population of particles (pions, specifically) that they start to overlap, and lose their individuality. By thinking in terms of quarks and gluons, it is not hard to anticipate what should happen. When there are many quarks and antiquarks of all colors, no very special behavior is required to achieve color neutrality, So the complications of confinement melt away. Similarly, the gain in energy from pairing off quarks and antiquarks, which drives chiral symmetry breaking, is overcome by the entropic advantage in letting these particles run free. So the complications of chiral symmetry breaking also melt away. At high temperatures, then, the asymptotically free quarks, antiquarks and gluons should be liberated, and stand revealed in their pristine glory.

Numerical simulations corroborate these expectations.[1] They indicate a drastic change in the structure of matter in a narrow range of temperatures around 1.5×10^{12} K, or, in more appropriate units, ~ 150 MeV. To anyone familiar with the spectrum of elementary particles, this will seem quite a low temperature. It is well below the mass of all the strongly interacting particles, except for pions. Thereby hangs a tale. At temperatures slightly below 150 MeV, we have a rather dilute gas of pions. Since pions have spin zero, π^+, π^- and π^0 simply correspond to 3 degrees of freedom. Yet at temperatures a bit above 150 MeV we have a quark–gluon plasma containing three flavors of quarks, each with three colors and two spins, and the corresponding antiquarks, plus eight kinds of gluons each with two possible helicities, for altogether 52 degrees of freedom. A drastic change indeed! It shows up directly as a vastly enhanced specific heat, for example.

What about experiments? In relativistic heavy ion collisions, one can realistically hope to produce a fireball that briefly, and in a limited region of space, achieves something close to equilibrium at temperatures approaching 300 MeV. The study of such collisions is in its infancy, though the baby is growing fast and should mature rapidly after the opening of the Relativistic Heavy Ion Collider at Brookhaven National Laboratory in the summer of 1999. Already, experimenters at CERN have found one intriguing indication that confinement comes undone in a narrow range of energies (or effective temperatures).[2] They have found a rather sudden drop J/ψ production, which they interpret as a collapse in the tendency for charm–anticharm mates, which are always produced close together, to stay paired.

For QCD at high temperature, then, we have a rich picture that can be viewed from several perspectives: theoretical, numerical, experimental. Some of its major outlines are already clear, while others, and many interesting details, are just coming into focus. Symmetry breaking is also a central ingredient of the electroweak part of the Standard Model, and of so-called grand unified theories that unify the strong and electroweak interactions. These symmetries are expected, like chiral symmetry in QCD, to be restored at high temperatures. But for electroweak and grand unified symmetry restoration, the supportive knowledge base is much narrower.

A particularly promising question is whether electroweak symmetry restoration is a continuous (second-order) or discontinuous (first-order) transition. To answer it, we will need to get a better grip on electroweak symmetry breaking itself. In particular, we will need to find out whether there are one or several Higgs particles, and to pin down their masses and interactions. Theories of the origin of the asymmetry between matter and antimatter and of cosmic inflation in the early universe are closely tied up with these questions.

Thus far I have discussed several problems that, while difficult and at present poorly understood, probably have some more or less conventional solution. Now I would like to mention two puzzles that might fall outside this pleasing class.

The first has to do with supersymmetry. This postulated symmetry transforms bosons into fermions. Of course, supersymmetry cannot be literally correct — there is definitely not a boson with the same mass and charge as the electron, for example. But the idea that supersymmetry is spontaneously broken — that it is a true symmetry of the basic equations, though not of the particular solution in which we live — has come to dominate much of our thinking about the future of particle physics. There is even some (indirect) experimental evidence for it. (See "Unification of Couplings.")

Now normally we expect that spontaneously broken symmetries are restored at high temperature. That is certainly the case for rotational symmetry when it is broken by ordinary magnetism, and for the chiral and electroweak symmetries discussed above. It is not so clear for supersymmetry, however. Indeed, the statistical mechanics of fermions and bosons is different, no matter what the temperature. So my puzzle is this: Is there any precise sense in which supersymmetry is restored at high temperature? Could its restoration be associated with a true phase transition?

My second puzzle has to do with the limits of Big Bang cosmology. The central hypothesis of conventional Big Bang cosmology is that the early universe is described by matter in thermal equilibrium, inhabiting an appropriate space-time obeying the equations of general relativity. It seems (at first sight) very reasonable to assume thermal equilibrium as the default option, for two reasons. First, because thermal equilibrium is famous for being, in a precise sense, the most probable state of matter. Second — and perhaps more to the point — because one finds, self-consistently, that for states anywhere near thermal equilibrium, particle interaction rates are very rapid compared to the expansion of the universe. Therefore, all such states rapidly approach accurate thermal equilibrium.

The puzzle is that by setting things up in this usual way, with matter as one ingredient and space-time or gravity as another separate one, we are in a backhanded way being quite selective about assuming equilibrium. In fact, it is very far from true that gravity is in thermal equilibrium, either in the standard cosmological models or in the real world. Approaching equilibrium for gravity would mean a universe full of black holes. You can see this most simply by considering what the universe would look like if it had somewhat more than critical density, as it approached the Big Crunch. The stars and galaxies would be squeezed together into gigantic black holes. That is what gravity would like to happen — it is what results as one approaches equilibrium, when gravity has had plenty of time to act — and it is very far indeed from being the time reverse of the standard Big Bang.

So the apparently innocuous starting point is actually rather peculiar. We assume perfect equilibrium for the non-gravitational forces, but essentially maximal disequilibrium for gravity. For anyone who believes that physics should strive for a unified theory of all the forces, having gravity fully integrated with the rest, this is quite annoying.

There is much more to be said about this problem, and indeed about most of the others raised above. But I think I have made my point, that thinking about temperature is a rewarding exercise, even for die-hard reductionists.

Note added, January 2006: The RHIC program has been very successful, and now we know much more about matter at high temperatures. The basic idea that quarks, antiquarks, and gluons get liberated at high temperatures has been confirmed, but there has been a surprise, too: the state they form appears to resemble a liquid more than a gas, at least at the temperatures RHIC achieves. Once again, reality is the ultimate teacher.

References

1. F. Karsch, *Nucl. Phys.* **A590**, 367c (1995).
2. M. C. Abreu *et al.*, *Phys. Lett.* **B410**, 327; *ibid.*, 337 (1997).

Breathless at the Heights

This series of three Reference Frame columns considers, and I think very probably answers, one of the great classic questions of physics:

> Why is gravity so feeble?

In "Scaling Mount Planck 1: A View from the Bottom" I argue that from a modern perspective the question might be better posed as:

> Why is the proton so light?

Indeed, in general relativity gravity becomes the theory of space-time's dynamical behavior. Its role is so basic that we probably should be expressing other things (like the proton's mass) in terms of gravitational parameters, rather than the other way around. In other words, we should just accept that gravity is what it is, and try to explain why it *appears* feeble to *us*. And since we're basically made of protons, and the force of gravity is proportional to mass, it comes down to why protons are so light.

That's a fortunate rephrasing, because QCD gives us a profound understanding of why protons weigh what they do, as explained in "The Origin of Mass." If we track it down, we find that the ridiculously small value of the proton's mass, as viewed by gravity, arises from a profound (Nobel Prize winning) dynamical phenomenon: asymptotic freedom. This amazing connection is spelled out in detail in "Scaling Mount Planck 1: A View from the Bottom."

For variety, now I'll sketch a slightly different way to look at it. The strong and gravitational interactions have equal strengths at very short distances (10^{-33} cm, the Planck length), where gravity is no longer feeble. As we go out to relatively larger distances, gravity gets feeble very fast, while the strong interaction gains power only slowly — logarithmically — according to the formulas of asymptotic freedom. When we finally reach distances like 10^{-13} cm, at which the strong interaction has acquired the power to bind quarks, and thereby make protons, gravity has become feeble indeed.

This wonderfully logical and coherent explanation of the feebleness of gravity depends, to be sure, on extrapolating the known laws of physics down to distances far smaller than any that have been probed experimentally. How will we tell if the extrapolation is sound?

We'll probably never be entirely certain, but two comments are in order:

Our bold extrapolation of quantum field theory to ultra-short distances has other consequences. It predicts that a stunningly beautiful unification of the strong, electromagnetic, and weak interactions will be possible, but only *if* a new class of particles, whose existence is suggested by other theoretical considerations (supersymmetry), really do exist. We'll find out whether they exist soon after the Large Hadron Collider (LHC) begins operations in late 2007. If they do, our extrapolation will be looking good! This line of thought is spelled out in "Scaling Mount Planck 2: Base Camp."

Our straightforward extrapolation of gravity might also be questioned. After heavy promotion of commotion about crises and revolutions in quantum gravity, there's a common misconception that everything about gravity and quantum mechanics is up in the air — maybe if we look a little more carefully we'll see that things actually fall sideways, or into new dimensions ... Well, maybe, but I think the crisis is vastly overblown, and the speculations largely gratuitous. Straightforward extrapolation is not only possible, but it works remarkably well numerically, in that the powers of all the interactions including gravity become equal at the Planck scale. That line of thought is spelled out in "Scaling Mount Planck 3: Is That All There Is?"

Some of the issues in "Scaling Mount Planck 2 and 3" also arose in "Unification of Couplings," where there was more room to discuss them.

Scaling Mount Planck 1: A View from the Bottom

Gravity dominates the large-scale structure of the universe, but only by default, so to speak. Matter arranges itself to cancel electromagnetism, and the strong and weak forces are intrinsically short range. At a more fundamental level, gravity is extravagantly feeble. Acting between protons, gravitational attraction is about 10^{36} times weaker than electrical repulsion. Where does this outlandish disparity come from? What does it mean?

These questions greatly disturbed Richard Feynman. His famous paper on quantizing general relativity,[1] in which he first described his discovery of the "ghost particles" that eventually played a crucial role in understanding modern gauge field theories, begins with a discussion of the smallness of gravitational effects on subatomic scales, after which he concludes,

> There's a certain irrationality to any work on [quantum] gravitation, so it's hard to explain why you do any of it. ... It is therefore clear that the problem we [are] working on is not the correct problem; the correct problem is: What determines the size of gravitation?

The same question drove Paul Dirac[2] to consider the radical idea that the fundamental "constants" of Nature are time dependent, so that the weakness of gravity could be related to the great age of the universe, through the following numerology: The observed expansion rate of the universe suggests that it began with a bang roughly 10^{17} seconds ago. On the other hand, the time it takes light to traverse the diameter of a proton is roughly 10^{-24} seconds. Squinting through rose-colored glasses, we can see that the ratio, 10^{-41}, is not so far from our mysterious 10^{-36}. (For what it's worth, the numbers agree better if we compare gravitational attraction versus electrical repulsion for electrons, instead of protons.) But the age of the universe, of course, changes with time. So if the numerological coincidence is to abide, something else — the relative strength of gravity, or the size of protons — will have to change in proportion. There are powerful experimental constraints on such effects, and Dirac's idea is not easy to reconcile with our standard modern theories of cosmology and fundamental interactions, which are tremendously successful.

In this column, I show that today it is natural to see the problem of gravity's feebleness in a new way — upside down and through a distorting lens compared to its superficial appearance. When viewed this way, the feebleness of gravity comes to seem much less enigmatic. In a sequel, I'll make a case that we're getting close to understanding it.

First let's quantify the problem. The mass of ordinary matter is dominated by protons (and neutrons), and the force of gravity is proportional to mass squared. Using Newton's constant, the proton mass, and fundamental constants, we can form the pure dimensionless number

$$N = G_N m_p^2 / \hbar c,$$

where G_N is Newton's constant, m_p is the proton mass, $\hbar$ is Planck's constant, and c is the speed of light. Substituting the measured values, we obtain

$$N \approx 3 \times 10^{-39}.$$

This is what we mean, quantitatively, when we say that gravity is extravagantly feeble.

We can interpret N directly in physical terms, too. Since the proton's geometrical size R is roughly the same as its Compton radius, $\hbar / m_p c$, the gravitational binding energy of a proton is roughly $G_N m_p^2 / R \approx N m_p c^2$. So N is the fractional contribution of gravitational binding energy to the proton's rest mass!

Soon after Max Planck introduced his constant $\hbar$ in the course of a phenomenological fit to the blackbody radiation spectrum, he pointed out the possibility[3] of building a system of units based on the three fundamental constants $\hbar$, c, and G_N. Indeed, from these three we can define a unit of mass $(\hbar c / G_N)^{1/2}$, a unit of length $(\hbar G_N / c^3)^{1/2}$, and a unit of time $(\hbar G_N / c^5)^{1/2}$ — what we now call the Planck mass, length, and time, respectively. Planck's proposal for a system of units based on fundamental physical constants was, when it was made, formally correct but rather thinly rooted in fundamental physics. Over the course of the 20th century, however, his proposal became compelling. Now there are profound reasons to regard c as the fundamental unit of velocity and $\hbar$ as the fundamental unit of action. In the special theory of relativity, there are symmetries relating space and time — and c serves as a conversion factor between the units in which space intervals and time intervals are measured. In quantum theory, the energy of a state is proportional to the frequency of its oscillations — and $\hbar$ is the conversion factor. Thus c and $\hbar$ appear directly as primary units of measurement in the basic laws of these two great theories. Finally, in general relativity theory, space-time curvature is proportional to the density of energy — and G_N (actually $1/G_N c^4$) is the conversion factor.

If we accept that G_N is a primary quantity, together with $\hbar$ and c, then the enigma of N's smallness looks quite different. We see that the question it poses is not, "Why is gravity so feeble?" but rather, "Why is the proton's mass so small?" For in natural (Planck) units, the strength of gravity simply is what it is, a primary quantity, while the proton's mass is the tiny number $\sqrt{N}$.

That's a provocative and fruitful way to invert the question, because we've attained quite a deep understanding of the origin of the proton's mass, as I discussed in an earlier column ("Mass without Mass 1"). The lion's share of the proton's mass can be accounted for in an approximation to quantum chromodynamics (QCD), where all the relevant particles — gluons, and up and down quarks — are taken to be massless. In that earlier column, I discussed this in conceptual terms; now let's look under the hood.

The key dynamical phenomenon is the running of the coupling (see "QCD Made Simple"). Looking at the classical equations of QCD, one would expect an attractive force between quarks that varies with the distance as g^2/r^2, where g is the coupling constant. This result is modified, however, by the effects of quantum fluctuations. The omnipresent evanescence of virtual particles renders empty space into a dynamical medium, whose response alters the force law.

In QCD, the antiscreening effect of virtual color gluons (asymptotic freedom) enhances the strength of the attraction by a factor that grows with the distance. This effect can be captured by defining an effective coupling, $g(r)$, that grows with distance. The attractive interaction among quarks wants to bind them together, but the potential energy to be gained by bringing quarks together must be weighed against its cost in kinetic energy. In a more familiar application, just this sort of competition between Coulomb attraction and localization energy is responsible for the stability and finite size of atoms.[4] Here, quantum-mechanical uncertainty implies that quark wavefunctions localized in space must contain a substantial admixture of high momentum, which translates directly, for a relativistic particle, into energy. If the attraction followed Coulomb's law, with a small coupling, the energetic price for staying localized would always outweigh the profit from attraction, and the quarks would not form a bound state. But the running coupling of QCD grows with distance, and that tips the balance. The quarks finally get reined in at distances where $g(r)$ becomes large.

The mechanism leading to this binding dynamics explains the "coincidence," noted above, that the geometric size of the proton is close to its Compton radius. That is because a substantial portion of its formation energy, from which its mass arises as $m_p = E/c^2$, is associated with the momenta of order $p \approx \hbar/R$, required for the quarks' localization, through $E = pc$. Simple algebra then yields $R \approx \hbar/m_p c$. But really no detailed calculation was required to reach this conclusion. Since only the broadest principles of special relativity and quantum mechanics come into the dynamics, the relationship follows by dimensional analysis.

Thus the proton mass is determined by the distance at which the running QCD coupling becomes strong. Let's call this the QCD distance. Our question, "Why is the proton mass so small?" has been transformed into the question, "Why is the QCD distance much larger than the Planck length?" To close our circle of ideas, we need to explain how, if only the Planck length is truly fundamental, this vastly different length can arise naturally.

This last elucidation, profound and beautiful, is worthy of the problem. It has to do with how the coupling runs. When the QCD coupling is weak, "running" is a bit of a misnomer. Actually the coupling creeps along, like a wounded snail. To be precise (and we can in fact calculate the behavior precisely, following the rules of quantum field theory, and even check it out experimentally[5]), the inverse coupling varies logarithmically with distance. In other words, the distance will need to change by many orders of magnitude for a moderately weak coupling to evolve into a strong one. So, finally, all we require to generate our large QCD distance dynamically is that, at the Planck length, the QCD coupling is moderately small (between a third and a half of what it is observed to be at 10^{-15} cm). From this modest and innocuous starting point, by following our logical flow upstream, we arrive at the tiny value of N, which at first sight seemed so absurd.

I've explained how the ridiculously feeble appearance of gravity is consistent with the idea that this force sets the scale for a fundamental theory of Nature. But does it? Stay tuned.

References

1. R. P. Feynman, *Acta Physica Polonica* **24**, 697 (1963).
2. P. A. M. Dirac, *Proc. Roy. Soc. London* **A165**, 199 (1939).
3. M. Planck, *Sitzungsber. Dtsch. Akad. Wiss. Berlin, Math-Phys. Tech. Kl.*, **440** (1899).
4. R. P. Feynman, R. B. Leighton and M. Sands, *The Feynman Lectures on Physics*, vol. 1, Addison-Wesley, Reading, Mass. (1963), pp. 38–4.
5. Summarized in *The Review of Particle Physics*, D. Groom *et al.* (Particle Data Group), *European Physical Journal* **C15**, 1 (2000), "Quantum Chromodynamics" entry.

Scaling Mount Planck 2: Base Camp

As I explained in the preceding column, if we believe that gravity will be a primary element within a unified theory of fundamental physics, then the classic question Why is gravity so weak? is much better posed as Why are protons so light? And since modern quantum chromodynamics (QCD) gives us a detailed and powerful understanding of the mass of the proton, I was able to give a detailed and powerful answer to this rephrased question. What I didn't do was present any serious evidence that my answer is the same as Nature's.

Let me recall very briefly the central points from that previous discussion. Quantum fluctuations in the quark and gluon fields turn empty space into a dynamical medium, which, under different circumstances, can either screen out or enhance the power of a source of color charge. Thus the strength of this coupling depends on the energy scale at which it is measured, an effect we call the running of couplings. Operationally, it means that the probability for radiation of color gluons that carry away large amounts of energy and momentum should decrease as the energy in question increases, or equivalently to increase as the energy shrinks. This behavior has been verified, with quantitative precision, in many experiments. Now because the coupling strength changes only logarithmically with the energy scale, only a big factor in this scale can make a significant change in the effective coupling strength. Thus a moderate color coupling at the enormously high energy scale that gravity suggests is fundamental (the Planck scale, about 10^{19} proton masses) will become large and capable of binding quarks together only at some much lower energy scale, where quarks bind into protons, and the building blocks of matter take shape. Following this scenario, it is straightforward to imagine that the running of the QCD coupling produces a relatively tiny mass for protons starting from a much larger fundamental scale, thereby putting the feebleness of gravity in satisfying perspective.

That's a fine sketch, as far as it goes. It's coherent, largely taken from life, and I think quite pretty. But I'd blush to call it a scientific trail guide for an ascent of Mount Planck. It's lacking in context and detail. And, beyond the bare observation that the logarithm of a very large number need not be very large, it lacks quantitative focus. To go further, we need to draw in other parts of physics besides QCD and gravity proper. Do they fill in our sketch? In particular, do they provide specific pointers toward our sought-for peak — the remote Planck scale? Or do they bring out remaining gaps? The answers, broadly speaking, are Yes, Yes — and Yes again. Overall, our sketch will grow much stronger and more coherent. Though holes do appear, I'll save their description for next time. Here let's savor some nice fancy doughnuts that surround them.

To begin, let's appreciate that the running of couplings is a general phenomenon of quantum field theory, not restricted to QCD. Indeed, its historical antecedents can be found in the prehistory of modern quantum field theory. When hints of a deviation from Paul Dirac's prediction for the spectrum of hydrogen were in the air, but before Willis Lamb's accurate measurements, Edwin Uehling in 1935 calculated a correction to the Coulomb potential due to the screening effect of virtual particles, the so-called vacuum polarization. It turns out that vacuum polarization provides a relatively small part of the Lamb shift. But it definitely must be included in order for theory to agree with experiment. This agreement provided early, direct evidence for what we today would call the running of the effective coupling for quantum electrodynamics. In recent times, this effect, and its weak interaction analog, has been richly documented in precision experiments at CERN's Large Electron-Positron Collider (LEP). Whereas the fine-structure constant α that we observe at vanishingly small energies is very nearly $\alpha(0) = 1/137.03599976(50)$, the value governing high-energy radiation in Z-boson production and decay is measured to be $\alpha(M_Z) = 1/127.934(27)$. The numerical value of the fine-structure constant, a conceptual gold nugget ardently pursued by Arthur Eddington and Wolfgang Pauli, has lost its luster. (There is a shiny gold nugget, nonetheless, whose nature will emerge in my next column.)

With both strong and weak couplings in play, it makes sense to ask if there is an energy scale where they equalize. The answer involves extrapolating many orders of magnitude beyond what has been accessed experimentally. Following the credo "It is more blessed to ask forgiveness than permission," let's try it. Certainly, nothing in the internal logic of quantum field theory forbids the extrapolation. Indeed, by its indication that fundamental dynamics evolves on a logarithmic scale, quantum field theory encourages us to think big. Carrying out the extrapolation, we find a most remarkable and encouraging result. The strong and weak couplings equalize — at roughly the Planck scale! Planck, of course, knew nothing of the strong or the weak interaction, nor of quantum field theory and running couplings. The Planck scale's reappearance in this entirely new context confirms his intuition about its fundamental significance.

The plot thickens when the remaining fundamental interaction, electromagnetism, is added to the mix. Its coupling also runs, as I've already mentioned. But in comparing this coupling with the strong or weak coupling, a new issue arises. The mediators of the strong and weak interactions are themselves strongly and weakly interacting particles; they have nonvanishing strong or weak charges. This property reflects the nonabelian character of the strong and weak gauge symmetries. The photon, however, is electrically neutral. So whereas for the strong and weak interactions there is a unique, natural unit of charge, for electromagnetism the natural unit is obscure. Should we use the charge of the u quark ($2/3e$), the d quark ($-1/3e$), the electron ($-e$), or something else? Since, as I've emphasized, small changes in the couplings correspond to big changes in energy scale, our answer for the unification scale is quite sensitive to this ambiguity.

Resolving it requires considerations of another order. Up to this point, I have been able to be rather vague about what actually happens at the Planck scale. We've seen that the strong coupling is not very strong up there, that it equalizes with the weak coupling, and both sorts of interactions become roughly comparable in strength to gravity. But it has not been necessary to speculate about the dynamics of a specific unified theory. Now it's unavoidable.

Unified gauge symmetry, which includes strong, weak, and electromagnetic interactions in a single structure, makes precise comparisons possible. Of course, our answer depends on what unifying symmetry is assumed. The original, simplest, and most natural possibilities were identified by Jogesh Pati and Abdus Salam and by Howard Georgi and Sheldon Glashow. On adopting either of those possibilities, we can make our calculation. And we find that the strong and electromagnetic couplings equalize, again, at roughly the Planck scale! And so do the weak and electromagnetic couplings.

A usably specific unified theory including gravity remains elusive. In the absence of such a theory, comparisons between the strength of gravity and that of the other forces cannot be precise. Thus the physical significance of exactly where these various unifications occur — whether it is at precisely the Planck scale rather than, say, the Planck scale times $1/8\pi^2$ — is correspondingly murky. I'll say more about it in my next column.

The question whether strong, weak, and electromagnetic couplings unite *with each other* at a common scale, by contrast, is ripe. If we take into account the virtual effects of only the particles in the Standard Model, the answer is, "Almost, but not quite." But if we make the further hypothesis, attractive on other grounds, that the known particles have heavier supersymmetric partners with masses in the neighborhood of 10^3 proton masses, then there is striking quantitative agreement.

Both the strength and the limitation of the running-of-couplings calculation lie in its insensitivity to details. Because the running depends only weakly (logarithmically) on the masses of the virtual particles involved, we can't use the success of our calculation to discriminate finely among detailed models of unified gauge symmetry, or supersymmetry, or the way in which these symmetries are broken.

Were there an abundance of independent evidence for these ideas, we might lament this robustness as a lost opportunity. But at present, independent evidence is extremely thin, especially for low-energy supersymmetry.

Our robust, successful calculation therefore provides a mercifully stable beacon amidst foggy mist. It reveals what appears to be a path with sound footing up Mount Planck, leading us to a dizzyingly high base camp. The indicated path ascends, following the tried and tested physics of quantum field theory, on a gentle logarithmic slope. In calibrated steps, it guides us directly from subnuclear phenomena of the strong and weak interactions to the heights that encode the feebleness of gravity. It is redoubled with unified gauge symmetry and dovetailed with low-energy supersymmetry.

As we approach such ethereal heights, the air grows thin, and even the normally sober can become giddy. Hallucinations are to be expected. How can we demonstrate to skeptics the reality of the places revealed in our visions? The traditional, and ultimately the only convincing, way is to bring back trophies. Unified gauge symmetry powerfully suggests, and almost requires, tiny violations of the laws of lepton and baryon and number conservation. These violations can be observed as small neutrino masses and as proton instability, respectively. Neutrino masses of appropriate magnitude have now been observed through oscillations (though questions remain, even here). Proton decay remains elusive. Low-energy supersymmetry predicts a whole new world of particles, several of which must be accessible to future accelerators including, specifically, the CERN Large Hadron Collider, which is scheduled to begin operation in 2007.

In the long view, then, there are genuine, tremendously exciting prospects for bringing this circle of ideas to fruition — or demolition. Unfortunately, the requisite experiments are difficult, slow, and expensive. The main ideas have been in place — ever promising, fundamentally unshaken, but mostly unfulfilled — for 20 years or more. The tempo, unfortunately, is poorly matched to news cycles or even to the timescales for academic promotions. It is tempting to hope that shortcuts will appear or even that the long, hard path, though beautiful, is illusory and that the peak is actually nearby. Personally, I prefer to anticipate that, here, beauty foretells truth, humbly accepting that, as Spinoza wrote, "All things great are as difficult as they are rare."

Scaling Mount Planck 3: Is That All There Is?

Let's quickly review the main points of the two earlier columns in this series. Gravity appears extravagantly feeble on atomic and laboratory scales, ultimately because the proton's mass m_p is *much* smaller than the Planck mass $M_{Planck} = (\hbar c / G_N)^{1/2}$, where $\hbar$ is Planck's quantum of action, c is the speed of light, and G_N, is Newton's gravitational constant. Numerically, $m_p / M_{Planck} \approx 10^{-18}$. If we aspire, in line with Planck's original vision and with modern ambitions for the unification of physics, to use the natural (Planck) system of units constructed from c, $\hbar$, and G_N (see "Scaling Mount Planck 1: A View from the Bottom," and if we agree that the proton is a natural object, then the very small ratio appears at first blush to pose a very big embarrassment. It mocks the central tenet of dimensional analysis, which is that natural quantities expressed in natural units should have numerical values close to unity.

Fortunately, we have a deep dynamical understanding of the origin of the proton's mass, thanks to quantum chromodynamics. The value of the proton's mass is determined by the scale Λ_{QCD}, at which the interaction between quarks — parameterized by the energy-dependent "running" QCD coupling constant $g_s(E)$ — starts to dominate their quantum-mechanical resistance to localization (see "Scaling Mount Planck 2: Base Camp"). More precisely, the criterion for the dominance of the quark-binding interaction is that the QCD analog of the fine structure constant $\alpha_s(E) \equiv g_s(E)^2/(4\pi\hbar c)$ becomes of order unity: $\alpha_s(\Lambda_{QCD}) \approx 1$. Because the energy dependence of $\alpha_s(E)$ is very mild, a long run in E is required to change its value significantly. Indeed, we find in this way that our QCD-based estimate of the proton mass, using $\alpha_s(m_p c^2) \approx 1$, corresponds to $g_s(M_{\text{Planck}} c^2) \approx 1/2$. So the extravagantly small value of m_p / M_{Planck} does not contradict the idea that M_{Planck} is the "natural" fundamental unit of mass, after all. Whereas naive analysis founders on the value of m_p, deeper understanding aims instead at $g_s(M_{\text{Planck}} c^2)$ — the basic coupling at the basic energy — as the primary quantity, from which m_p is derived. And $g_s(M_{\text{Planck}} c^2)$ *is* of order unity!

A conceptually independent line of evidence likewise points to $M_{\text{Planck}} c^2$ as a fundamental energy scale. By postulating the existence of an encompassing symmetry at that scale, and weaving the separate gauge symmetries SU(3)×SU(2)×U(1) of the standard model into a larger whole, we can elucidate a few basic features of the standard model that would otherwise remain cryptic. The scattered multiplets of fermions and their peculiar hypercharge assignments click together like pieces of a disassembled watch. And, most impressively, the disparate coupling strengths we observe at low energy are derived quantitatively from a single coupling — none other than our friend $g_s(M_{\text{Planck}} c^2)$ — at the basic scale.

In all those previous considerations, gravity itself has figured only passively, as a numerical backdrop. It has supplied us with the numerical value of G_N, but that's all. Now, in this concluding column, I examine how (and to what extent) gravity, as a dynamical theory, fits within this circle of ideas.

A lot of portentous drivel has been written about the quantum theory of gravity, so I'd like to begin by making a fundamental observation about it that tends to be obfuscated. *There is a perfectly well-defined quantum theory of gravity that agrees accurately with all available experimental data.* (I have heard two grand masters of theoretical physics, Richard Feynman and J. D. Bjorken, emphasize this point on public occasions.)

Here it is. Take classical general relativity as it stands: the Einstein–Hilbert action for gravity, with minimal coupling to the standard model of matter. Expand the metric field in small fluctuations around flat space, and pass from the classical to the quantum theory following the canonical procedure. This is just what we do for any other field. It is, for example, how we produce *quantum* chromodynamics from classical gauge theory. Applied to general relativity, this approach gives you a theory of gravitons interacting with matter.

More specifically, this procedure generates a set of rules for Feynman graphs, which you can use to compute physical processes. All the classic consequences of general relativity, including the derivation of Newton's law as a first approximation, the advance of Mercury's perihelion, the decay of binary pulsar orbits due to gravitational radiation, and so forth, follow from straightforward application of these rules within a framework in which the principles of quantum mechanics are fully respected.

To define the rules algorithmically, we need to specify how to deal with ill-defined integrals that arise in higher orders of perturbation theory. The same problem already exists in the standard model, even before gravity is included. There we deal with ill-defined integrals using renormalization theory. We can do the same here. In renormalization theory, we specify by hand the values of some physical parameters, and thereby fix the otherwise ill-defined integrals. A salient difference between how renormalization theory functions in the standard model and how it extends to include gravity is that, whereas in the standard model by itself we need only specify a finite number of parameters to fix all the integrals, after we include gravity we need an infinite number. But that's all right. By setting all but a very few of those parameters equal to zero, we arrive at an adequate — indeed, a spectacularly successful — theory. It is just this theory that practicing physicists always use, tacitly, when they do cosmology and astrophysics. (For the experts: The prescription is to put the coefficients of all nonminimal coupling terms to zero at some reference energy scale, call it ε, well below the Planck scale. The necessity to choose an ε introduces an ambiguity in the theory, but the consequences of that ambiguity are both far below the limits of observation and well beyond our practical ability to calculate corrections expected from mundane, non-gravitational interactions.)

Of course the theory just described, despite its practical success, has serious shortcomings. Any theory of gravity that fails to explain why our richly structured vacuum, full of symmetry-breaking condensates and virtual particles, does not weigh much more than it does is a profoundly incomplete theory. This stricture applies equally to the most erudite developments in string and M theory and to the humble bottom-up approach used here. This gaping hole in our understanding of Nature is the notorious problem of the cosmological term. Perhaps less pressing, but still annoying, is that the above-mentioned ambiguity in the theory of gravity at ultralarge energy–momentum makes it difficult to address questions about what happens in ultraextreme conditions, including such interesting situations as the earliest moments of the Big Bang and the endpoints of gravitational collapse.

Nevertheless it makes good sense to take our working theory of gravity at face value and to see whether it fits into the attractive picture of unification we have built for the strong, weak, and electromagnetic interactions. Again, a crucial question is the apparent disparity between the coupling strengths. For the standard model interactions, logarithmic running of couplings with energy was a subtle quantum-mechanical phenomenon, caused by the screening or anti-screening effect of virtual particles. With gravity, the main effect is much simpler — and much bigger. Gravity, in general relativity, responds directly to energy–momentum. So the effective strength of the gravitational interaction, when measured by probes carrying larger energy–momentum, appears larger. That is a classical effect, and it goes as a power, not a logarithm, of the energy.

Now on laboratory scales, gravity is *much* weaker than the other interactions — roughly a factor 10^{-40}. But we've seen that unification of the standard model couplings occurs at a very large energy scale, precisely because their running is logarithmic. And at this energy scale, we find that gravity, which runs faster, has almost caught up to the other interactions! Since the mathematical form of the interactions is not precisely the same, we cannot make a completely rigorous comparison, but simple comparisons of forces or scattering amplitudes give numbers like 10^{-2}. Gravity is still weaker, but not absurdly so. Given the immensity of the original disparity, and the audacity of our extrapolations, this minor discrepancy qualifies, if not quite as full success in achieving, at least as further encouragement toward trusting, the ideal of unification.

Let me summarize. Planck observed in 1900 that one could construct a system of units based on c, $\hbar$, G_N. Subsequent developments displayed those quantities as conversion factors in profound physical theories. Now we find that Planck's units, although preposterous for everyday laboratory work, are very suitable for expressing the deep structure of what I consider our best working model of Nature, as sketched in this three-part series of columns. Planck proposed, implicitly, that the mountain of theoretical physics would be built to purely conceptual specifications, using just those units. Now we've taken the measure of Mount Planck from several different vantage points: from QCD, from unified gauge theories, from gravity itself — and found a consistent altitude. It therefore comes to seem that Planck's magic mountain, born in fantasy and numerology, may well correspond to physical reality. If so, then reductionist physics begins to face the awesome question, compounded of fulfillment and yearning, that heads this column.

At Sea in the Depths

When I was a small boy I fantasized about becoming a magician. But when I learned the tricks I discovered that they relied more on deception than on special powers, and then becoming a magician didn't seem so appealing anymore. But later, to my surprise, I found myself becoming a magician after all. I learned quantum mechanics.

What is magic, after all? It's something that works, but you don't understand why. As "What is Quantum Theory?" exposes, I don't understand quantum mechanics properly (and none of the readers of *Physics Today* has helped me out). I conjure with it, and so far mostly good things have happened. But it's disconcerting, not to mention embarrassing, for an enlightened rationalist like me to rely on frequent consultations with a mysterious oracle, who rarely gives straight answers, instead quoting odds. Do I really understand her pronouncements? Can she be trusted? I'd really like to find out what makes her tick.

I didn't major in physics, and I learned it in quirky ways. When I was a small boy I heard that there was this great magician Albert Einstein, and I wanted to learn his tricks. After reading some popularizations (I remember especially Lincoln Barnett's *The Universe and Doctor Einstein*, and Bertrand Russell's *ABC of Relativity*) I wanted to get to the source.

After many attempts and much backtracking, by the time I was in college I was able to read Einstein's great 1915 paper establishing general relativity. I was deeply impressed by its style. As described in "Total Relativity: Mach 2004," the paper begins with an extended philosophical discussion of relative motion (including the phrase "epistemologically unsatisfactory", which you won't find in many physics papers) and Mach's principle, including a memorable thought experiment. Then Einstein settles down to business, and few words are wasted as the new theory of space, time and gravity is revealed with admirable brevity.

At the time I didn't understand exactly what the first part of the paper had to do with the rest of it, but there was a lot I didn't understand very well, and I didn't agonize over the issue. I had the general impression that the philosophical discussion was so deep and penetrating, that once you grasped it the mathematics and physics just fell out as technical bagatelles. Einstein!!

My conceptual disconnect lay dormant for many years until, inspired by the cosmological term problem, I started thinking about maybe modifying the theory of gravity myself. I went back to Einstein and Mach for inspiration, and was amazed at what I found: the second half of Einstein's paper actually *contradicts* the first part. Yet, in the spirit of Walt Whitman's jovial admission

Do I contradict myself?

Very well then I contradict myself.

I contain multitudes.

the first part, featuring Mach's principle, still appears extremely interesting and provocative. And, since it really hasn't been done justice, it might yet stimulate new fundamental physics.

I'm not sure exactly where "Life's Parameters" came from. I've been intensely interested in neurobiology since my student days, and under Stan Leibler's tutelage I got a serious schooling in modern cell biology (and read the big Alberts *et al.* text). But the immediate impetus for the column was thinking through and writing about why the proton has a ridiculously small mass in Planck units (Items 15–17). It was natural to ask, why do the *practical* units of physics have the values they do?

What makes them practical? They're practical because they're well adapted to human life, of course. And so, the question moves to biology.

A more subterranean stream of inspiration came was Feynman's famous after-dinner talk, *There's Plenty of Room at the Bottom*. (Also here

I should mention von Neumann's *The Computer and the Brain*, which has some of the same spirit, and which made a deep impression on me.) In that talk, Feynman explored the limits of physical technology for handling information in time and space. Now that we understand so much about fundamental biology, and are beginning to understand how to engineer with it, it will be fruitful, I'm suggesting, to do some comparable exercises for biological technology. How fast can creations based on that technology compute? How small can they be? How large? How long can they last? How will all that change in space, or in the atmospheres and gravity of other planets, or in environments of our design? I'm planning to think a lot more about these questions, and maybe give some after-dinner talks …

What Is Quantum Theory?

Over the period 1885–1889, Heinrich Hertz[1] discovered that electromagnetic waves can travel through empty space, and demonstrated experimentally that these waves move at the speed of light and are transversely polarized. His work confirmed predictions James Clerk Maxwell had made 20 years earlier, in 1864. Hertz's major papers were collected in a book, *Electric Waves*, for which he wrote an extensive introduction. In that introduction occurs his famous, extraordinary statement: "To the question: What is Maxwell's theory? I know of no shorter or more definite answer than the following: 'Maxwell's theory is Maxwell's system of equations.'"

Superficially, this statement might appear innocent — even ingenuous — but it goes deep, and in its time it caused a sensation. There was, at the time, a rival tradition of electromagnetic theories, especially strong in Germany, which advocated action-at-a-distance formulations in preference to fields. These theories had the advantage of continuing the tremendously successful Newtonian tradition, and of using familiar, highly developed mathematical methods. They also had enormous flexibility. With velocity-dependent force laws, most of the previously known facts about electricity and magnetism could readily be described using action-at-a-distance. Arnold Sommerfeld wrote[2] of his student days (1887–1889) in Koenigsberg, "The total picture of electrodynamics thus presented to us was awkward, incoherent, and by no means self-contained."

Perhaps some modification would also describe Hertz's new results. (Indeed, we know now that by using retarded potentials one *can* reproduce the Maxwell equations from an action-at-a-distance theory, rather elegantly in fact.) So Hertz sought to forestall unproductive debates between rival theories with identical physical content by focusing on the bottom-line content. Sommerfeld continues, "When I read Hertz's great paper, it was as though scales fell from my eyes."

Furthermore, Hertz wanted to purify Maxwell's work. The point is that Maxwell reached his equations through a complex process of constructing and modifying mechanical models of the ether and, according to Hertz, "… when Maxwell composed his great treatise, the accumulated hypotheses of his earlier mode of conception no longer suited him, or else he discovered contradictions in them and so abandoned them. But he did not eliminate them completely …"

Yet a modern physicist, while not contradicting it, could not rest satisfied with Hertz's answer to his question. Maxwell's theory is much more than Maxwell's equations. Or, to put it differently, merely writing down Maxwell's equations, and doing them justice, are two quite different things.

Indeed, a modern physicist, asked what is Maxwell's theory, might be more inclined to answer that it is special relativity plus gauge invariance. While not altering Maxwell's equations, in a real sense these concepts tell us *why* that superficially complicated system of partial differential equations must take precisely the form it does, what its essential nature is, and how it might be generalized. This last feature bears abundant fruit in the modern Standard Model. The core of the Standard Model is a mighty generalization of gauge invariance, which provides successful descriptions of physical phenomena far beyond anything Maxwell or Hertz could have imagined.

With this history as background, let us return to this column's analogous question: What is Quantum Theory? At one level, we can answer along the lines of Hertz. Quantum theory is the theory you find written down in textbooks of quantum theory. Perhaps its definitive exposition is Dirac's book.[3] Conversely, you can find, in the early parts of Dirac's book, statements very much in the Hertzian spirit:

> The new scheme becomes a precise physical theory when all the axioms and rules of manipulation governing the mathematical quantities are specified and when in addition certain laws are laid down connecting physical facts with the mathematical formalism, so that from any given physical conditions equations between the mathematical quantities may be inferred and vice versa.

Of course, the equations of quantum theory are notoriously harder to interpret than Maxwell's equations. The leading interpretations of quantum theory introduce concepts that are extrinsic to its equations ("observers"), or even contradict them ("collapse of the wave function"). The relevant literature is famously contentious and obscure. I believe it will remain so until someone constructs, within the formalism of quantum mechanics, an "observer," that is, a model entity whose states correspond to a recognizable caricature of conscious awareness, and demonstrates that the perceived interaction of this entity with the physical world, following the equations of quantum theory, accords with our experience. That is a formidable project, extending well beyond what is conventionally considered physics. Like most working physicists, I expect (perhaps naively) that this project can be accomplished, and that the equations will survive its completion unscathed. In any case, only after its successful completion might one legitimately claim that quantum theory is defined by the equations of quantum theory.

Stepping now toward firmer ground, let us consider the equations themselves. The heart of quantum theory, which plays for it the central role analogous to the role of Maxwell's equations in electrodynamics, lies in the commutation relations among dynamical variables. Specifically, it is in these commutation relations — and, ultimately, only here — that Planck's constant appears. The most familiar commutation relation, $[p,q] = -i\hbar$, conjoins linear momentum with position, but there are also different ones between spins, or between fermion fields. In formulating these commutation relations, the founders of quantum theory were guided by analogy, by aesthetics, and — ultimately — by a complex dialogue with Nature, through experiment. Here is how Dirac describes the crucial step:[4]

> The problem of finding quantum conditions is not of such a general character. ... It is instead a special problem which presents itself with each particular dynamical system one is called upon to study. ... a fairly general method of obtaining quantum conditions ... is the method of *classical analogy* [original italics].

I think it is fair to say that one does not find here a profound guiding principle comparable to the equivalence of different observers (that inspires both relativity theories), or of different potentials (that inspires gauge theories).

Those profound guiding principles of physics are statements of *symmetry*. Is it possible to phrase the equations of quantum theory as statements of symmetry? An interesting but brief and inconclusive discussion of this occurs in Herman Weyl's singular text, where he proposes [the original is entirely italics!]:[5] "The kinematical structure of a physical system is expressed by an irreducible unitary projective representation of abelian rotations in Hilbert space."

Naturally, I won't be able to unpack this formulation here, but three comments do seem appropriate. First, Weyl shows that his formulation contains the Heisenberg algebra of quantum mechanics and the quantization of boson and fermion fields as special cases, but also allows additional possibilities. Second, the sort of symmetry he proposes — abelian — is the simplest possible kind. Third, his symmetry of quantum kinematics is entirely separate and independent from the other symmetries of physics.

The next level in understanding may come when an overarching symmetry is found, melding the conventional symmetries and Weyl's symmetry of quantum kinematics (made more specific, and possibly modified) into an organic whole. Perhaps Weyl himself anticipated this possibility, when he signed off his pioneering discussion with: "It seems more probable that the scheme of quantum kinematics will share the fate of the general scheme of quantum mechanics: to be submerged in the concrete physical laws of the only existing physical structure, the actual world."

To summarize, I feel that after seventy-five years — and innumerable successful applications — we are still two big steps away from understanding quantum theory properly.

References

1. An excellent recent brief biography of Hertz, including an extensive selection of his original papers end contemporary commentary, is J. Mulligan, *Heinrich Rudolf Hertz*, Garland (New York), 1994.
2. A. Sommerfeld, *Electrodynamics*, Academic (London), 1964, p. 2.
3. P. A. M. Dirac, *Quantum Mechanics*, 4th revised edition, Oxford (London), 1967, p. 15.
4. *Ibid.*, p. 84.
5. H. Weyl, *The Theory of Groups and Quantum Mechanics*, Dover (New York), 1950, p. 272.

Total Relativity: Mach 2004

Like the child hero of *The Emperor's New Clothes*, Ernst Mach (1838–1916) unsettled conventional wisdom by making simple observations, obvious in retrospect. Mach's close critical analysis of the empirical value of physical concepts and his insistence that they must justify their use helped produce the atmosphere in which special and general relativity, and later quantum theory, could be conceived.

Mach's masterpiece is *The Science of Mechanics*.[1] It is fascinating to read, even today, and every physicist ought to have that pleasure. In an annotated narrative, Mach dissects the conceptual innovations and presuppositions that marked the history of the science of motion, from its prescientific roots through the late 19th century. He was especially critical of Newton's concepts of absolute time and space:

> Absolute time can be measured by comparison with no motion; it has therefore neither a practical nor a scientific value; and no one is justified in saying that he knows aught about it. It is an idle metaphysical conception.[1]

Here's what Albert Einstein, in his self-styled "obituary," said about Mach's book:

> Even [James Clerk] Maxwell and [Heinrich] Hertz, who in retrospect appear as those who demolished the faith in mechanics as the final basis of all physical thinking, in their conscious thinking adhered throughout to mechanics as the secured basis of physics. It was Ernst Mach who, in his history of mechanics

(Geschichte der Mechanik), shook this dogmatic faith. This book exercised a profound influence upon me in this regard while I was a student. I see Mach's greatness in his incorruptible skepticism and independence.[2]

Special relativity puts all space-time frames that move with respect to one another at constant velocity on an equal footing. It thereby renders moot the notion of a unique "preferred" value for any single object's velocity. Mach's deconstruction of motion, however, went much further. It culminated in a concept of total relativity, Mach's principle, which remains provocative to this day.

Here is Isaac Newton's original formulation of his concept of absolute space:

> If a bucket, suspended by a long cord, is so often turned about that finally the cord is strongly twisted, then is filled with water, and held at rest together with the water; and afterwards by the acceleration of a second force, it is suddenly set whirling about the contrary way, and continues, while the cord is untwisting itself, for some time in this motion; the surface of the water will at first be level, just as it was before the vessel began to move; but, subsequently, the vessel, by gradually communicating its motion to the water, will make it begin sensibly to rotate, and the water will recede little by little from the middle and rise up at the sides of the vessel, its surface assuming a concave form. (This experiment I have made myself.) ... when the relative motion of the water had decreased, the rising of the water at the side of the vessel indicated an endeavor to recede from the axis; and this endeavor revealed the real motion of the water.[3]

Mach insisted that the relative motion of bucket and distant stars is responsible for the observed concave surface. In Mach's own words:

> Newton's experiment with the rotating vessel simply informs us, that the relative motion of the water with respect to the sides of the vessel produces no noticeable centrifugal forces, but that such forces are produced by its relative rotation with respect to the mass of the Earth and the other celestial bodies. No one is competent to say how the experiment would turn out if the sides of the vessel increased in thickness and mass until they were ultimately several leagues thick.[1]

An Ideal Unrealized

A remarkable invocation of Mach's principle occurs near the beginning of Einstein's great foundational paper on general relativity:

> In classical mechanics, and no less in the special theory of relativity, there is an inherent epistemological defect which was, perhaps for the first time, clearly pointed out by Ernst Mach. We will elucidate it by the following example: Two fluid bodies of the same size and nature hover freely in space at so great a distance from each other and from all other masses that only those gravitational forces need be taken into account which arise from the interaction of different parts of the same body.
>
> Let either mass, as judged by an observer at rest relative to the other mass, rotate with constant angular velocity about the line joining the masses. This is a verifiable relative motion of the two bodies. Now let us imagine that each of the bodies has been surveyed by means of measuring instruments at rest relatively to itself, and let the surface of S_1 prove to be a sphere, and that of S_2 an ellipsoid of revolution. Thereupon we put the question — What is the reason for this difference between the two bodies? No answer can be admitted as epistemologically satisfactory, unless the reason given is an *observable fact of experience* ... .
>
> Newtonian mechanics does not give a satisfactory answer to this question ... .
>
> We have to take it that the general laws of motion, which in particular determine the shapes of S_1 and S_2, must be such that the mechanical behavior of S_1 and S_2 is partly conditioned, in quite essential respects, by distant masses which we have not included in the system under consideration.[4]

The preceding quotation, part of a lengthy methodological discussion that constitutes a significant fraction of this otherwise terse paper, makes it clear, as Einstein acknowledged on many occasions, that as he worked toward constructing general relativity theory his thinking was guided by Mach's principle. Ironically, however, Einstein's general relativity as finally formulated does *not* embody Mach's total relativity principle. If one analyzes the thought-experiment that Einstein outlined using general relativity, the result is just the same as in Newtonian mechanics! Einstein must have realized this, but he does not mention it in the paper.

Though Mach's principle is not an automatic consequence of the equations of general relativity, Einstein attempted in later work to impose it as a criterion to select out acceptable (that is, "epistemologically satisfactory") solutions. To rule out the troublesome behavior realized in his thought-experiment with two isolated bodies, he postulated that in reality there is no such entity as an isolated body! The universe must be spatially closed, with no boundary, and on large scales uniformly filled with matter. Although those ideas have been extremely stimulating for cosmology, it is (to say the least) not clear that they are true. Modern inflationary cosmology, for example, is agnostic regarding ultimate closure and definitely suggests nonuniformity on the largest scales. That suggestion may not be the last word on the subject, but it's become harder than ever to regard Einstein's cosmological implementation of Mach's principle as anything but an ad hoc patch.

Mechanistic variations on the theme, more in line with Mach's original idea that the distant stars cause inertia, appear in the literature. The Lense–Thirring effect demonstrates "frame-dragging" by a rotating shell: Inertial frames inside the shell rotate, in the same sense as the shell, relative to the inertial frames at infinity. Dennis Sciama has argued that a gravitational analog of electromagnetic induction would react against relative acceleration, and could act at such large distances as to provide the effect of inertia.

The Perspective of Symmetry

But these quasi-cosmological implementations of Mach's principle (which, when examined in detail, both have serious technical problems), raise questions of symmetry, just as Einstein's did. If inertia depends on the distribution of matter at large distances, why, in a lumpy universe, should inertia be accurately isotropic? Why, in an expanding universe, should it be constant? To derive these properties from the distant stars would appear to require fine adjustment of their influence. To put it bluntly, it leaves us at the mercy of astrology.

Einstein's principle of equivalence, as embodied in general relativity, appears to represent a deeper insight. It states that in any small region of space-time, there exist systems of coordinates — inertial frames — in which the laws of special relativity are valid. (And it states further that a gravitational field is equivalent, over a small region, to the use of a frame that is accelerated with respect to the local inertial frames.) This principle can be phrased as a symmetry principle: local Lorentz invariance.

Mach's principle, or total relativity, goes beyond the principle of equivalence. Total relativity can also be stated as a symmetry principle. It instructs us that in the primary equations (in other words, before their solution reveals the crucial influence of distant bodies!) we should put *all* motions on an equal footing, not just those that correspond to constant relative velocity. It claims that the choice of coordinates is entirely a matter of convention, and requires that we remove all intrinsic structure from space-time. On that basis any choice of coordinates should be on equal footing, since the labels implementing the coordinates could be undergoing arbitrary motions. But in general relativity, space-time is not without structure, and it is not true that all coordinate systems are equally good (notwithstanding contrary statements that pervade the literature — starting, as we've seen, with Einstein's original paper). General relativity includes a metric field, which tells us how to assign numerical measures to intervals of time and space. It's convenient to choose frames in which the metric field takes its simplest possible form, because in such frames the laws of physics assume their simplest form.

Posing the issue, Einstein *versus* Mach, as a question of symmetry brings it within a circle of ideas that are central to modern fundamental physics. In the standard electroweak model, we have a Higgs field that breaks local gauge symmetries of the primary equations; in quantum chromodynamics, we have a quark-antiquark condensate that breaks both those symmetries and others; and in unification schemes, generalizations of the symmetry-breaking idea are used freely.

The symmetry perspective suggests further questions that could prove fruitful for the future of physics. It invites us to contemplate theories enjoying larger symmetries than are realized in the equivalence principle of general relativity. Mach's principle, from this perspective, is the hypothesis that a larger, primary theory should include total relativity — that is, physical equivalence among all different coordinate systems. (A different generalization appears in Kaluza–Klein theory and its modern descendants: In the process of compactifying extra dimensions, the higher-dimensional equivalence principle is broken down to the smaller symmetry of its 3 + 1 dimensional version.) This primary symmetry of the equations must, of course, be badly broken in the particular solution that describes the world we observe. Nevertheless, its conceptual influence would be felt through restrictions it imposes on the equations of the physics. As we struggle with the problem of the cosmological "constant," constructive suggestions for augmenting the equivalence principle could prove most welcome.

Mach's austere empiricism is a disinfectant that, taken too far, can induce sterility. Mach himself never accepted special relativity. He also denounced atomism and harassed his great contemporary Ludwig Boltzmann over it.[5] In private correspondence (quoted in reference 5), Einstein wrote that Mach's approach to science "cannot give birth to anything living, it can only exterminate vermin." Yet in this sharp statement, I believe Einstein meant to be judicious. Exterminating vermin is a necessary and sometimes challenging task, even if it is not so transcendent as giving birth. In the world of ideas, as opposed to the world of events, we can choose what to retain. The good that men do lives after them, the evil is oft interred with their bones.

References

1. E. Mach, *The Science of Mechanics: A Critical Historical Account of Its Development*, Open Court, La Salle, IL (1893).
2. A. Einstein, in *Albert Einstein: Philosopher-Scientist*, P. Schilpp, ed., Harper and Row, New York (1949), p. 1.
3. I. Newton, *Mathematical Principles of Natural Philosophy*, I. B. Cohen, A. Whitman, trans., U. of Calif. Press, Berkeley (1999).
4. A. Einstein, *Annalen der Physik* **49**, 769 (1916); Eng. trans. in *The Principle of Relativity*, Dover, New York (1952), p. 111.
5. D. Lindley, *Boltzmann's Atom: The Great Debate That Launched a Revolution in Physics*, Free Press, New York (2001).

Life's Parameters

Planck's units — 10^{-6} gram, 10^{-33} centimeter, 10^{-44} second — are derived from fundamental parameters that appear in the most basic theories of physics. They are constructed from suitable combinations of the speed of light c, the quantum of action h, and the Newtonian gravitational constant G. These quantities are the avatars of Lorentz symmetry, wave–particle duality, and bending of space-time by matter, respectively.

The mismatch between Planck's units and the practical units of mass, length, and time — to wit, 1 g, 1 cm, 1 s — is so enormous as to be grotesque. (Our discussion will be smoother using these, rather than the "standard" SI units.) Quantitative disparities of this order pose qualitative challenges for our understanding of the world. Why do we find it helpful to use units that are so far removed from the fundamentals?

The central mission of my recent trilogy of Reference Frame columns, "Scaling Mount Planck," was to explain how the value of the proton mass, a skimpy 10^{-18} Planck units, might emerge from a theory in which Planck units are basic. Superficially, the appearance of such a small number violates the guiding principle of dimensional analysis: that natural quantities in natural units should be expressed as numbers of order unity. But a profound and well-established dynamical effect, the logarithmic running of couplings, together with the basic understanding of protons provided by quantum chromodynamics, lets us understand where the small number comes from.

From a conventional, reductionist perspective this calculation of the proton mass solves the main problem involved in relating Planck's units to mundane reality. The business of fundamental physics, from that perspective, is to understand the basic building blocks. To put it crudely: Having understood protons, you're entitled to declare victory. (Strictly speaking, electrons count for something too.) But for a broader vision, such insight, although important, poses a new challenge. We want to understand in a more detailed and comprehensive way how the texture of our everyday world, the macrocosm, relates to the fundamentals.

Along that road, an important step is to understand the basis for the practical shorthand we use in describing that world. We need to deconstruct the macros that we use to construct the macrocosm. Which brings us back to our problem: Why do we find it helpful to use grams, centimeters, and seconds — CGS units?

As the words "we" and "helpful" suggest, this is not a conventional question in pure physics. It has very much to do with what we are, as physical beings, and how we interact with the physical world.

Planck Mass and Practical Mass

Let's start with mass. We've started with a mismatch of 10^6 between the gram and Planck's unit of mass. Our earlier triumph in "Scaling Mount Planck", which took us from the Planck mass to the proton mass, aggravates this mismatch by a factor of 10^{18}. This leaves us a factor of 10^{24} to explain.

That number, of course, is essentially Avogadro's number, the number of protons in a gram. Why is that number so big? Well, we find it convenient to use grams, because we are — very roughly — gram-sized. It takes of order a million protons and neutrons to make a functional protein or macro-molecule, a billion such building blocks (with their aqueous environment) to make a functional cell, and a billion cells to make a simple tissue fragment. Multiply these factors, and there's your 10^{24}. Biology, as it has evolved on Earth, requires that kind of hierarchy of structures to build up architecture complicated enough to support beings capable of doing physics.

This quick explication does not do justice to the very interesting question of why each stage of the hierarchy requires large numbers of basic units from the previous stage nor to the specific values of the large number. Certainly, at the first stage ensuring stability of function against quantum and thermal fluctuations is of key importance. The complexity of metabolism, which requires many different catalytic agents and organelles, is a key at the second stage.

The third stage, passing from cells to intelligent creatures, appears more contingent. Indeed, the emergence of multicellular life is a relatively recent evolutionary event, and to this day single-cell forms remain common. And among multicellular forms we find a vast range of masses, including some very unintelligent giants such as apatosaurs and trees. Altogether, intelligence seems to be a biological epiphenomenon. Not all species evolve toward it as a function of time, nor at any one time do even the most massive creatures necessarily accommodate it.

For better or worse, we have only one (semi-)convincing example of evolved intelligence to look at: the human brain. And while our understanding of that structure is advancing rapidly, it is still primitive. We can't understand why the human brain is the size it is when we still don't know how that brain works. There are, however, two observations that suggest that human intelligence could not be supported with a brain of significantly smaller mass. First is the extremely suggestive historical fact that rich cultural artifacts emerged simultaneously with a vast increase in the size of human brains, both occurring on timescales that are very short by evolutionary standards. Second is the fact that human childbirth is made difficult by the size of neonatal brains (and neonates are far from finished products). Together, these observations suggest that sheer brain mass is crucial to the emergence of intelligence, and that, consistent with functionality, there has been substantial evolutionary pressure to keep the mass as small as possible.

Planck Length and Practical Length

In any case, given an explanation of the disparity in mass units, the disparity in length units is a straightforward consequence. A centimeter is roughly 10^8 Bohr radii, or atomic sizes, and so a cubic centimeter is just what encompasses those same 10^{24} atoms that make a gram.

I hasten to confess that the Bohr radius, which is undoubtedly the key thing here, itself has a problematic value when expressed in Planck units. The Bohr radius is most naturally expressed as $r_B = h^2/m_e e^2$, where h is Planck's quantum of action, m_e is the electron mass, and e is the electron charge. Alternatively, we can write it as $r_B = h/c \times 1/\alpha \times 1/m_e$, where c is the speed of light and α is the fine-structure constant. Now h and c are, of course, both just unity in Planck units, and α does not pose a major difficulty. (In the framework of unified gauge theories, its observed value corresponds to a near-unity value of the unified coupling at the Planck scale.) But m_e, at about 10^{-22} times the Planck mass, is so very small as to be a very big embarrassment. The best that can be said is that this embarrassment is not a new one.

Planck Time and Practical Time

In these derivations of the practical mass and length scales, I used simple arguments and crude estimates. I'm confident, though, that with some work they could be firmed up and enriched considerably. By comparison, when it comes to the unit of time, I'm somewhat at a loss.

Indeed, hints that biological timescales are highly negotiable seem to be all around us. When watching trees adapt to their environment, we run out of patience and must resort to time-lapse photography, while flies elude our swats, and to follow the beating of their wings, we need to watch slow-motion movies.

So why does it take about a second to have a thought? At a mechanistic level, this timescale is tied up with the diffusion rate of signal molecules across synapses, opening and closing of receptors, the capacitance and conductivity of the fatty membranes that facilitate nerve impulses, reaction rates for secondary messengers, and possibly other factors. From the point of view of physics, these are complicated phenomena, and it seems extremely difficult to tie them to the fundamentals — or, therefore, to perceive fundamental constraints on their values. It is far from obvious that evolution has been driven to optimize the rate of thought. We might even anticipate that this rate could be drastically modulated by means of physio-chemistry ("speed" worthy of the name!), or altered by genetic engineering. These observations also suggest that in seeking the deep source of the second a bottom-up approach from microphysics is doomed, and we must consider environmental and possibly historical (evolutionary) factors.

A possible clue is that the value g of the acceleration due to near-Earth gravity comes out to be of order unity in practical CGS units. Since g sets the tempo for purposeful motion near Earth's surface, creatures working at this tempo will adopt as the natural unit of time, accommodating the gram and centimeter, something close to $(g\, cm^2/g)^{1/2}$. And indeed we do.

By way of contrast, it is striking that the practical unit of temperature, the degree (regarded as a unit of energy), corresponds to the strange value

°C ~ 10^{-16} g cm^2/s^2. This temperature provides some rough measure of available energy sources and heat sinks, and so it is very relevant to the physics of computation. But if we try to find the deep source of the time unit in (g cm^2/°C)$^{1/2}$, we'd wind up with 10^8 s! Obviously this does not correspond to the speed of thought; but then again we do not think with macroscopic units (like Tinkertoy^{TM} computers), but at a molecular level. If instead we translate the degree directly into an atomic timescale, using Planck's constant, we find the unit $h/°C \sim 10^{-11}$ s. This is so much smaller than the practical unit tailored to our thought processes as to strongly suggest, again, that we operate far from the physical limit.

There is direct evidence for that conclusion. The impressively tiny — and still shrinking — size and timescales of our artificial thinking progeny, electronic computers, are much closer to fundamental physics. They were designed that way! Careful use of the laws of physics makes possible a higher density of more rapid thought than did biological evolution. When the coming quick-witted second-generation silicon physicists define their own practical units, they'll use different ones, and they'll have a much easier time understanding where they came from.

Once and Future History

"The Dirac Equation" and "Fermi and the Elucidation of Matter" are fraternal twins. "The Dirac Equation" was a homework assignment from Graham Farmelo, for his collection of essays on great equations, *It Must Be Beautiful*. "Fermi and the Elucidation of Matter" was a homework assignment from Jim Cronin, for a collection of tributes to Fermi, *Fermi Remembered*, to commemorate the 100th anniversary of his birth.

Superficially, matter and light seem to be utterly different sorts of things. And not only superficially: until well into the 20th century, the best scientific understandings of these two elements of reality used entirely different and contradictory concepts. Discrete particles were the ingredients for matter, but continuous fields described light. Matter was permanent, but light was transient. It was Dirac and Fermi, primarily, who transcended these dualities. Their work initiated modern quantum field theory. So these pieces provide a gentle introduction to "Quantum Field Theory."

There's a lot more in these two pieces than I'll mention here; let me only call special attention to the concluding section of "The Dirac Equation," which has some sharp observations on mathematical creativity.

A general observation: The more I've learned about most "great" historical figures like Alexander, Napoleon, or many others I won't name, the less admirable they've appeared to me. There are exceptions, notably Benjamin Franklin and Abraham Lincoln, but they are rare. In physics I've found the proportions reversed. The more I learn about Dirac and Fermi — and also Bohr (Item 33) — the more admirable, in different ways, they seem.

"The Standard Model Transcended" is a report, for *Nature* News and Views, on the historic first announcement that neutrinos have mass. I was present at the conference in Osaka, and the atmosphere of release and fulfillment there was tangible. People were glowing like newlyweds.

What does it mean? As I explain in "The Standard Model Transcended," a levitating pyramid now rests on one point.

"Masses and Molasses" and "In Search of Symmetry Lost" are half-sibs. Both are devoted to the same subject, namely the exotic superconductivity of what we see as empty space. (That phenomenon is commonly called the Higgs phenomenon, a name that manages to be both uninformative and misleading.) I know I'm the father of "Masses and Molasses" (never mind how), but it got its looks from somebody at *New Scientist*. Note especially the penetrance of the dominant short-sentence trait. Anyway, switching metaphors, the two pieces go together naturally: "Masses and Molasses" forms an appetizer to "In Search of Symmetry Lost," the entrée.

"From 'Not Wrong' to (Maybe) Right" and "Unification of Couplings" are another pair. Both deal with a remarkable, quantitative evidence for the validity of ambitious speculations about the validity of quantum field theory at ultra-small distances, ultimate unification of the fundamental forces of nature, and supersymmetry. The Large Hadron Collider (LHC) will bring forth abundant testimonial fruit. It's a thrilling prospect, in the abstract, though I have to admit that after more than 20 years of waiting I've learned how to control my excitement and get some sleep. "From 'Not Wrong' to (Maybe) Right" tells the inside personal story of the key calculation, while "Unification of Couplings" explains it scientifically.

The Dirac Equation

One cannot escape the feeling that these mathematical formulae have an independent existence and an intelligence of their own, that they are wiser than we are, wiser even than their discoverers, that we get more out of them than was originally put into them.

— H. Hertz, on Maxwell's equations for electromagnetism

A great deal of my work is just playing with equations and seeing what they give.

— P. A. M. Dirac

It gave just the properties one needed for an electron. That was really an unexpected bonus for me, completely unexpected.

— P. A. M. Dirac, on the Dirac equation

Of all the equations of physics, perhaps the most "magical" is the Dirac equation. It is the most freely invented, the least conditioned by experiment, the one with the strangest and most startling consequences.

In early 1928 (the receipt date on the original paper is January 2), Paul Adrien Maurice Dirac (1902–1984), a 25-year-old recent convert from electrical engineering to theoretical physics, produced a remarkable equation, forever to be known as the Dirac equation. Dirac's goal was quite concrete, and quite topical. He wanted to produce an equation that would describe the behavior of electrons more accurately than previous equations. Those equations incorporated either special relativity or quantum mechanics, but not both. Several other more prominent and experienced physicists were working on the same problem.

Unlike those other physicists, and unlike the great classics of physics, Newton and Maxwell, Dirac did not proceed from a minute study of experimental facts. Instead he guided his search using a few basic facts and perceived theoretical imperatives, some of which we now know to be wrong. Dirac sought to embody these principles in an economical, mathematically consistent scheme. By "playing with equations," as he put it, he hit upon a uniquely simple, elegant solution. This is, of course, the equation we now call the Dirac equation.

Some consequences of Dirac's equation could be compared with existing experimental observations. They worked quite well, and explained results that were otherwise quite mysterious. Specifically, as I'll describe below, Dirac's equation successfully predicts that electrons are always spinning and that they act as little bar magnets. It even predicts the rate of the spin and the strength of the magnetism. But other consequences appeared inconsistent with obvious facts. Notably, Dirac's equation contains solutions that seem to describe a way for ordinary atoms to wink out into bursts of light, spontaneously, in a fraction of a second.

For several years Dirac and other physicists struggled with an extraordinary paradox. How can an equation be "obviously right" since it accounts accurately for many precise experimental results, and achingly beautiful to boot — and yet manifestly, catastrophically wrong?

The Dirac equation became the fulcrum on which fundamental physics pivoted. While keeping faith in its mathematical form, physicists were forced to reexamine the meaning of the symbols it contains. It was in this confused, intellectually painful re-examination — during which Werner Heisenberg wrote to his friend Wolfgang Pauli, "The saddest chapter of modern physics is and remains the Dirac theory" and "In order not to be irritated with Dirac I have decided to do something else for a change..." — that truly modern physics began.

A spectacular result was the prediction of *antimatter* — more precisely, that there should be a new particle with the same mass as the electron, but the opposite electric charge, and capable of annihilating an electron into pure energy. Particles of just this type were promptly identified, through painstaking scrutiny of cosmic ray tracks, by Carl Anderson in 1932.

The more profound, encompassing result was a complete reworking of the foundations of our description of matter. In this new physics, particles are mere ephemera. They are freely created and destroyed; indeed, their fleeting existence and exchange is the source of all interactions. The truly fundamental objects are universal, transformative ethers: quantum fields.

These are the concepts that underlie our modern, wonderfully successful Theory of Matter (usually called, quite unpoetically, the Standard Model). And the Dirac equation itself, drastically reinterpreted and vastly generalized, but never abandoned, remains a central pillar in our understanding of Nature.

1. Dirac's Problem and the Unity of Nature

The immediate occasion for Dirac's discovery, and the way he himself thought about it, was the need to reconcile two successful, advanced theories of physics that had gotten slightly out of synch. By 1928 Einstein's special theory of relativity was already over two decades old, well digested, and fully established. (The general theory, which describes gravitation, is not part of our story here. Gravity is negligibly weak on atomic scales.) On the other hand, the new quantum mechanics of Heisenberg and Schrödinger, although quite a young theory, had already provided brilliant insight into the structure of atoms, and successfully explained a host of previously mysterious phenomena. Clearly, it captured essential features of the dynamics of electrons in atoms. The difficulty was that the equations developed by Heisenberg and Schrödinger did not take off from Einstein's relativistic mechanics, but from the old mechanics of Newton. Newtonian mechanics can be an excellent approximation for systems in which all velocities are much smaller than the speed of light, and this includes many cases of interest in atomic physics and chemistry. But the experimental data on atomic spectra, which one could address with the new quantum theory, was so accurate that small deviations from the Heisenberg–Schrödinger predictions could be observed. So there was a strong "practical" motivation to search for a more accurate electron equation, based on relativistic mechanics. Not only young Dirac, but also several other major physicists, were after such an equation.

In hindsight we can discern that much more ancient and fundamental dichotomies were in play: light versus matter; continuous versus discrete. These dichotomies present tremendous barriers to the goal of achieving a unified description of Nature. Of the theories Dirac and his contemporaries sought to reconcile, relativity was the child of light and the continuum, while quantum theory was the child of matter and the discrete. After Dirac's revolution had run its course, all were reconciled, in the mind-stretching conceptual amalgam we call a quantum field.

The dichotomies light/matter and continuous/discrete go deep. The earliest sentient proto-humans noticed them. The ancient Greeks articulated them clearly, and debated them inconclusively. Specifically, Aristotle distinguished Fire and Earth as primary elements — light versus matter. And he argued, against the Atomists, in favor of a fundamental plenum ("Nature abhors a vacuum") — upholding the continuous, against the discrete.

These dichotomies were not relieved by the triumphs of classical physics; indeed, they were sharpened.

Newton's mechanics is best adapted to describing the motion of rigid bodies through empty space. While Newton himself in various places speculated on the possible primacy of either side of both dichotomies, Newton's followers emphasized his "hard, massy, impenetrable" atoms as the fundamental building blocks of Nature. Even light was modeled in terms of particles.

Early in the nineteenth century a very different picture of light, according to which it consists of waves, scored brilliant successes. Physicists accepted that there must be a continuous, space-filling ether to support these waves. The discoveries of Faraday and Maxwell, assimilating light to the play of electric and magnetic fields, which are themselves continuous entities filling all space, refined and reinforced this idea.

Yet Maxwell himself succeeded, as did Ludwig Boltzmann, in showing that the observed properties of gases, including many surprising details, could be explained if the gases were composed of many small, discrete, well-separated atoms moving through otherwise empty space. Furthermore J. J. Thomson experimentally, and Hendrik Lorentz theoretically, established the existence of electrons as building blocks of matter. Electrons appear to be indestructible particles, of the sort that Newton would have appreciated.

Thus as the twentieth century opened, physics featured two quite different sorts of theories, living together in uneasy peace. Maxwell's electrodynamics is a continuum theory of electric and magnetic fields, and of light, that makes no mention of mass. Newton's mechanics is a theory of discrete particles, whose *only* mandatory properties are mass and electric charge.[a]

---

[a] That is, to predict the motion of a particle you need to know its charge and its mass: no more, no less. The value of the charge can be zero; then the particle will have only gravitational interactions.

Early quantum theory developed along two main branches, following the fork of our dichotomies, but with hints of convergence.

One branch, beginning with Planck's work on radiation theory, and reaching a climax in Einstein's theory of photons, dealt with light. Its central result is that light comes in indivisible minimal units, photons, with energy and momentum proportional to the frequency of the light. This, of course, established a particle-like aspect of light.

The second branch, beginning with Bohr's atomic theory and reaching a climax in Schrödinger's wave equation, dealt with electrons. It established that the stable configurations of electrons around atomic nuclei were associated with regular patterns of wave vibrations. This established a wave-like property of matter.

Thus the fundamental dichotomies softened. Light is a bit like particles, and electrons are a bit like waves. But sharp contrasts remained. Two differences, in particular, appeared to distinguish light from matter sharply.

First, if light is to be made of particles, then they must be very peculiar particles, with internal structure, for light can be polarized. To do justice to this property of light, its particles must have some corresponding property. There can't be an adequate description of a light beam specifying only that it is composed of so-and-so many photons with such-and-such energies; those facts will tell us how bright the beam is, and what colors it contains, but not how it is polarized. To get a complete description, one must also be able to say which way the beam is polarized, and this means that each photon must somehow carry around an arrow that allows it to keep a record of the light's polarity. This would seem to take us away from the traditional ideal of elementary particles. If there's an arrow, what's *it* made of? And why can't it be separated from the particle?

Second, and more profound, photons are evanescent. Light can be radiated, as when you turn on a flashlight, or absorbed, as when you cover it with your hand. Therefore particles of light can be created or destroyed. This basic, familiar property of light and photons takes us far away from the traditional ideal of elementary particles. The stability of matter would seem to require indestructible building blocks, with properties fundamentally different from evanescent photons.

The Dirac equation, and the crisis it provoked, forced physicists, finally, to transcend all these dichotomies. The consequence is a unified concept of substance, that is surely one of humanity's greatest intellectual achievements.

2. The Early Payoff: Spin

Dirac was working to reconcile the quantum mechanics of electrons with special relativity. He thought — mistakenly, we now know — that quantum theory required equations of a particularly simple kind, the kind mathematicians call first-order. Never mind why he thought so, or precisely what first-order means; the point is that he wanted an equation that is, in a certain very precise sense, of the simplest possible kind. Tension arises because it is not easy to find an equation that is both simple in this sense and also consistent with the requirements of special relativity. To construct such an equation, Dirac had to expand the terms of the discussion. He found he could not get by with a single first-order equation — he needed a system of four intricately related ones, and it is actually this system we refer to as "the" Dirac equation.

Two equations were quite welcome. Four, initially, were a big problem.

First, the good news.

Although the Bohr theory gave a good rough account of atomic spectra, there were many discrepant details. Some of the discrepancies concerned the number of electrons that could occupy each orbit, others involved the response of atoms to magnetic fields, as revealed in the movement of their spectral lines. Wolfgang Pauli had shown, through detailed analysis of the experimental evidence, that Bohr's model could describe complex atoms, even roughly, only if there were a tight restriction on how many electrons could occupy any given orbit. This is the origin of the famous Pauli exclusion principle. Today we learn this principle in the form "only one electron can occupy a given state." But Pauli's original proposal was not so neat; it came with some disturbing fine print. For the number of electrons that could occupy a given Bohr orbital was not one, but two. Pauli spoke obscurely of a "classically non-describable duplexity," but — needless to say — did not describe any reason for it.

In 1925 two Dutch graduate students, Samuel Goudsmit and George Uhlenbeck, devised a possible explanation of the magnetic response problems. If electrons were actually tiny magnets, they showed, the discrepancies would disappear. Their model's success required that all electrons must have the same magnetic strength, which they could calculate. They went on to propose a mechanism for the electron's magnetism. Electrons, of course, are electrically charged particles. Electric charge in circular motion generates magnetic fields. Thus, if for some reason electrons were always rotating about their own axis, their magnetism might be explained.

This intrinsic *spin* of electrons would have an additional virtue. If the rate of spin were the minimum allowed by quantum mechanics,[b] then Pauli's "duplexity" would be explained. For the spin would have no freedom to vary in magnitude, but only the option to point either up or down. Many eminent physicists were quite skeptical of Goudsmit and Uhlenbeck. Pauli himself tried to dissuade them from publishing their work. For one thing, their model seemed to require the electron to rotate at an extraordinarily rapid rate, at its surface probably faster than the speed of light. For another, they gave no account of what holds an electron together. If it is an extended distribution of electric charge, all of the same sign, it will want to fly apart — and rotation, by introducing centrifugal forces, only makes the problem worse. Finally, there was a quantitative mismatch between their requirements for the strength of the electron's magnetism and the amount of its spin. The ratio of these two quantities is governed by a factor called the gyromagnetic ratio, written g. Classical mechanics predicts $g = 1$, whereas to fit the data Goudsmit and Uhlenbeck postulated $g = 2$. But despite these quite reasonable objections, their model stubbornly continued to agree with experimental results!

Enter Dirac. His system of equations allowed a class of solutions, for small velocities, in which only two of the four functions appearing in his equations are appreciable. This was duplexity, but with a difference. Here it fell out automatically as a consequence of implementing general principles, and most definitely did not have to be introduced *ad hoc*. Better yet, using his equation Dirac could calculate the magnetism of electrons, also without further assumptions. He got $g = 2$. Dirac's great paper of 1928 wastes no words. Upon demonstrating this result, he says simply

> The magnetic moment is just that assumed in the spinning electron model.

And a few pages later, after working out the consequences, he concludes laconically

> The present theory will thus, in the first approximation, lead to the same energy levels as those obtained by [C. G.] Darwin, which are in agreement with experiment.

---

[b]In quantum mechanics, only certain values of the discrete spin are allowed. This is closely related to the restriction on allowed Bohr orbitals.

His results spoke loudly for themselves, with no need for amplification. From then on, there was no escaping Dirac's equation. Whatever difficulties arose — and there were some big and obvious ones — they would be occasions for struggle, not desertion. Such gleaming jewels of insight would be defended at all costs.

Although his intellectual starting point, as I mentioned, was quite different and more abstract, Dirac begins his paper by referring to Goudsmit, Uhlenbeck, and the experimental success of their model. Only in the second paragraph does he reveal his hand. What he says is quite pertinent to the themes I emphasized above.

> The question remains as to why Nature should have chosen this particular model for the electron instead of being satisfied with a point-charge. One would like to find some incompleteness in the previous methods of applying quantum mechanics to the point-charge such that, when removed, the whole of the duplexity phenomena follow without arbitrary assumptions.

Thus Dirac is not offering a new model of electrons, as such. Rather, he is defining a new *irreducible* property of matter, inherent in the nature of things, specifically in the consistent implementation of relativity and quantum theory, that arises even in the simplest possible case of structureless point particles. Electrons happen to be embodiments of this simplest possible form of matter. Dirac retains the valuable properties of Goudsmit and Uhlenbeck's "spin," specifically its fixed magnitude and its magnetic action, which aid in the description of observed realities, but bases them on a much deeper foundation. The arbitrary and unsatisfactory features of their model are bypassed.

We were looking for an arrow that would be a necessary and inseparable part of elementary bits of matter, like polarization for photons. Well, there it is!

The spin of the electron has many practical consequences. It is responsible for ferromagnetism, and for the enhancement of magnetic fields in the core of electric coils, that forms the heart of modern power technology (motors and dynamos). By manipulating electron spins, we can store and retrieve a great deal of information in a very small volume (magnetic tape, disk drives). Even the much smaller and more inaccessible spin of atomic nuclei plays a big role in modern technology. Manipulating such spins with radio and magnetic fields, and sensing their response, is the basis of the magnetic resonance imaging (MRI) so useful in medicine.

This application, among many others, would be inconceivable (literally!) without the exquisite control of matter that only fundamental understanding can bring.

Spin in general, and Dirac's prediction for the magnetic moment in particular, has also played a seminal role in the subsequent development of fundamental physics. Small deviations from Dirac's $g=2$ were discovered by Polykarp Kusch and collaborators in the 1940s. They provided some of the first quantitative evidence for the effects of virtual particles, a deep and characteristic property of quantum field theory. Very large deviations from $g=2$ were observed for protons and neutrons in the 1930s. This was an early indication that protons and neutrons are not fundamental particles in the same sense that electrons are. But I'm getting ahead of the story ...

3. The Dramatic Surprise: Antimatter

Now for the 'bad' news.

Dirac's equation consists of four components. That is, it contains four separate wave functions to describe electrons. Two components have an attractive and immediately successful interpretation, as we just discussed, describing the two possible directions of an electron's spin. The extra doubling, by contrast, appeared at first to be quite problematic.

In fact, the extra equations contain solutions with *negative* energy (and either direction of spin). In classical (non-quantum) physics the existence of extra solutions would be embarrassing, but not necessarily catastrophic. For in classical physics, you can simply choose not to use these solutions. Of course that begs the question why *Nature* chooses not to use them, but it is a logically consistent procedure. In quantum mechanics, even this option is not available. In quantum physics, generally "that which is not forbidden is mandatory." In the specific case at hand, we can be quite specific and precise about this. All solutions of the electron's wave equation represent possible behaviors of the electron, that will arise in the right circumstances. Assuming Dirac's equation, if you start with an electron in one of the positive-energy solutions, you can calculate the rate for it to emit a photon and transition into one of the negative-energy solutions. Energy must be conserved overall, but that is not a problem here — it just means that the energy of the emitted photon would be *more* than that of the electron which emitted it! Anyway, the rate turns out to be ridiculously fast, a small fraction of a second. So you can't ignore the negative-energy solutions for long. And since an electron has never been observed to do something so peculiar as radiating more energy than it starts with, there was, on the fact of it, a terrible problem with the quantum mechanics of Dirac's equation.

Dirac was well aware of this problem. In his original paper, he simply acknowledged

> For this second class of solutions W [the energy] has a negative value. One gets over the difficulty on the classical theory by arbitrarily excluding those solutions that have a negative W. One cannot do this on the quantum theory, since in general a perturbation will cause transitions from states with W positive to states with W negative. ... The resulting theory is therefore still only an approximation, but it appears to be good enough to account for all the duplexity phenomena without arbitrary assumptions.

and left it at that. This was the situation that provoked Heisenberg's outbursts to Pauli, quoted earlier.

By the end of 1929 — not quite two years later — Dirac made a proposal to address the problem. It exploited the Pauli exclusion principle, according to which no two electrons obey the same solution of the wave equation. What Dirac proposed was a radically new conception of empty space. He proposed that what we consider 'empty' space is in reality chock-a-block with negative-energy electrons. In fact, according to Dirac, *'empty' space actually contains electrons obeying all the negative energy solutions.* The great virtue of this proposal is that it explains away the troublesome transitions from positive to negative solutions. A positive-energy electron can't go to a negative-energy solution, because there's always another electron already there and the Pauli exclusion principle won't allow a second one to join it.

It sounds outrageous, on first hearing, to be told that what we perceive as empty space is actually quite full of stuff. But, on reflection, why not? We have been sculpted by evolution to perceive aspects of the world that are somehow useful for our survival and reproductive success. Since unchanging aspects of the world, upon which we can have little influence, are not useful in this way, it should not seem terribly peculiar that they would escape our untutored perception. In any case, we have no warrant to expect that naive intuitions about what is weird or unlikely provide reliable guidance for constructing models of fundamental structure in the microworld, because these intuitions derive from an entirely different realm of phenomena. We must take it as it comes. The validity of a model must be judged according to the fruitfulness and accuracy of its consequences.

So Dirac was quite fearless about outraging common sense. He focused, quite properly, on the observable consequences of his proposal.

Since we are considering the idea that the ordinary state of "empty" space is far from empty, it is helpful to have a different, more noncommittal word for it. The one physicists like to use is "vacuum."

In Dirac's proposal, the vacuum is full of negative-energy electrons. This makes the vacuum a medium, with dynamical properties of its own. For example, photons can interact with the vacuum. One thing that can happen is that if you shine light on the vacuum, providing photons with enough energy, then a negative-energy electron can absorb one of these photons, and go into a positive-energy solution. The positive-energy solution would be observed as an ordinary electron, of course. But in the final state there is also a *hole* in the vacuum, because the solution originally occupied by the negative-energy electron is no longer occupied.

The idea of holes was, in the context of a dynamical vacuum, startlingly original, but it was not quite unprecedented. Dirac drew on an analogy with the theory of heavy atoms, which contain many electrons. Within such atoms, some of the electrons correspond to solutions of the wave equation that reside nearby the highly charged nucleus, and are very tightly bound. It takes a lot of energy to break such electrons free, and so under normal conditions they present an unchanging aspect of the atom. But if one of these electrons absorbs a high-energy photon (an X-ray) and is ejected from the atom, the change in the normal aspect of the atom is marked by its *absence*. The absence of an electron, which would have supplied negative charge, by contrast looks like a positive charge. The positive effective charge follows the orbit of the missing electron, so it has the properties of a positively charged particle.

Based on this analogy and other hand-waving arguments — the paper is quite short, and almost devoid of equations — Dirac proposed that holes in the vacuum are positively charged particles. The process where a photon excites a negative-energy electron in the vacuum to a positive energy is then interpreted as the photon creating an electron and a positively charged particle (the hole). Conversely, if there is a preexisting hole, then a positive-energy electron can emit a photon and occupy the vacant negative-energy solution. This is interpreted as the annihilation of an electron and a hole into pure energy. I referred to a photon being emitted, but this is only one possibility. Several photons might be emitted, or any other form of radiation that carries away the liberated energy.

Dirac's first hole theory paper was entitled "A Theory of Electrons and Protons." At the time protons were the only known positively charged particles. It was therefore natural to try to identify the hypothetical holes as protons. But this identification soon raised severe difficulties. Specifically, the two sorts of process we just discussed — production of electron-proton pairs, and annihilation of electron-proton pairs — have never been observed. The second is even more problematic, because it predicts that hydrogen atoms spontaneously self-destruct in microseconds — which, thankfully, they do not.

The identification of holes with protons also involved a logical difficulty. Based on the symmetry of the equations, one could demonstrate that the holes must have the same mass as the electrons. But a proton has, of course, a much larger mass than an electron.

In 1931 Dirac withdrew his earlier identification of holes with protons, and accepted the logical outcome of his own equation and the dynamical vacuum it required:

> A hole, if there was one, would be a new kind of elementary particle, unknown to experimental physics, having the same mass and opposite charge of the electron.

On August 2, 1932, Carl Anderson, an American experimentalist studying photographs of the tracks left by cosmic rays in a cloud chamber, noticed some tracks that lost energy as expected for electrons, but were bent in the opposite direction by the magnetic field. He interpreted this as indicating the existence of a new particle, now known as the anti-electron or positron, with the same mass as the electron but the opposite electric charge. Ironically, Anderson was completely unaware of Dirac's prediction.

Thousands of miles away from his rooms at Saint John's, Dirac's holes — the product of his theoretical vision and revision — had been found, descending from the skies of Pasadena. So in the long run the "bad" news turned out to be "even better" news. Negative-energy frogs became positronic princes.

Today positrons are no longer a marvel, but a tool. A notable use is to take pictures of the brain in action — PET scans, for positron-electron tomography. How do positrons get into your head? They are snuck in by injecting molecules containing atoms whose radioactive nuclei decay with positrons as one of their products. These positrons do not go very far before they annihilate against some nearly electron, usually producing two photons, which escape your skull, and can be detected. Then you can reconstruct where the original molecule went, to map out metabolism, and you can also study energy loss of the photons on the way out, to get a density profile, and ultimately an image, of the brain tissue.

Another notable application is to fundamental physics. You can accelerate positrons to high energy, as you can of course electrons, and bring the beams together. Then the positrons and electrons will annihilate, producing a highly concentrated form of "pure energy." Much of the progress in fundamental physics over the past half century has been based on studies of this type, at a series of great accelerators all over the world, the latest and greatest being the LEP (large electron-positron) collider at CERN, outside Geneva. I'll be discussing a stunning highlight of this physics a little later.

The physical ideas of Dirac's hole theory, which as I mentioned had some of its roots in the earlier study of heavy atoms, fed back in a big way into solid-state physics. In solids one has a reference or ground configuration of electrons, with the lowest possible energy, in which electrons occupy all the available states up to a certain level. This ground configuration is the analog of the vacuum in hole theory. There are also configurations of higher energy, wherein some of the low-energy states are not used by any electron. In these configurations there are vacancies or "holes" — that's what they're called, technically — where an electron would ordinarily be. Such holes behave in many respects like positively charged particles. Solid-state diodes and transistors are based on clever manipulation of holes and electron densities at junctions between different materials. One also has the beautiful possibility to direct electrons and holes to a place where they can combine (annihilate). This allows you to design a source of photons that you can control quite precisely, and leads to such mainstays of modern technology as LEDs (light-emitting diodes) and solid-state lasers.

In the years since 1932 many additional examples of anti-particles have been observed. In fact, for every particle that has ever been discovered, a corresponding anti-particle has also been found. There are antineutrons, antiprotons, antimuons (the muon itself is a particle very similar to the electron, but heavier), antiquarks of various sorts, even antineutrinos, and anti-π mesons, anti-K mesons...[c] Many of these particles do not obey the Dirac equation, and some of them do not even obey the Pauli exclusion principle. So the physical reason for the existence of antimatter must be very general — much more general than the arguments that first led Dirac to predict the existence of positrons.

In fact, there is a very general argument that if you implement both quantum mechanics and special relativity, every particle must have a corresponding antiparticle. A proper presentation of the argument requires either sophisticated mathematical background or lots of patience. Here I'll be content with a rough demonstration of why antimatter is a plausible consequence of implementing both relativity and quantum mechanics.

Consider a particle, let's say a shmoo, to give it a name (while emphasizing that it could be *anything*), moving east at very nearly the speed of light. According to quantum mechanics, there is actually some uncertainty in its position. So there's some probability, if you measure it, that you will find that the shmoo is slightly west of its expected mean position at an initial time, and slightly east of its expected mean position at a later time. So it has traveled further than you might have expected during this interval — which means it was traveling more quickly. But since the expected velocity was essentially the speed of light, the faster speeds required to accommodate uncertainty threaten to violate special relativity, which requires that particles cannot move faster than the speed of light. It's a paradox.

With antiparticles, you can escape the paradox. It requires orchestrating a symphony of weird ideas, but it's the only way people have figured out how to do it, and it seems to be Nature's way. The central idea is that, yes, uncertainty does mean that you can find a shmoo where special relativity tells you your shmoo can't be — but the shmoo you observe is not necessarily the same as the one you were looking for! For it's also possible that at the later time there are two shmoos, the original one and a new one. To make this consistent there must also be an anti-shmoo, to balance the charge, and to cancel out any other conserved quantities that might be associated with the additional shmoo. What about the energy balance — aren't we getting out more than we put in? Here, as often in quantum theory, to avoid contradictions you must be specific and concrete in thinking about what it means to measure something. One

---

[c]An interesting case is the photon, which is its own antiparticle. This is not possible for a charged particle, but the photon is electrically neutral.

way to measure the shmoo's position would be to shine light on it. But to measure the position of a fast-moving shmoo accurately we have to use high-energy photons, and there's also then the possibility such a photon will create a shmoo-anti-shmoo pair. And in that case — closing the circle — when you report the result of your position measurement, you might be talking about the wrong shmoo!

4. The Deepest Meanings: Quantum Field Theory

Dirac's hole theory is brilliantly clever, but Nature goes deeper. Although hole theory is internally consistent, and can cover a wide range of applications, there are several important considerations that force us to go beyond it.

First, there are particles that do not have spin, and do not obey the Dirac equation, and yet have antiparticles. This is no accident: the existence of antiparticles is a general consequence of combining quantum mechanics and special relativity, as I just discussed. Specifically, for example, positively charged π^+ mesons (discovered in 1947) or W^+ bosons (discovered in 1983) are quite important players in elementary particle physics, and they do have antiparticles π^- and W^-. But we can't use Dirac's hole theory to make sense of these antiparticles, because π^+ and W^+ particles don't obey the Pauli exclusion principle. So there is no possibility of interpreting their antiparticles as holes in a filled sea of negative-energy solution. If there are negative-energy solutions, whatever equation they satisfy,[d] occupying them with one particle will not prevent another particle from entering the same state. Thus catastrophic transitions into negative-energy states, which Dirac's hole theory prevents for electrons, must be avoided in a different way.

Second, there are processes in which the number of electrons minus the number of positrons changes. An example is the decay of a neutron into a proton, an electron, and an antineutrino. In hole theory the excitation of a negative-energy electron into a positive-energy state is interpreted as creation of a positron-electron pair, and de-excitation of a positive-energy electron into an unoccupied negative-energy state is interpreted as annihilation of an electron-positron pair. In neither case does the difference between the number of electrons and the number of

---

[d]In fact these particles obey wave equations that do have negative-energy solutions.

positrons change. Hole theory cannot accommodate changes in this difference. So there are definitely important processes in Nature, even ones specifically involving electrons, that do not fit easily into Dirac's hole theory.

The third and final reason harks back to our initial discussion. We were looking to break down the great dichotomies light/matter and continuous/discrete. Relativity and quantum mechanics, separately, brought us close to success, and the Dirac equation, with its implication of spin, brought us closer still. But so far we haven't quite got there. Photons are evanescent, electrons ... well, they're evanescent too, as a matter of experimental fact, as I just mentioned, but we haven't yet adequately fit that feature into our theoretical discussion. In hole theory electrons can come and go, but only as positrons go and come.

These are not so much contradictions as indications of missed opportunity. They indicate that there ought to be some alternative to hole theory that covers all forms of matter, and that treats the creation and destruction of particles as a primary phenomenon.

Ironically, Dirac himself had earlier constructed the prototype of such a theory. In 1927, he applied the principles of the new quantum mechanics to Maxwell's equations of classical electrodynamics. He showed that Einstein's revolutionary postulate that light comes in particles — photons — was a consequence of the logical application of these principles, and that the properties of photons were correctly accounted for. Few observations are so common as that light can be created from non-light, say by a flashlight, or absorbed and annihilated, say by a black cat. But translated into the language of photons, this means that the quantum theory of Maxwell's equations is a theory of the creation and destruction of particles (photons). Indeed, the electromagnetic field appears, in Dirac's quantum theory of electromagnetism, primarily as an agent of creation and destruction. Photons arise as excitations of this field, which is the primary object. Photons come and go, but the field abides. The full significance of this development seems to have escaped Dirac and all of his contemporaries for some time, perhaps precisely because of the apparent specialness of light (dichotomy!). But it is a general construction, which can be applied to the object that appears in Dirac's equation — the electron field — as well.

The result of a logical application of the principles of quantum mechanics to Dirac's equation is an object similar to what he found for Maxwell's equations. It is an object that destroys electrons, and creates positrons.[e] Both are examples of *quantum fields*. When the object that appears in Dirac's equation is interpreted as a quantum field, the negative-energy solutions take on a completely different meaning, with no problematic aspects. The positive-energy solutions multiply electron destruction operators, while the negative-energy solutions multiply positron creation operators. In this framework, the difference between the two kinds of solution is that negative energy represents the energy you need to borrow to make a positron, while positive energy is what you gain by destroying an electron. The possibility of negative numbers is no more paradoxical here than in your bank balance.

With the development of quantum field theory, the opportunities that Dirac's equation and hole theory made evident, but did not quite fulfill, were finally met. The description of light and matter was put, at last, on a common footing. Dirac said, with understandable satisfaction, that with the emergence of quantum electrodynamics physicists had attained foundational equations adequate to describe "all of chemistry, and most of physics."

In 1932 Enrico Fermi constructed a successful theory of radioactive decays (beta decays), including the neutron decay I mentioned before, by exporting the concepts of quantum field theory far from their origin. Since these processes involve the creation and destruction of protons — the epitome of 'stable' matter — the old dichotomies had finally been transcended. Both particles and light are epiphenomena, surface manifestations of the deeper and abiding realities, quantum fields. These fields fill all of space, and in this sense they are continuous. But the excitations they create, whether we recognize them as particles of matter or as particles of light, are discrete.

In hole theory we had a picture of the vacuum as filled with a sea of negative-energy electrons. In quantum field theory, the picture is quite different. But there is no returning to innocence. The new picture of the vacuum differs even more radically from naive "empty space." Quantum uncertainty, combined with the possibility of processes of creation and destruction, implies a vacuum teeming with activity. Pairs of particles and antiparticles fleetingly come to be and pass away. I once wrote a sonnet about virtual particles, and here it comes:

---

[e]There is also a closely related object, the Hermitean conjugate, that creates electrons and destroys positrons.

Beware of thinking nothing's there —
Remove what you can; despite your care
Behind remains a restless seething
Of mindless clones beyond conceiving.

They come in a wink, and dance about;
Whatever they touch is seized by doubt:
What am I doing here? What should I weigh?
Such thoughts often lead to rapid decay.

Fear not! The terminology's misleading;
Decay is virtual particle breeding
And seething, though mindless, can serve noble ends,
The clone-stuff, exchanged, makes a bond between friends.

To be or not? The choice seems clear enough,
But Hamlet oscillated. So does this stuff.

5. Aftermaths

With the genesis of quantum field theory, we reach a natural intellectual boundary for our discussion of the Dirac equation. By the mid-1930s the immediate paradoxes this equation raised had been resolved, and its initial promise had been amply fulfilled. Dirac received the Nobel Prize in 1933, Anderson in 1935.

In later years the understanding of quantum field theory deepened, and its applications broadened. Using it, physicists have constructed (and established with an astonishing degree of rigor and beyond all reasonable doubt) what will stand for the foreseeable future — perhaps for all time — as the working Theory of Matter. How this happened, and the nature of the theory, is an epic story involving many other ideas, in which the Dirac equation as such plays a distinguished but not a dominant role. But some later developments are so closely linked to our main themes, and so pretty in themselves, that they deserve mention here.

There is another sense in which the genesis of quantum field theory marks a natural boundary. It is the limit beyond which Dirac himself did not progress. Like Einstein, in his later years Dirac took a separate path. He paid no attention to most of the work of other physicists, and dissented from the rest. In the marvelous developments that his work commenced, Dirac's own participation was peripheral.

5.1. QED and Magnetic Moments

Interaction with the ever-present dynamical vacuum of quantum field theory modifies the observed properties of particles. We do not see the hypothetical properties of the "bare" particles, but rather the physical particles, "dressed" by their interaction with the quantum fluctuations in the dynamical vacuum.

In particular, the physical electron is not the bare electron, and it does not quite satisfy Dirac's $g = 2$. When Polykarp Kusch made very accurate measurements, in 1947, he found that g is larger than 2 by a factor 1.00119. Now this is not a very large correction, quantitatively, but it was a great stimulus to theoretical physics, because it provided a very concrete challenge. At that time there were so many loose ends in fundamental physics — a plethora of unexpected, newly discovered particles including muons, π mesons, and others, no satisfactory theory explaining what force holds atomic nuclei together, fragmentary and undigested results about radioactive decays, anomalies in high-energy cosmic rays — that it was hard to know where to focus. In fact, there was a basic philosophical conflict about strategy.

Most of the older generation, the founders of quantum theory, including Einstein, Schrödinger, Bohr, Heisenberg, and Pauli, were prepared for another revolution. They thought it was fruitless to spend time trying to carry out more accurate calculations in quantum electrodynamics, since this theory was surely incomplete and probably just wrong. It did not help that the calculations required to get more accurate results are very difficult, and that they seemed to give senseless (infinite) answers. So the old master were searching for a different kind of theory, unfortunately with no clear direction.

Ironically, it was a younger generation of theorists — Schwinger, Feynman, Dyson, and Tomonaga in Japan — who played a conservative role.[f] They found a way to perform the more accurate calculations, and get meaningful finite results, without changing the underlying theory. The theory they used, in fact, was just the one Dirac had constructed in the 20s and 30s. The result of an epochal calculation by Schwinger, including the effects of the dynamic vacuum, was a small correction to Dirac's $g = 2$. It too was reported in 1947, and it agreed spectacularly well with

---

[f] Seminal contributions were also made by the slightly older theorists Kramers and Bethe, and by the theorist-turned-experimentalist Lamb.

Kusch's contemporary measurements. Many other triumphs followed. Kusch received the Nobel Prize in 1955; Schwinger, Feynman, and Tomonaga jointly in 1965 (the delay is hard to understand!).

Strangely, Dirac did not accept the new procedures. Caution was perhaps justified in the early days, when the mathematical methods being used were unfamiliar and sketchy and involved a certain amount of inspired guesswork. But the technical difficulties were cleaned up in due course.[g]

Feynman called QED "the jewel of physics — our proudest possession." But in 1951 Dirac wrote

> Recent work by Lamb, Schwinger and Feynman and others has been very successful ... but the resulting theory is an ugly and incomplete one.

And in his last paper, in 1984,

> These rules of renormalization give surprisingly good agreement with experiments. Most physicists say that these working rules are, therefore, correct. I feel that this is not an adequate reason. Just because the results happen to be in agreement with experiment does not prove that one's theory is correct.

You might notice a certain contrast in tone between the young Dirac, who clung to his equation like a barnacle because it explained experimental results, and the older inhabitant of the same body.

Today the experimental determination of the magnetic moment of the electron is

$$(g/2)_{\text{experiment}} = 1.001\ 159\ 652\ 188\ 4\ (43)$$

while the theoretical prediction, firmly based on QED, calculated to high accuracy, is

$$(g/2)_{\text{theory}} = 1.001\ 159\ 652\ 187\ 9\ (43)$$

---

[g]Although QED does have problems of principle, if it is regarded (unrealistically!) as a completely closed theory, they are problems at a different level than what troubled Dirac, and they are very plausibly solved by embedding QED into a larger, asymptotically free theory — see below. This has scant practical effect on most of its predictions.

where the uncertainty in the last two digits is indicated. It is the toughest, most accurate confrontation between intricate — but precisely defined! — theoretical calculations and delicate — but precisely controlled! — experiments in all of science. That's what Feynman meant by "our proudest possession."

Ever more accurate determination of the magnetic moment of the electron, and of its kindred particle the muon, remains an important frontier of experimental physics. With the accuracies now achievable, the results will be sensitive to effects of quantum fluctuations due to hypothetical new heavy particles — in particular, those predicted by supersymmetry.

5.2. QCD and the Theory of Matter

The magnetic moment of the proton does not satisfy Dirac's $g = 2$, but instead has $g \approx 5.6$. For neutrons it is worse. Neutrons are electrically neutral, so the simple Dirac equation for neutrons predicts no magnetic moment at all. In fact the neutron has a magnetic moment about 2/3 as large as that of a proton, and with the opposite orientation relative to spin. That corresponds to an infinite value of g, since the neutron is electrically neutral. The discrepant values of these magnetic moments were the earliest indication that protons and neutrons are more complicated objects than electrons.

With further study, many more complications appeared. The forces among protons and neutrons were found to be very complicated. They depend in a bewildering way not only on the distance between them, but also on their velocities, and spin orientations, and all combinations of these together. In fact, it soon appeared that they are not "forces" in the traditional sense at all. To have a force between protons, in the traditional sense, would mean that the motion of one proton can be affected by the presence of another, so that when you shoot one proton by another, it swerves. What you actually observe is that when one proton collides with another, typically many particles emerge, most of which are highly unstable. There are π mesons, K mesons, ρ mesons, Λ and Σ baryons, their antiparticles, and many more. All these particles interact very powerfully with each other. And so the problem of nuclear forces, a frontier of physics starting in the 1930s, became the problem understanding a vast new world of particles and reactions, the most powerful in Nature. Even the terminology changed. Physicists no longer refer to nuclear forces, but to the strong interaction.

Now we know that all the complexities of the strong interaction can be described, at a fundamental level, by a theory called quantum chromodynamics, or QCD, a vast generalization of QED. The elementary building blocks of QCD are quarks and gluons. There are six different kinds, or 'flavors' of quarks: u, d, s, c, b, t (up, down, strange, charm, bottom, top). The quarks are very similar to one another, differing mainly in their mass. Only the lightest ones, u and d, are found in ordinary matter. Making an analogy to the building blocks of QED, quarks play roughly the role of electrons, and gluons play roughly the role of photons. The big difference is that whereas in QED there is just one type of charge, and one photon, in QCD there are three types of charge, called colors, and eight gluons. Some gluons respond to color charges, similarly to the way photons respond to electric charge. Others mediate transitions between one color and another. Thus (say) a u quark with blue charge can radiate a gluon and turn into a u quark and green charge. Since all the charges overall must be conserved, this particular gluon must have blue charge $+1$, green charge -1. Since gluons themselves carry unbalanced color charge, in QCD there are elementary processes where gluons radiate other gluons. There is nothing like that in QED. Photons are electrically neutral, and to a very good approximation they do not interact with other photons. Much of the richness and complexity of QCD arises because of this new feature.

Described thus baldly and verbally, without grounding in concepts or phenomena, QCD might seem both arbitrary and fantastic. Unfortunately, I won't be able to do justice to its symmetry and mathematical beauty here. But some brief explications are in order. ...

How did we arrive at such a theory? And how do we know it's right? In the case of QCD, these are two very different questions. The historical path to its discovery was tortuous, with many false trails and blind alleys. But in retrospect, it didn't have to be that way. If the right kind of ultrahigh-energy accelerators had come on line earlier, QCD would have stared us in the face.[h] This *gedanken*-history brings together most of the ideas I've discussed in this article, and forms a fitting conclusion to its physical part.

---

[h] Up to a couple of profound but well-posed and solvable problems, as I'll shortly discuss.

When electrons and positrons are accelerated to ultrahigh energy and then made to collide, two kinds of events are observed. In one kind of event the particles in the final states are leptons and photons. For this class of events, usually the final state is just a lepton and its anti-lepton; but in about 1% of the events there is also a photon, and in about 0.01% of the events there are also two photons. The probability for these sorts of events, and for the various particles to come out at various angles with different energies, can all be computed using QED, and it all works out very nicely. Conversely, if you hadn't known about QED, you could have figured out the basic rules for the fundamental interaction of QED — that is, the emission of a photon by an electron — just by studying these events. The fundamental interaction of light with matter is laid out right before your eyes.

In the other kind of event, you see something rather different. Instead of just two or at most a handful of particles coming out, there are many. And they are different kinds of particles. The particles you see in this second class of events are things like π mesons, K mesons, protons, neutrons, and their antiparticles — all particles that, unlike photons and leptons, have strong interactions. The angular distribution of these particles is very structured. They do not come out independently, every which way. Rather, they emerge in just a few directions, making narrow sprays or (as they're usually called) "jets." About 90% of the time there are just two jets, in opposite directions; roughly 90% of the time there are three jets, .9% four jets — you can guess the pattern.

Now if you squint a little, and don't resolve the individual particles, but just follow the flow of energy and momentum, then the two kinds of events — the QED 'particle' events, and the 'jetty' events with strongly interacting particles — look just the same!

So (in this imaginary history) it would have been hard to resist the temptation to treat the jets as if they are particles, and propose rules for the likelihood of different radiation patterns, with different numbers, angles, and energies of the jet-particles, in direct analogy to the procedures that work for QED. And this would work out very nicely, because rules quite similar to those for QED actually do describe the observations. Of course, the rules that work are precisely those of QCD, including the new processes where glue radiates glue. All these rules — the foundational elements of the entire theory — could have been derived directly from the data. "Quarks" and "gluons" would be words with direct and precise operational definitions, in terms of jets.

Still, there would have been two big conceptual puzzles. Why do the experiments show 'quarks' and 'gluons' instead of just quarks and gluons — that is, jets, instead of just particles? And how do you connect the theoretical concepts that directly and successfully describe the high-energy events to all the other phenomena of the strong interaction? The connection between the supposedly foundational theory and the mundane observations is, to say the least, not obvious. For example, you would like to construct protons out of the 'quarks' and 'gluons' that appear in the fundamental theory. But this looks hopeless, since the jets in terms of which 'quarks' and 'gluons' are operationally defined often contain, among other things, protons.

There is an elegant solution to these problems. It is the phenomenon of *asymptotic freedom* in QCD. According to asymptotic freedom, radiation events that involve large changes in the flow of energy and momentum are rare, while radiation events that involve only small changes in energy and momentum are very common. Asymptotic freedom is not a separate assumption, but a deep mathematical consequence of the structure of QCD.

Asymptotic freedom neatly explains why there are jets in electron-positron annihilations at high energies, in the class of events containing strongly interacting particles. Immediately after the electron and positron annihilate, you have a quark and an antiquark emerging. They are moving rapidly, in opposite directions. They quickly radiate gluons, and the gluons themselves radiate, and a complicated cascade develops, with many particles. But despite all this commotion the overall flow of energy and momentum is not significantly disturbed. Radiations that disturb the flow of energy and momentum are rare, according to asymptotic freedom. So there is a large multiplicity of particles all moving in the same direction, the direction originally staked out by the quark or antiquark. In a word, we've produced a jet. When one of those rare radiations that disturbs the flow of energy and momentum takes place, the radiated gluon starts a jet of its own. Then we have a three-jet event. And so forth.

Asymptotic freedom also indicates why the description of protons (and the other strongly interacting particles) that we actually observe as individual stable, or quasi-stable, entities are complicated objects. For such particles are, more or less by definition, configurations of quarks, antiquarks, and gluons that have a reasonable degree of stability. But since the quarks, antiquarks, and gluons all have a very high probability for radiating, no simple configuration will have this property. The only possibility for stability involves dynamic equilibrium, in which the emission of radiation in one part of the system is balanced by its absorption somewhere else.

As things actually happened, asymptotic freedom was discovered theoretically (by Davd Gross and me, and independently by David Politzer) and QCD was proposed as the theory of the strong interaction (by Gross and me) in 1973, based on much less direct evidence. The existence of jets was anticipated, and their properties were *predicted* theoretically, in considerable detail, before their experimental observation. Based on these experiments, and many others, today QCD is accepted as the fundamental theory of the strong interaction, on a par with QED as the description of the electromagnetic interaction.

There has also been enormous progress in using QCD to describe the properties of protons, neutrons, and the other strongly interacting particles. This involves very demanding numerical work, using the most powerful computers, but the results are worth it. One highlight is that we can calculate from first principles, with no important free parameters, the masses of protons and neutrons. As I explained, from a fundamental point of view these particles are quite complicated dynamical equilibria of quarks, antiquarks, and gluons. Most of their mass — and therefore most of the mass of matter, including human brains and bodies — arises from the pure energy of these objects, themselves essentially massless, in motion, according to $m = E/c^2$. At this level, at least, we are ethereal creatures.

Dirac said that QED described "most of physics, and all of chemistry." Indeed, it is the fundamental theory of the outer structure of atoms (and much more). In the same sense, QCD is the fundamental theory of atomic nuclei (and much more). Together, they constitute a remarkably complete, well tested, fruitful and economical Theory of Matter.

6. The Fertility of Reason

I've now discussed in some detail how "playing with equations" led Dirac to an equation laden with consequences that he did not anticipate, and that in many ways he resisted, but that proved to be true and enormously fruitful. How could such a thing happen? Can mathematics be truly creative? Is it really possible, by logical processing or calculation, to arrive at essentially new insights — to get out more than you put in?

This question is especially timely today, since it lies at the heart of debates regarding the nature of machine intelligence — whether it may develop into a species of mind on a par with human intelligence, or even its eventual superior.

At first sight, the arguments against appear compelling.

Most powerful, at least psychologically, is the argument from introspection. Reflecting on our own thought processes, we can hardly avoid an unshakeable intuition that they do not consist exclusively, or even primarily, of rule-based symbol manipulation. It just doesn't feel that way. We normally think in images and emotions, not just symbols. And our streams of thought are constantly stimulated and redirected by interactions with the external world, and by internal drives, in ways that don't seem to resemble at all the unfolding of mathematical algorithms.

Another argument derives from our experience with modern digital computers. For these are, in a sense, ideal mathematicians. They follow precise rules (axioms) with a relentlessness, speed, and freedom from error that far surpasses what is possible for humans. And in many specialized, essentially mathematical tasks, such as arranging airline flight plans or oil delivery schedules to maximize profits, they far surpass human performance. Yet by common, reasonable standards even the most powerful modern computers remains fragile, limited, and just plain dopey. A trivial programming mistake, a few lines of virus code, or a memory flaw can bring a powerful machine to a halt, or send it into an orgy of self-destruction. Communication requires a rigidly controlled format, supporting none of the richness of natural language. Absurd output can, and often does, emerge uncensored and unremarked.

Upon closer scrutiny however, these arguments raise questions and doubts. Although the nature of the map from patterns of electrical signals in nerve cells to processes of human thought remains deeply mysterious in many respects, quite a bit is known, especially about the early stages of sensory processing. Nothing that has been discovered so far suggests that anything more exotic than electric and chemical signaling, following well-established physical laws, is involved. The vast majority of scientists accept as a working hypothesis that a map from patterns of electric signals to patterns of thought must and does exist. The pattern of photons impinging on our retina is broken up and parsed out into elementary units, fed into a bewildering series of different channels, processed, and (somehow) reassembled to give us the deceptively simple "picture of the world," organized into objects in space, that we easily take for granted.

The fact is we do not have the slightest idea how we accomplish most of what we do, even — perhaps especially — our most basic mental feats. People who've attempted to construct machines that can recognize objects appearing in pictures, or that can walk around and explore the world like a toddler, have had a very frustrating time, even though they can do these things very easily themselves. They can't teach others how they do these things because they themselves don't know. Thus it seems clear that introspection is an unreliable guide to the deep structure of thought, both as regards what is known and what is unknown.

Turning to experience with computers, any negative verdict is surely premature, since they are evolving rapidly. One recent benchmark is Deep Blue's over the great world chess champion Garry Kasparov in a brief match. No one competent to judge would deny that play at this level would be judged a profoundly creative accomplishment, if it were performed by a human. Yet such success in a limited domain only sharpens the question: What is missing, that prevents the emergence of creativity from pure calculation over a broad front? In thinking about this tremendous question, I believe case studies can help us.

In modern physics, and perhaps in the whole of intellectual history, no episode better illustrates the profoundly creative nature of mathematical reasoning than the history of the Dirac equation. In hindsight, we know that what Dirac was trying to do is strictly impossible. The rules of quantum mechanics, as they were understood in 1928, cannot be made consistent with special relativity. Yet from inconsistent assumptions Dirac was led to an equation that remains a cornerstone of physics to this day.

So here we are presented with a specific, significant, well-documented example of how mathematical reasoning about the physical world, culminating in a specific equation, led to results that came as a complete surprise to the thinker himself. Seemingly in defiance of some law of conservation, he got out much more than he put in. How was such a leap possible? Why did Dirac, in particular, achieve it? What drove Dirac and his contemporaries to persist in clinging to his equation, when it led them out to sea?[i]

---

[i] Much later, in the 1960s, Heisenberg recalled "Up till that time [1928] I had the impression that, in quantum theory, we had come into the harbor, into the port. Dirac's paper threw us out into the sea again."

Insights emerge from two of Dirac's own remarks. In his characteristically terse essay "My Life as a Physicist" he pays extended tribute to the value of his training as an engineer, including:

> The engineering course influenced me very strongly. ... I've learned that, in the description of Nature, one has to tolerate approximations, and that even work with approximations can be interesting and can sometimes be beautiful.

Along this line, one source of Dirac's (and others') early faith in his equation, which allowed him to overlook its apparent flaws, was simply that he could find approximate solutions of it that agreed brilliantly with experimental data on the spectrum of hydrogen. In his earliest papers he was content to mention, without claiming to solve, the difficulty that there were other solutions, apparently equally valid mathematically, that had no reasonable physical interpretation.

Along what might superficially seem to be a very different line, Dirac often paid tribute to the heuristic power of mathematical beauty:

> The research worker, in his efforts to express the fundamental laws of Nature in mathematical form, should strive mainly for mathematical beauty.

This was another source of early faith in Dirac's equation. It was (and is) extraordinarily beautiful.

Unfortunately, it is hard to make precise, and all but impossible to convey to a lay reader, the nature of mathematical beauty. But we can draw some analogies with other sorts of beauty. One feature that can make a piece of music, a novel, or a play beautiful is the accumulation of tension between important, well-developed themes, which is then resolved in a surprising and convincing way. One feature that can make a work of architecture or sculpture beautiful is symmetry — balance of proportions, intricacy toward a purpose. The Dirac equation possesses both these features to the highest degree.

Recall that Dirac was working to reconcile the quantum mechanics of electrons with special relativity. It is quite beautiful to see how the tension between conflicting demands of simplicity and relativity can be harmonized, and to find that there is essentially only one way to do it. That is one aspect of the mathematical beauty of the Dirac equation. Another aspect, its symmetry and balance, is almost sensual. Space and time, energy and momentum, appear on an equal footing. The different terms in the system of equations must be choreographed to the music of relativity, and the pattern of 0s and 1s (and *i*s) dances before your eyes.

The lines converge when the needs of physics lead to mathematical beauty, or — in rare and magical moments — when the requirements of mathematics lead to physical truth. Dirac searched for a mathematical equation satisfying physically motivated hypotheses. He found that to do so he actually needed a system of equations, with four components. This was a surprise. Two components were most welcome, as they clearly represented the two possible directions of an electron's spin. But the extra doubling at first had no convincing physical interpretation. Indeed, it undermined the assumed meaning of the equation. Yet the equation had taken on a life of its own, transcending the ideas that gave birth to it, and before very long the two extra components were recognized to portend the spinning positron, as we saw.

With this convergence, I think, we reach the heart of Dirac's method in reaching the Dirac equation, which was likewise Maxwell's in reaching the Maxwell equations, and Einstein's in reaching both the special and the general theories of relativity. They proceed by *experimental logic*. That concept is an oxymoron only on the surface. In experimental logic, one formulates hypotheses in equations, and experiments with those equations. That is, one tries to improve the equations from the point of view of beauty and consistency, and then checks whether the "improved" equations elucidate some feature of Nature. Mathematicians recognize the technique of "proof by contradiction": To prove A, you assume the opposite of A, and reach a contradiction. Experimental logic is "validation by fruitfulness": To validate A, assume it, and show that it leads to fruitful consequences. Relative to routine deductive logic, experimental logic abides by the Jesuit credo "It is more blessed to ask forgiveness than permission." Indeed, as we have seen, experimental logic does not regard inconsistency as an irremediable catastrophe. If a line of investigation has some success, and is fruitful, it should not be abandoned on account of its inconsistency, or its approximate nature. Rather, we should look for a way to make it true.

With all this in mind, let us return to the question of the creativity of mathematical reasoning. I said before that modern digital computers are, in a sense, ideal mathematicians. Within any reasonable, precisely axiomatized domain of mathematics, we know how to program a computer so it will systematically prove all the valid theorems.[j] A modern machine of this sort could churn through its program, and output valid theorems, much faster and more reliably than any human mathematician could. But running such a program to do advanced mathematics would be no better than setting the proverbial horde of monkeys to typing, hoping to reproduce Shakespeare. You'd get a lot of true theorems, but essentially all of them would be trivial, with the gems hopelessly buried amidst the rubbish. In practice, if you peruse journals of mathematics or mathematical physics, not to speak of literary magazines, you won't find much work submitted by computers. Attempts to teach computers to do "real." creative mathematics, like the attempts to teach them to recognize real objects or navigate the real world, have had very limited success. Now we begin to see that these are closely related problems. Creative mathematics and physics rely not on perfect logic, but rather on an experimental logic. Experimental logic involves noticing patterns, playing with them, making assumptions to explain them, and — especially — recognizing beauty. And creative physics requires more: abilities to sense and cherish patterns in the world, and to value not only logical consistency, but also (approximate!) fidelity to the world as observed.

So, returning to the central question: Can purely mathematical reasoning be creative? Undoubtedly, if it is used à la Dirac, in concert with the abilities to tolerate approximations, to recognize beauty, and to learn by interacting with the real world. Each of these factors has played a role in all the great episodes of progress in physics. The question returns, as a challenge to ground those abilities in specific mechanisms.

---

[j]This is a consequence of Godel's completeness theorem for first-order predicate logic. Sophisticated readers may wonder how this result, that all valid theorems can be proved in mechanical fashion, can be consistent with Godel's famous incompleteness theorem. (It's not a misprint: Godel proved both completeness and incompleteness theorems.) To make a long story short, Godel's incompleteness theorem shows that in any rich mathematical system you will be able to formulate meaningful statements such that neither the statement nor its denial is a theorem. Such "incompleteness" does not contradict the possibility of systematically enumerating all the theorems.

Acknowledgments

My work is supported in part by funds provided by the U.S. Department of Energy (D.O.E.) under cooperative research agreement #DF-FC02-94ER40818. This presentation is adapted from my chapter "A Piece of Magic: The Dirac Equation" in the book *It Must Be Beautiful, The Great Equations of Modern Science*, ed. G. Farmelo (Granta Books, 2002).

References

1. For background material on atomic physics and quantum theory, including excerpts from important original sources, I highly recommend H. Boorse and L. Motz, *The World of the Atom* (Basic Books, 1966). Of course, some of its more "timely" parts appear somewhat dated today.
2. Dirac's classic is *The Principles of Quantum Mechanics* (Fourth Edition, Cambridge, 1958).
3. A demanding but honest and beautiful treatment of the principles of quantum electrodynamics, with no mathematical prerequisites, is R. P. Feynman, *QED: The Strange Theory of Light and Matter* (Princeton, 1985).
4. For a brief account of QCD, easily accessible after Feynman's book, with no mathematical prerequisites, see F. Wilczek, "QCD Made Simple", *Physics Today*, **53N8** 22–28, (2000). I'm at work on a full account, to be called simply *QCD* (Princeton).
5. For a conceptual review of quantum field theory, see my article "Quantum Field Theory" in the American Physical Society Centenary issue of *Rev. Mod. Phys.* **71**, S85–S95, (1999); this issue is also published as *More Things in Heaven and Earth — A Celebration of Physics at the Millennium*, B. Bederson, ed. (Springer-Verlag, New York, 1999). It contains several other reflective articles that touch on many of our themes.

Addendum: The Dirac Equation, Spelled Out

And if he is compelled to look straight at the light, will he not have a pain in his eyes which will make him turn away to take refuge in the objects of vision which he can see, and which he will conceive to be in reality clearer than the things which are now being shown to him? (Plato)

I've described the Dirac equation, but I didn't write it down. If you're ready to dare the dazzling light of its mathematical formulation, continue.

Let's start simple, with the proper form of Einstein's famous

$$E = mc^2. \tag{1}$$

That gives the correct equation for the energy of a particle of mass m *when it is at rest*. The more general version, that covers moving particles, is

$$E = \sqrt{p^2 c^2 + m^2 c^4} \tag{2}$$

where p is the momentum of the particle. (In the case $p = 0$, this goes back to $E = mc^2$.)

According to Pythagoras the square of the momentum, p^2, is equal to the sum of the squares of the momentum in the x, y and z directions, which we write as $p^2 = p_x^2 + p_y^2 + p_z^2$. I'll need this fact momentarily.

Now if you work with ordinary numbers, there's nothing you can do to simplify that expression for E. Dirac didn't like the square root, so he introduced new kinds of "numbers" that allow you to get rid of it. Dirac's version is

$$E = (p_x \eta_x + p_y \eta_y + p_z \eta_z)c + \rho m c^2 \tag{3}$$

On the face of it this looks quite different from Einstein's equation (in its general form). But the proof of the pudding is in the squaring. The square of Einstein's equation is

$$E^2 = p^2 c^2 + m^2 c^4 = (p_x^2 + p_y^2 + p_z^2)c^2 + m^2 c^4 . \tag{4}$$

When we square Dirac's version, we find the same result *if* the new quantities η and ρ satisfy the equations

$$\eta_x^2 = \eta_y^2 = \eta_z^2 = \rho^2 = 1 \tag{5}$$

and

$$\eta_x \eta_y + \eta_y \eta_x = \eta_x \eta_z + \eta_z \eta_x = \eta_y \eta_z + \eta_z \eta_y = \rho \eta_x + \eta_x \rho$$
$$= \rho \eta_y + \eta_y \rho = \rho \eta_z + \eta_z \rho = 0 . \tag{6}$$

It's easy to see that these equations can't be solved with ordinary numbers. When you multiply ordinary numbers the order of multiplying doesn't matter, so if η_x and η_y were ordinary numbers we'd have $\eta_x \eta_y + \eta_y \eta_x = 2\eta_x \eta_y$, and the only way this could vanish would be for one of the factors vanished, but that contradicts $\eta_x^2 = \eta_y^2 = 1$. We can, however, get a solution if we use 4×4 *matrices*. (If you don't know what matrices are, you not going to learn it here. Sorry.) Here it comes:

$$\eta_x = \begin{pmatrix} 0 & 1 & 0 & 0 \\ 1 & 0 & 0 & 0 \\ 0 & 0 & 0 & -1 \\ 0 & 0 & -1 & 0 \end{pmatrix} \tag{7}$$

$$\eta_y = \begin{pmatrix} 0 & -i & 0 & 0 \\ i & 0 & 0 & 0 \\ 0 & 0 & 0 & i \\ 0 & 0 & -i & 0 \end{pmatrix} \tag{8}$$

$$\eta_z = \begin{pmatrix} 1 & 0 & 0 & 0 \\ 0 & -1 & 0 & 0 \\ 0 & 0 & -1 & 0 \\ 0 & 0 & 0 & 1 \end{pmatrix} \tag{9}$$

$$\rho = \begin{pmatrix} 0 & 0 & 1 & 0 \\ 0 & 0 & 0 & 1 \\ 1 & 0 & 0 & 0 \\ 0 & 1 & 0 & 0 \end{pmatrix} \tag{10}$$

Now we're almost home. The remaining step follows Schrödinger's prescription for passing from classical to quantum mechanics. That is, we replace the energy by a time derivative (times $i\hbar$) and the momenta by spatial derivatives (times $-i\hbar$), and let the whole thing act on a wave function. Thus we arrive at the Dirac equation proper:

$$i\hbar \frac{\partial}{\partial t} \psi(x,y,z,t) = \left[-i\hbar c \left(\frac{\partial}{\partial x} \eta_x + \frac{\partial}{\partial y} \eta_y + \frac{\partial}{\partial z} \eta_z \right) + mc^2 \rho \right]$$
$$\times \psi(x,y,z,t). \tag{11}$$

Since the ηs and ρ are 4×4 matrices, the wave function $\psi(x,y,z,t)$ has four components.

Fermi and the Elucidation of Matter

Fermi helped establish a new framework for understanding matter, based on quantum theory. This framework refines and improves traditional atomism in two crucial respects. First, the elementary constituents of matter belong to a very small number of classes, and all objects of a given class (e.g., all electrons) are rigorously identical, indeed indistinguishable. This profound identity is demonstrated empirically through the phenomena of quantum statistics, and is explained by the principles of free quantum field theory. Second, objects of one class can mutate into objects of other classes. Such mutability can be understood as manifesting interacting quantum field theory. Fermi contributed to establishing theoretical foundations for the new viewpoint, through his work on quantum statistics and quantum field theory, and to its fruitful application and empirical validation, through his work on beta decay, nuclear transmutation, and primeval strong interaction theory.

I feel privileged to return to my alma mater to pay tribute to the memory of Enrico Fermi on the occasion of the 100th anniversary of his birth. Fermi has always been one of my heroes, as well as my scientific "great-grandfather" (in the line Fermi → Chew → Gross → FW). It was also a joy and an inspiration, in preparing, to browse through his papers.

I was asked to speak for one-half hour on "Fermi's Contribution to Modern Physics." This task requires, of course, severe selection. Later speakers will discuss Fermi's remarkable achievements as teacher and scientific statesman, so clearly my job is to focus on his direct contributions to the scientific literature. That still leaves far too much material, if I'm to do anything but catalogue. Cataloguing would be silly, as well as tedious, since Fermi's collected works, with important commentaries from his associates, are readily available [1, 2]. What I decided to do, instead, was to identify and follow a unifying thread that could tie together Fermi's most important work. Though Fermi's papers are extraordinarily varied and always focused on specific problems, such a thread was not difficult to discern. Fermi was a prolific contributor to what I feel is the most important achievement of twentieth-century physics: an accurate and, for practical purposes, completely adequate theory of matter, based on extremely nonintuitive concepts, but codified in precise and useable equations.

1. Atomism Transformed

Since the days of Galileo and Newton, the goal of physics — rarely articulated, but tacit in its practice — was to derive dynamical equations so that, given the configuration of a material system at one time, its configuration at other times could be predicted. The description of the Solar System, based on Newtonian celestial mechanics, realized this goal. This description excellently accounts for Kepler's laws of planetary motion, the tides, the precession of the equinoxes, and much else, but it gives no *a priori* predictions for such things as the number of planets and their moons, their relative sizes, or the dimensions of their orbits. Indeed, we now know that other stars support quite different kinds of planetary systems. Similarly, the great eighteenth- and nineteenth-century discoveries in electricity, magnetism, and optics, synthesized in Maxwell's dynamical equations of electromagnetism, provided a rich description of the behavior of given distributions of charges, currents, and electric and magnetic fields, but no explanation of why there should be specific reproducible forms of matter.

More specifically, nothing in classical physics explains the existence of elementary building blocks with definite sizes and properties. Yet one of the most basic facts about the physical world is that matter is built up from a few fundamental building blocks (e.g., electrons, quarks, photons, gluons), each occurring in vast numbers of identical copies. Were this not true there could be no lawful chemistry, because every atom would have its own quirky properties. In Nature, by contrast, we find accurate uniformity of properties, even across cosmic scales. The patterns of spectral lines emitted by atoms in the atmospheres of stars in distant galaxies match those we observe in terrestrial laboratories.

Qualitative and semiquantitative evidence for some form of atomism has been recognized for centuries [3]. Lucretius gave poetic expression to ancient atomism and Newton endorsed it in his famous Query 31, beginning

> It seems probable to me, that God in the beginning formed matter in solid, massy, hard, impenetrable, moveable particles, of such sizes and figures, and with such other properties, and in such proportions to space, as most conduced to the ends for which He formed them; and that these primitive particles being solids, are incomparably harder than any porous bodies compounded of them, even so very hard, as never to wear or break in pieces; no ordinary power being able to divide what God Himself made one in the first creation.

In the nineteenth century Dalton, by identifying the law of definite proportions, made atomism the basis of scientific chemistry. Clausius, Maxwell, and Boltzmann used it to construct successful quantitative theories of the behavior of gases. But in these developments the properties of atoms themselves were not derived, but assumed, and their theory was not much advanced beyond Newton's formulation. In particular, two fundamental questions went begging.

Question 1: Why is matter built from vast numbers of particles that fall into a small number of classes? And why do all particles of a given class rigorously display the same properties?

The indistinguishability of particles is so familiar and so fundamental to all of modern physical science that we could easily take it for granted. Yet it is by no means obvious. For example, it directly contradicts one of the pillars of Leibniz's metaphysics, his "principle of the identity of indiscernibles," according to which two objects cannot differ solely in number, but will always exhibit some distinguishing features. Maxwell thought the similarity of different molecules so remarkable that he devoted the last part of his *Encyclopedia Britannica* entry on Atoms — well over a thousand words — to discussing it. He concluded

> the formation of a molecule is therefore an event not belonging to that order of Nature in which we live ... it must be referred to the epoch, not of the formation of the earth or the solar system ... but of the establishment of the existing order of Nature ... .

Question 2: Why are there different classes of particles? And why do they exist in the proportions they do?

As we have just seen, both Newton and Maxwell considered this question, but thought that its answer laid beyond the horizon of physics.

By the end of the twentieth century, physics had made decisive progress on these questions. From a science of "how," it had expanded into a science of "what" that supported a much deeper appreciation of "why". A radically new model of matter had been constructed. The elementary building blocks had been inventoried, the equations for their behavior had been precisely defined. The building blocks of our transformed atomism are, paradoxically, both far more reproducible and far more fluid in their properties than classically conceived atoms. Fermi was a chief architect of this construction, contributing to the design at many levels, as I shall now discuss.

2. The Identical and the Indistinguishable

From the perspective of classical physics the indistinguishability of electrons (or other elementary building blocks) is both inessential and surprising. If electrons were nearly but not quite precisely identical, for example, if their masses varied over a range of a few parts per billion, then according to the laws of classical physics different specimens would behave in nearly but not quite the same way. And since the possible behavior is continuously graded, we could not preclude the possibility that future observations, attaining greater accuracy than is available today, might discover small differences among electrons. Indeed, it would seem reasonable to expect that differences would arise, since over a long lifetime each electron might wear down, or get bent, according to its individual history.

The first evidence that the similarity of like particles is quite precise, and goes deeper than mere resemblance, emerged from a simple but profound reflection by Josiah Willard Gibbs in his work on the foundations of statistical mechanics. It is known as "Gibbs' paradox," and goes as follows. Suppose that we have a box separated into two equal compartments A and B, both filled with equal densities of hydrogen gas at the same temperature. Suppose further that there is a shutter separating the compartments, and consider what happens if we open the shutter and allow the gas to settle into equilibrium. The molecules originally confined to A (or B) might then be located anywhere in A + B. Thus, since there appear to be many more distinct possibilities for distributing the molecules, it would seem that the entropy of the gas, which measures the number of possible microstates, will increase. On the other hand, one might have the contrary intuition, based on everyday experience, that the properties of gases in equilibrium are fully characterized by their volume, temperature, and density. If that intuition is correct, then in our thought experiment the act of opening the shutter makes no change in the state of the gas, and so of course it generates no entropy. In fact, this result is what one finds in actual experiments.

The experimental verdict on Gibbs' paradox has profound implications. If we could keep track of every molecule we would certainly have the extra entropy, the so-called entropy of mixing. Indeed, when gases of *different* types are mixed, say hydrogen and helium, entropy is generated. Since entropy of mixing is not observed for (superficially) similar gases, there can be no method, *even in principle*, to tell their molecules apart. Thus we cannot make a rigorous statement of the kind, "Molecule 1 is in A, molecule 2 is in A, ... , molecule n is in A," but only a much weaker statement, of the kind, "There are n molecules in A." In this precise sense, hydrogen molecules are not only similar, nor even only identical, but beyond that, indistinguishable.

Fermi motivated his concept of state-counting [4] through a different, though related, difficulty of classical statistical mechanics, namely its failure to account for Nernst's law. This law, abstracted from empirical data, implied that the entropy of a material vanishes at the absolute zero of temperature. Like Gibbs's paradox it indicates that there are far fewer states than appear classically, and in particular no entropy of mixing. Fermi therefore proposed to generalize Pauli's exclusion principle from its spectroscopic origins to a universal principle, not only for electrons but as a candidate to describe matter in general:

> We will therefore assume in the following that, at most, one molecule with given quantum numbers can exist in our gas: as quantum numbers we must take into account not only those that determine the internal motions of the molecule but also the numbers that determine its translational motion.

It is remarkable, and perhaps characteristic, that Fermi uses the tested methods of the old quantum theory in evaluating the properties of his ideal gas. He places his molecules in an imaginary shallow harmonic well, identifies the single-particle energy levels the levels by appeal to the Bohr–Sommerfeld quantization rules, and assigns one state of the system to every distribution of particles over these levels. (Of course, he does not fail to remark that the results should not depend on the details of this procedure.) All the standard results on the ideal Fermi–Dirac gas are then derived in a few strokes, by clever combinatorics.

Indeed, after a few introductory paragraphs, the paper becomes a series of equations interrupted only by minimal text of the general form, "now let us calculate … ." So the following interpolation, though brief, commands attention:

> At the absolute zero point, our gas molecules arrange themselves in a kind of shell-like structure which has a certain analogy to the arrangement of electrons in an atom with many electrons.

We can see in this the germ of the Thomas-Fermi model of atoms [5], which grows very directly out of Fermi's treatment of the quantum gas, and gives wonderful insight into the properties of matter [6]. Also, of course, the general scheme of fermions in a harmonic well is the starting point for the *nuclear* shell model — of which more below.

The successful adaptation of Fermi's ideal gas theory to describe electrons in metals, starting with Sommerfeld and Bethe, and many other applications, vindicated his approach.

3. The Primacy of Quantum Fields 1: Free Fields

As I have emphasized, Fermi's state-counting logically requires the radical indistinguishability of the underlying particles. It does not explain that fact, however, so its success only sharpens the fundamental problem posed in our Question 1. Deeper insight into this question requires considerations of another order. It comes from the synthesis of quantum mechanics and special relativity into quantum field theory.

The field concept came to dominate physics starting with the work of Faraday in the mid-nineteenth century. Its conceptual advantage over the earlier Newtonian program of physics, to formulate the fundamental laws in terms of forces among atomic particles, emerges when we take into account the circumstance, unknown to Newton (or, for that matter, Faraday) but fundamental in special relativity, that physical influences travel no faster than a finite limiting speed. For this implies that the force on a given particle at a given time cannot be inferred from the positions of other particles at that time, but must be deduced in a complicated way from their previous positions. Faraday's intuition that the fundamental laws of electromagnetism could be expressed most simply in terms of fields filling space and time was of course brilliantly vindicated by Maxwell's mathematical theory.

The concept of locality, in the crude form that one can predict the behavior of nearby objects without reference to distant ones, is basic to scientific practice. Practical experimenters — if not astrologers — confidently expect, on the basis of much successful experience, that after reasonable (generally quite modest) precautions isolating their experiments, they will obtain reproducible results.

The deep and ancient historic roots of the field and locality concepts provide no guarantee that these concepts remain relevant or valid when extrapolated far beyond their origins in experience, into the subatomic and quantum domain. This extrapolation must be judged by its fruits. Remarkably, the first consequences of relativistic quantum field theory supply the answer to our Question 1, in its sharp form including Fermi's quantum state-counting.

In quantum field theory, particles are not the primary reality. Relativity and locality demand that fields, not particles, are the primary reality. According to quantum theory, the excitations of these fields come in discrete lumps. These lumps are what we recognize as particles. In this way, particles are derived from fields. Indeed, what we call particles are simply the form in which low-energy excitations of quantum fields appear. Thus all electrons are precisely alike because all are excitations of the same underlying Ur-stuff, the electron field. The same logic, of course, applies to photons or quarks, or even to composite objects such as atomic nuclei, atoms, or molecules.

Given the indistinguishability of a class of elementary particles, including complete invariance of their interactions under interchange, the general principles of quantum mechanics teach us that solutions forming any representation of the permutation symmetry group retain that property in time. But they do not constrain which representations are realized. Quantum field theory not only explains the existence of indistinguishable particles and the invariance of their interactions under interchange, but also constrains the symmetry of the solutions. There are two possibilities, bosons or fermions. For bosons, only the identity representation is physical (symmetric wave functions); for fermions, only the one-dimensional odd representation is physical (antisymmetric wave functions). One also has the spin-statistics theorem, according to which objects with integer spin are bosons, whereas objects with half odd-integer spin are fermions. Fermions, of course, obey Fermi's state-counting procedure. Examples are electrons, protons, neutrons, quarks, and other charged leptons and neutrinos.

It would not be appropriate to review here the rudiments of quantum field theory, which justify the assertions of the preceding paragraph. But a brief heuristic discussion seems in order.

In classical physics particles have definite trajectories, and there is no limit to the accuracy with which we can follow their paths. Thus, in principle, we could always keep tab on who's who. Thus classical physics is inconsistent with the rigorous concept of indistinguishable particles, and it comes out on the wrong side of Gibbs' paradox.

In the quantum theory of indistinguishable particles the situation is quite different. The possible positions of particles are described by waves (i.e., their wave-functions). Waves can overlap and blur. Related to this, there is a limit to the precision with which we can follow their trajectories, according to Heisenberg's uncertainty principle. So when we calculate the quantum-mechanical amplitude for a physical process to take place, we must sum contributions from all ways in which it might have occurred. Thus, to calculate the amplitude that a state with two indistinguishable particles of a given sort — call them quants — at positions x_1, x_2 at time t_i, will evolve into a state with two quants at x_3, x_4 at time t_f, we must sum contributions from all possible trajectories for the quants at intermediate times. These trajectories fall into two distinct categories. In one category, the quant initially at x_1 moves to x_3, and the quant initially at x_2 moves to x_4. In the other category, the quant initially at x_1 moves to x_4, and the quant initially at x_2 moves to x_3. Because (by hypothesis) quants are indistinguishable, the final states are the same for both categories. Therefore, according to the general principles of quantum mechanics, we must add the amplitudes for these two categories. We say there are "direct" and "exchange" contributions to the process. Similarly, if we have more than two quants, we must add contributions involving arbitrary permutations of the original particles.

Since the trajectories fall into different classes, we may also imagine adding the amplitudes with relative factors. Mathematical consistency severely limits our freedom, however. We must demand that the rule for multiplying amplitudes, when we sum over states at an intermediate time, is consistent with the rule for the overall amplitude. Since the final result of double exchange is the same as no exchange at all, we must assign factors in such a way that direct × direct = exchange × exchange, and the only consistent possibilities are direct/exchange = ±1. These correspond to bosons (+) and fermions (−), respectively. The choice of sign determines how the interference term between direct and exchange contributions contributes to the square of the amplitude, that is the probability for the overall process. This choice is vitally important for calculating the elementary interactions even of short-lived, confined, or elusive particles, such as quarks and gluons, whose equilibrium statistical mechanics is moot [7].

4. The Centerpiece: Beta Decay

The preceding consequences of quantum field theory follow from its basic "kinematic" structure, independent of any specific dynamical equations. They justified Fermi's state-counting, and rooted it in a much more comprehensive framework. But Fermi's own major contributions to quantum field theory came at the next stage, in understanding its dynamical implications.

Fermi absorbed quantum electrodynamics by working through many examples and using them in teaching. In so doing, he assimilated Dirac's original, rather abstract formulation of the theory to his own more concrete way of thinking. His review article [8] is a masterpiece, instructive and refreshing to read even today. It begins

> Dirac's theory of radiation is based on a very simple idea; instead of considering an atom and the radiation field with which it interacts as two distinct systems, he treats them as a single system whose energy is the sum of three terms: one representing the energy of the atom, a second representing the electromagnetic energy of the radiation field, and a small term representing the coupling energy of the atom and the radiation field … .

and soon continues

> A very simple example will explain these relations. Let us consider a pendulum which corresponds to the atom, and an oscillating string in the neighborhood of the pendulum which represents the

radiation field … . To obtain a mechanical representation of this [interaction] term, let us tie the mass M of the pendulum to a point A of the string by means of a very thin and elastic thread a … [I]f a period of the string is equal to the period of the pendulum, there is resonance and the amplitude of vibration of the pendulum becomes considerable after a certain time. This process corresponds to the absorption of radiation by the atom.

Everything is done from scratch, starting with harmonic oscillators. Fully elaborated examples of how the formalism reproduces concrete experimental arrangements in space-time are presented, including the Doppler effect and Lippmann fringes, in addition to the "S-matrix" type scattering processes that dominate modern textbooks.

Ironically, in view of what was about to happen, Fermi's review article does not consider systematic quantization of the electron field. Various processes involving positrons are discussed on an *ad hoc* basis, essentially following Dirac's hole theory.

With hindsight, it appears obvious that Chadwick's discovery of the neutron in early 1932 marks the transition between the ancient and the classic eras of nuclear physics. (The dominant earlier idea, consonant with an application of Occam's razor that in retrospect seems reckless, was that nuclei were made from protons and tightly bound electrons. This idea is economical of particles, of course, but it begs the question of dynamics, and it also has problems with quantum statistics — N^{14} would be 21 particles, and a fermion, whereas molecular spectroscopy shows it is a boson.) At the time, however, there was much confusion [9].

The worst difficulties, which brought into question the validity of quantum mechanics and energy conservation in the nuclear domain, concerned β decay. On the face of it, the observations seemed to indicate a process $n \to p + e^-$, wherein a neutron decays into a proton plus electron. However, this would indicate boson statistics for the neutron, and it reintroduces the spectroscopic problems mentioned above. Moreover, electrons are observed to be emitted with a nontrivial spectrum of energies, which is impossible for a two-body decay if energy and momentum are conserved. Bohr in this context suggested abandoning conservation of energy. However, Pauli suggested that the general principles of quantum theory could be respected and the conservation laws could be maintained if the decay were in reality $n \to p + e^- + \bar{\nu}$ with a neutral particle $\bar{\nu}$ escaping detection.

Fermi invented the term "neutrino" for Pauli's particle. This started as a little joke in conversation, *neutrino* being the Italian diminutive of *neutrone*, suggesting the contrast of a little neutral one versus a heavy neutral one. It stuck, of course. (Actually, what appears in neutron decay is what we call today an antineutrino.)

More importantly, Fermi took Pauli's idea seriously and literally, and attempted to bring it fully into line with special relativity and quantum mechanics. This meant constructing an appropriate quantum field theory. Having mastered the concepts and technique of quantum electrodynamics, and after absorbing the Jordan–Wigner technique for quantizing fermion fields, he was ready to construct a quantum field theory for β decay. He chose a Hamiltonian of the simplest possible type, involving a local coupling of the four fields involved, one for each particle created or destroyed. There are various possibilities for combining the spinors into a Lorentz-invariant object, as Fermi discusses. He then calculates the electron emission spectrum, including the possible effect of nonzero neutrino mass.

It soon became apparent that these ideas successfully organize a vast wealth of data on nuclear β decays in general. They provided, for forty years thereafter, the ultimate foundation of weak interaction theory, and still remain the most useful working description of a big chapter of nuclear and particle physics. Major refinements including the concept of universality, parity violation, V-A theory — which together made the transition to a deeper foundation, based on the gauge principle, compelling — not to mention the experimental investigation of neutrinos themselves, were all based firmly on the foundation supplied by Fermi.

5. The Primacy of Quantum Fields 2: From Local Interactions to Particles/Forces (Real and Virtual)

Although Fermi's work on β decay was typically specific and sharply focused, by implication it set a much broader agenda for quantum field theory. It emphasized the very direct connection between the abstract principles of interacting quantum field theory and a most fundamental aspect of Nature, *the ubiquity of particle creation and destruction processes.*

Local interactions involve products of field operators at a point. When the fields are expanded into creation and annihilation operators multiplying modes, we see that these interactions correspond to processes wherein particles can be created, annihilated, or changed into different kinds of particles. This possibility arises, of course, in the primeval quantum field theory, quantum electrodynamics, where the primary interaction arises from a product of the electron field, its Hermitean conjugate, and the photon field. Processes of radiation and absorption of photons by electrons (or positrons), as well as electron-positron pair creation, are encoded in this product. But because the emission and absorption of light is such a common experience, and electrodynamics such a special and familiar classical field theory, this correspondence between formalism and reality did not initially make a big impression. The first conscious exploitation of the potential for quantum field theory to describe processes of transformation was Fermi's theory of beta decay. He turned the procedure around, inferring from the observed processes of particle transformation the nature of the underlying local interaction of fields. Fermi's theory involved creation and annihilation not of photons but of atomic nuclei and electrons (as well as neutrinos) — the ingredients of "matter." It thereby initiated the process whereby classical atomism, involving stable individual objects, was replaced by a more sophisticated and accurate picture. In this picture only the fields, and not the individual objects they create and destroy, are permanent.

This line of thought gains power from its association with a second general consequence of quantum field theory, *the association of forces and interactions with particle exchange*. When Maxwell completed the equations of electrodynamics, he found that they supported source-free electromagnetic waves. The classical electric and magnetic fields took on a life of their own. Electric and magnetic forces between charged particles are explained as due to one particle acting as a source for electric and magnetic fields, which then influence others. With the correspondence of fields and particles, as it arises in quantum field theory, Maxwell's discovery corresponds to the existence of photons, and the generation of forces by intermediary fields corresponds to the exchange of virtual photons.

The association of forces (or more generally, interactions) with particles is a general feature of quantum field theory. "Real" particles are field excitations that can be considered usefully as independent entities, typically because they have a reasonably long lifetime and can exist spatially separated from other excitations, so that we can associate transport of definite units of mass, energy, charge, etc., with them. But in quantum theory all excitations can also be produced as short-lived fluctuations.

These fluctuations are constantly taking place in what we regard as empty space, so that the physicists' notion of vacuum is very far removed from simple nothingness. The behavior of real particles is affected by their interaction with these virtual fluctuations. Indeed, according to quantum field theory that's all there is! So observed forces must be ascribed to fluctuations of quantum fields — but these fields will then also support genuine excitations, that is, real particles. Tangible particles and their "virtual" cousins are as inseparable as two sides of the same coin. This connection was used by Yukawa to infer the existence and mass of pions from the range of nuclear forces. (Yukawa began his work by considering whether the exchange of virtual electrons and neutrinos, in Fermi's β decay theory, might be responsible for the nuclear force! After showing that these virtual particles gave much too small a force, he was led to introduce a new particle.) More recently it has been used in electroweak theory to infer the existence, mass, and properties of W and Z bosons prior to their observation, and in QCD to infer the existence and properties of gluon jets prior to their observation.

This circle of ideas, which to me forms the crowning glory of twentieth-century physics, grew around Fermi's theory of β decay. There is a double meaning in my title, "Fermi and the Elucidation of Matter." For Fermi's most beautiful insight was, precisely, to realize the profound commonality of matter and light.

6. Nuclear Chemistry

The other big classes of nuclear transformations, of quite a different character from β decay, are those in which no leptons are involved. In modern language, these are processes mediated by the strong and electromagnetic interactions. They include the fragmentation of heavy nuclei (fission) and the joining of light nuclei (fusion). These processes are sometimes called nuclear chemistry, since they can be pictured as rearrangements of existing materials — protons and neutrons — in the same way that ordinary chemistry can be pictured as rearrangements of electrons and nuclei. In this terminology, it would be natural to call β decay nuclear alchemy.

Fermi discovered the technique that above all others opened up the experimental investigation of nuclear chemistry. This is the potent ability of slow neutrons to enter and stir up nuclear targets. Fermi regarded this as his greatest discovery. In an interview with S. Chandrasekhar, quoted in [2], he described it:

> I will tell you now how I came to make the discovery which I suppose is the most important one I have made. We were working very hard on the neutron-induced radioactivity and the results we were obtaining made no sense. One day, as I came to the laboratory, it occurred to me that I should examine the effect of placing a piece of lead before the incident neutrons. Instead of my usual custom, I took great pains to have the piece of lead precisely machined. I was clearly dissatisfied with something: I tried every excuse to postpone putting the piece of lead in its place. When finally, with some reluctance, I was going to put it in its place, I said to myself: "No, I do not want this piece of lead here; what I want is a piece of paraffin." It was just like that, with no advance warning, no conscious prior reasoning. I immediately took some odd piece of paraffin and placed it where the piece of lead was to have been.

There are wonderful accounts of how he mobilized his group in Rome to exploit, with extraordinary joy and energy, his serendipitous discovery [2].

I will not retell that story here, nor the epic saga of the nuclear technology commencing with the atomic bomb project [10]. We are still coming to terms with the destructive potential of nuclear weapons, and we have barely begun to exploit the resource of nuclear energy.

From the point of view of pure physics, the significance of Fermi's work in nuclear chemistry was above all to show that nuclei could be usefully described as composites wherein protons and neutrons, though in close contact, retain their individual identity and properties. This viewpoint reached its apex in the shell model of Mayer and Jensen [11]. Famously, Fermi helped inspire Mayer's work, and in particular suggested the importance of spin-orbit coupling, which proved crucial. The success of Mayer's independent-particle model for describing quantum systems in which the interactions are very strong stimulated deep work in the many-body problem. It also provided the intellectual background for the quark model.

7. Last Insights and Visions

With developments in nuclear chemistry making it clear that different nuclei are built up from protons and neutrons, and developments in beta-decay theory showing how protons and neutrons can transmute into one another, our Question 2, to understand the constitution of the world, came into sharp focus. Specifically, it became possible to pose the origin of the elements as a scientific question. Fermi was very much alive to this possibility that his work had opened up. With Turkevich he worked extensively on Gamow's "ylem" proposal that the elements are built up by successive neutron captures, starting with neutrons only in a hot, rapidly expanding Universe. They correctly identified the major difficulty with this idea, that is, the insurmountable gap at atomic number 5, where no suitably stable nucleus exists. Yet a rich and detailed account of the origin of elements can be constructed, very nearly along these lines. The Fermi–Turkevich gap at $A=5$ is reflected in the observed abundances, in that less than 1% of cosmic chemistry resides in such nuclei, the astronomers' "metals." Elements beyond $A=5$ (except for a tiny amount of Li^7) are produced in a different way, in the reactions powering stars, and are injected back into circulation through winds or, perhaps, with additional last-minute cooking, during supernova explosions. Also, the correct initial condition for Big Bang nucleosynthesis postulates thermal equilibrium at high temperatures, not all neutrons. Though he didn't quite get there, Fermi envisioned this promised land.

All this progress in observing, codifying, and even controlling nuclear processes was primarily based on experimental work. The models used to correlate the experimental data incorporated relativistic kinematics and basic principles of quantum mechanics, but were not derived from a closed set of underlying equations. The experimental studies revealed that the interaction between nucleons at low energy is extremely complex. It depends on distance, velocity, and spin in a completely entangled way. One could parametrize the experimental results in terms of a set of energy-dependent functions — "phase shifts" — but these functions displayed no obvious simplicity.

The lone triumph of high-energy theory was the discovery of Yukawa's pion. This particle, with the simple local coupling postulated by Yukawa's theory, could account semiquantitatively for the long-range tail of the force. Might it provide the complete answer? No one could tell for sure — the necessary calculations were too difficult.

Starting around 1950 Fermi's main focus was on the experimental investigation of pion-nucleon interactions. They might be expected to have only the square root of the difficulty in interpretation, so to speak, since they are closer to the core-element of Yukawa's theory. But pion-nucleon interactions turned out to be extremely complicated also.

With the growing complexity of the observed phenomena, Fermi began to doubt the adequacy of Yukawa's theory. No one could calculate the consequences of the theory accurately, but the richness of the observed phenomena undermined the basis for hypothesizing that with point-like protons, neutrons, and pions one had reached bottom in the understanding of strongly interacting matter. There were deeper reasons for doubt, arising from a decent respect for Question 2. Discovery of the μ particle, hints of more new particles in cosmic ray events (eventually evolving into our K mesons), together with the familiar nucleons, electrons, photons, plus neutrinos and the pions, indicated a proliferation of "elementary" particles. They all transformed among one another in complicated ways. Could one exploit the transformative aspect of quantum field theory to isolate a simple basis for this proliferation — fewer, more truly elementary building blocks?

In one of his late theoretical works, with Yang, Fermi proposed a radical alternative to Yukawa's theory, that might begin to economize particles. They proposed that the pion was not fundamental and elementary at all, but rather a composite particle, specifically a nucleon-antinucleon bound state. This was a big extrapolation of the idea behind the shell model of nuclei. Further, they proposed that the primary strong interaction was what we would call today a four-fermion coupling of the nucleon fields. The pion was to be produced as a consequence of this interaction, and Yukawa's theory as an approximation — what we would call today an effective field theory. The primary interaction in the Fermi–Yang theory is of the same form of interaction as appears in Fermi's theory of β decay, though of course the strength of the interaction and the identities of the fields involved are quite different. In his account of this work [12] Yang says

> As explicitly stated in the paper, we did not really have any illusions that what we suggested may actually correspond to reality .... Fermi said, however, that as a student one solves problems, but as a research worker one raises questions ....

Indeed, the details of their proposal do not correspond to our modern understanding. In particular, we have learned to be comfortable with a proliferation of particles, so long as their fields are related by symmetries. But some of the questions Fermi and Yang raised — or, I would say, the directions they implicitly suggested — were, in retrospect, fruitful ones. First, the whole paper is firmly set in the framework of relativistic quantum field theory. Its goal, in the spirit of quantum electrodynamics and Fermi's β decay theory, was to explore the possibilities of that framework, rather than to overthrow it. For example, at the time the existence of antinucleons had not yet been established experimentally. However, the existence of antiparticles is a general consequence of relativistic quantum field theory, and it is accepted with minimal comment. Second, building up light particles from much heavier constituents was a liberating notion. It is an extreme extrapolation of the lowering of mass through binding energy. Nowadays we go much further along the same line, binding infinitely heavy (confined) quarks and gluons into the observed strongly interacting particles, including both pions and nucleons. Third, and most profoundly, the possibility of deep similarity in the basic mechanism of the strong and the weak interaction, despite their very different superficial appearance, is anticipated. The existence of just such a common mechanism, rooted in concepts later discovered by Fermi's co-author — Yang–Mills theory — is a central feature of the modern theory of matter, the so-called Standard Model.

In another of his last works, with Pasta and Ulam [13], Fermi enthusiastically seized upon a new opportunity for exploration — the emerging capabilities of rapid machine computation. With his instinct for the border of the unknown and the accessible, he chose to revisit a classic, fundamental problem that had been the subject of one of his earliest papers, the problem of approach to equilibrium in a many-body system. The normal working hypothesis of statistical mechanics is that equilibrium is the default option and is rapidly attained in any complex system unless some simple conservation law forbids it. But proof is notoriously elusive, and Fermi thought to explore the situation by controlled *numerical* experiments, wherein the degree of complexity could be varied. Specifically, he coupled together various modest numbers of locally coupled nonlinear springs. A stunning surprise emerged: The approach to equilibrium is far from trivial; there are emergent structures, collective excitations that can persist indefinitely. The topic of solitons, which subsequently proved vast and fruitful, was foreshadowed in this work. And the profound but somewhat nebulous question, central to an adequate understanding of Nature, of how ordered structures emerge spontaneously from simple homogeneous laws and minimally structured initial conditions, spontaneously began to pose itself.

Not coincidentally, the same mathematical structure — locally coupled nonlinear oscillators — lies at the foundation of relativistic quantum field theory. Indeed, as we saw, this was the way Fermi approached the subject right from the start. In modern QCD the emergent structures are protons, pions, and other hadrons, which are well hidden in the quark and gluon field "springs." Numerical work of the kind Fermi pioneered remains our most reliable tool for studying these structures.

Clearly, Fermi was leading physics toward fertile new directions. When his life was cut short, it was a tremendous loss for our subject.

8. Fermi as Inspiration: Passion and Style

Surveying Fermi's output as a whole, one senses a special passion and style, unique in modern physics. Clearly, Fermi loved his dialogue with Nature. He might respond to her deepest puzzles with queries of his own, as in the explorations we have just discussed, or with patient gathering of facts, as in his almost brutally systematic experimental investigations of nuclear transmutations and pion physics. But, he also took great joy in solving, or simply explicating, her simpler riddles, as the many Fermi stories collected in this volume attest.

Fermi tackled ambitious problems at the frontier of knowledge, but always with realism and modesty. These aspects of his scientific style shine through one of Fermi's rare "methodological" reflections, written near the end of his life [14]:

> When the Yukawa theory first was proposed there was a legitimate hope that the particles involved, protons, neutrons, and pi-mesons, could be legitimately considered as elementary particles. This hope loses more and more its foundation as new elementary particles are rapidly being discovered.
>
> It is difficult to say what will be the future path. One can go back to the books on method (I doubt whether many physicists actually do this) where it will be learned that one must take experimental data, collect experimental data, organize experimental data, begin to make working hypotheses, try to correlate, and so on, until eventually a pattern springs to life and one has only to pick out the results. Perhaps the traditional scientific method of the text books may be the best guide, in the lack of anything better ... .

Of course, it may be that someone will come up soon with a solution to the problem of the meson, and that experimental results will confirm so many detailed features of the theory that it will be clear to everybody that it is the correct one. Such things have happened in the past. They may happen again. However, I do not believe that we can count on it, and I believe that we must be prepared for a long, hard pull.

Those of you familiar with the subsequent history of the strong-interaction problem will recognize that Fermi's prognosis was uncannily accurate. A long period of experimental exploration and pattern recognition provided the preconditions for a great intellectual leap and synthesis. The process would, I think, have gratified Fermi, but not surprised him.

The current situation in physics is quite different from what Fermi lived through or, at the end, described. After the triumphs of the twentieth century, it is easy to be ambitious. Ultimate questions about the closure of fundamental dynamical laws and the origin of the observed Universe begin to seem accessible. The potential of quantum engineering, and the challenge of understanding how one might orchestrate known fundamentals into complex systems, including powerful minds, beckon. Less easy, perhaps, is to remain realistic and (therefore) appropriately modest — in other words, to craft important subquestions that we can answer definitively, in dialogue with Nature. In this art Fermi was a natural grand master, and the worthy heir of Galileo.

References

[1] *Collected Works*, 2 volumes, E. Fermi (University of Chicago Press, 1962).
[2] Another indispensable source is the scientific biography *Enrico Fermi, Physicist*, E. Segre (University of Chicago Press, 1970).
[3] An accessible selection of original sources and useful commentaries is *The World of the Atom*, 2 volumes, eds. H. Boorse and L. Motz (Basic Books, 1966).
[4] Papers 30 and 31 in [1]. A portion is translated into English in [3].
[5] Papers 43–48 in [1].
[6] *Thomas-Fermi and Related Theories of Atoms and Molecules*, E. Lieb, in *Rev. Mod. Phys.* **53** No. 4, pt. 1, 603–641 (1981).
[7] Actually, the statistical mechanics of quarks and gluons may become accessible in ongoing experiments at RHIC.
[8] Paper 67 in [1].

[9] *Inward Bound: Of Matter and Forces in the Physical World*, A. Pais (Oxford, 1986).
[10] *The Making of the Atomic Bomb*, R. Rhodes (Simon and Schuster, 1986).
[11] *Elementary Theory of Nuclear Shell Structure*, M. Mayer (Wiley, 1955).
[12] C. N. Yang, introduction to paper 239 in [1].
[13] Paper 266 in [1].
[14] Paper 247 in [1].

The Standard Model Transcended

Prolonged applause met Professor Kajita's presentation last month* of results from the SuperKamiokande detector.[1] After decades of ardent searching, marked by numerous false alarms and several tantalizing but precarious hints, a physical phenomenon lying beyond the framework of the Standard Model of physics had been clearly identified. Neutrinos have mass.

The new observations concern the different types of neutrino (electron, muon and tau) produced by cosmic rays in the Earth's atmosphere, and they strengthen earlier indications[2] that there are fewer muon-type neutrinos than expected. In the details of its dependence on energy and angle, this anomaly is consistent with a very specific hypothesis: muon neutrinos are a mixture of two states with different masses. As time progresses, the theory goes, these two states fall out of phase; then the proportions of the mixture have changed, and no longer fit the specifications for a muon neutrino. The new mixture is part muon neutrino, part something else (probably a tau neutrino) — the muon neutrino has 'oscillated'.

The extent of the oscillation should depend in a very specific way on the difference between the masses, the original proportions of the mixture, the neutrino's energy and the distance it has travelled. The distance is related to the neutrino's arrival angle, as it originated in the atmosphere; and the energy is related to the energy released in the

---

*Neutrino '98, Takayama, Japan, 4–9 June 1998.

detector. The predicted dependence of the oscillation on angle and energy matches the observed anomalies quite well, so it seems appropriate — to the extent that it is ever possible — to claim that the observations 'rule in' the hypothesis. The neutrino mass difference needed to explain the observations is minuscule: a few hundredths of an electron volt, or less than 10^{-7} times the mass of the electron.

But the raw phenomena do not begin to convey the discovery's significance. It cannot be considered as a bizarre, tiny correction to our previous understanding; rather, it is a first step towards a more comprehensive, and much more beautiful, formulation of the fundamental laws of physics. For, as I shall explain, this neutrino mass confirms bold theoretical ideas about symmetry and unification of forces that arise from the deep structure of the Standard Model.

In the Standard Model, the quarks and leptons that are the building blocks of matter fall into five separate classes, or multiplets (Fig. 1). There are 'strong' interactions, mediated by colour gluons, which can transform particles within a multiplet; 'weak' interactions, mediated by W bosons, that can transform particles between multiplets; and 'hypercharge' interactions, mediated by a mixture of photon and Z boson, which leave the particles unchanged (and underlie the more familiar electromagnetic interaction). It would be difficult to overstate the power and practical success of the Standard Model, but one cannot deny the superficially graceless cast of Fig. 1. Because we are discussing the most basic laws of Nature, we have a right to expect better.

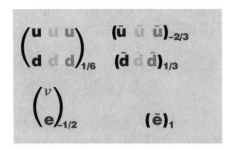

FIGURE 1. Multiplets of the Standard Model of particle physics. The up and down quarks, neutrino and electron shown here form only the first particle family (of three). The figure also omits right-handed particles, but the same pattern of five isolated multiplets is repeated for each family and handedness. The strong nuclear interaction transforms particles horizontally; the weak interaction, vertically. The subscript is hypercharge.

Fortunately, postulating a higher degree of symmetry produces a prettier picture.[3,4] One version, called SO(10) symmetry,[5] looks especially good in light of the new neutrino mass measurements. In this unified theory (see Box, overleaf), all the particles of the Standard Model appear

SO(10) marshals the particles

By unifying quarks and leptons into a single class, the most important properties of these particles can be summarized in a set of rules for dealing with symmetrical symbols. The appropriate mathematical structure (the 16-dimensional spinor representation of SO(10)) is formidable, but some appreciation of the scheme can be conveyed fairly simply.

The core of the Standard Model is the idea that the non-gravitational interactions of matter, the strong, weak and electromagnetic interactions, are all described by the responses of particles called gauge bosons (the photon, for example) to several types of charge. These charges are usually called 'colours'. The gauge bosons of the strong interaction respond to and change three colours, usually called red, white and blue. The gauge bosons of the weak interaction respond to and change two colours, let us call them green and purple. The remaining 'hypercharge' gauge boson, from which the ordinary photon is constructed, responds to a combination of charges: one-half the sum of green and purple, minus one-third the sum of red, white and blue. The Standard Model does not explain why this particular combination is picked out, but it is inevitable within unification schemes such as SO(10).

This follows from one of the formal rules, the 'bleaching rule', that an equal mixture of all available charges cancels out. Thus, for example, a particle carrying a unit of red plus a unit of white colour charge will look the same, as far as the colour gluons of the strong interaction are concerned, as if it carried a negative unit of blue charge: the sum of red, white and blue units cancels.

The bleaching rule eliminates from the strong interaction the gauge boson that responds to the sum of red, white and blue charges, and from the weak interaction the gauge boson that responds to the sum of green and purple charges. These are bogus particles: 'gauge bogons'. Likewise, it eliminates from the unified model, which treats all five colours together, the gauge boson that couples to the sum of all five colours. However, the difference between total weak and strong colour charges is not bleached out. The gauge boson that responds to this (the difference between the strong and weak gauge bogons) is precisely the hypercharge component of the Standard Model.

The multiplet shown here is constructed as a five-bit register, with entries + and −, subject to the rule that the total number of − signs is odd (in each case, changing all the signs gives the right-handed particle equivalent, not shown). Each entry corresponds to either plus or minus half a unit of the associated colour charge. So the first particle has half a unit each of green and purple charge; these cancel according to the bleaching rule, so this particle has no weak interaction. It has minus half a unit of red charge, and half a unit each of white and blue charge, equivalent to minus a full unit of red charge. These are precisely the properties of one of the particles in nature: the antiparticle of the right-handed red down quark. In all the rows up to the last, we find a precise match between the mathematical demands of this scheme and observed particles.

But the last row describes a totally bleached particle. It has neither strong, nor weak, nor electromagnetic interactions. It is none other than our new friend, the particle that according to this scheme gives mass to the neutrino: the N.

Colours					Hypercharge	Particle name
R	W	B	G	P	$-(R+W+B)/3 + (G+P)/2$	
−	+	+	+	+	1/3	$\bar d$
+	−	+	+	+	1/3	$\bar d$
+	+	−	+	+	1/3	$\bar d$
+	+	+	−	+	−1/2	e
+	+	+	+	−	−1/2	ν
−	−	−	+	+	1	$\bar e$
−	+	+	−	−	1/6	d
+	−	+	−	−	1/6	d
+	+	−	−	−	1/6	d
−	+	+	+	−	1/6	u
−	+	+	−	+	1/6	u
+	−	+	+	−	1/6	u
−	+	−	−	−	−2/3	$\bar u$
+	−	−	−	−	−2/3	$\bar u$
−	−	+	−	−	−2/3	$\bar u$
−	−	−	−	−	0	N

in one multiplet, and transformations within it include both the strong and weak interactions, now appearing on an equal footing.

The SO(10) unification of matter into one multiplet is a stunning achievement, but has always raised two nagging questions. First, completing the symmetry requires an additional particle, here called N. Where is it? Second, the postulated symmetry between the strong and weak interactions requires that they have equal power. They do not.

The predicted properties of N are peculiar. Considering only the interactions of the Standard Model (instead of the full SO(10)), N is completely neutral — it has no strong, weak or electromagnetic interactions. That makes it elusive. Related to this neutrality is a unique feature of N's mass.

Whereas all the other particles would be massless in empty space, and acquire mass only as a result of their interactions with the omnipresent so-called Higgs field, N has an independent mass. Indeed, rather than giving it mass the interaction of N with the Higgs field mixes it with an ordinary neutrino. All potentially observable consequences of the existence of N depend on the value of its mass. Fortunately, insight into this arises from our second nagging question — the inequality between strong and weak forces.

The coupling strength of forces in the Standard Model depends on distance in a precisely calculable way.[6,7] By extrapolating these calculations to much smaller distances — equivalently, much larger energies — we can test the idea that the strong, weak and electromagnetic interactions are different facets of one universal interaction.[8] Remarkably, it works. The various coupling strengths do meet, indicating that unification occurs at distances below 10^{-32} m, or energies above 10^{16} GeV. In comparison, direct measurements currently peter out at around 10^{-18} m, or 100 GeV.

Now I can show how the different strands knit together to produce neutrino masses.[9,10] N's mass is closely related to the unification scale — a very large mass scale by other standards. Ordinary neutrinos then acquire mass indirectly through their interaction with N: it is possible for a left-handed neutrino (ν) to change into a left-handed N, then a right-handed anti-N, and finally into a right-handed anti-ν (see the Box). This process requires a quantum fluctuation to account for the fleeting existence of N and anti-N. But quantum fluctuations to states of such large energy are very rare. So the process whereby ordinary neutrinos flip their handedness, which is a measure of their mass, is heavily suppressed. Quantitatively, this argument predicts neutrino masses of about the value reported by Kajita — which no other current theories do.

What's next? The theory of neutrino masses and mixings is far from complete.[2] The long-standing apparent shortage of neutrinos from the Sun might be explained by oscillations of electron neutrinos, as might a reported anomaly in accelerator neutrino experiments. And neutrinos of large enough mass — about a hundred times the mass difference that SuperKamiokande has measured — could provide a significant 'dark matter' component to the density of the Universe, and so affect the evolution of large-scale structure in galaxy clustering.

Some less direct ramifications may prove still more profound. Unification of couplings seems to require another extension of the Standard Model, to include approximate 'supersymmetry'.[11] One of the main goals of the forthcoming Large Hadron Collider will be to see whether this is correct — if it is, a whole new world of phenomena will open up, involving new heavy partners for all known particles. But supersymmetry, as it solves old problems, poses new ones. Supersymmetric theories provide new mechanisms whereby protons might decay,[12,13] and the predicted rates are already precariously close to violating experimental limits. More experiments could intensify the crisis — or begin to bring it to a satisfactory climax.

It might be said, with some justice, that I have erected here an enormous inverted pyramid of theory, supported on one point. But what an improvement this is, over no support at all! Suddenly, and at last, we begin to see the embodiment of long-anticipated, seductive dreams of pure reason.

References

1. http://www-sk.icrr.u-tokyo.ac.jp/doc/index.html
2. Wilczek, F., *Nature* **391**, 123–124 (1998).
3. Pati, J. & Salam, A., *Phys. Rev. Lett.* **31**, 661–664 (1973).
4. Georgi, H. & Glashow, S., *Phys. Rev. Lett.* **32**, 438–441 (1974).
5. Georgi, H., in *Particles and Fields 1974* (ed. Carlson, C.) 575–582 (Am. Inst. Phys., New York, 1975).
6. Gross, D. & Wilczek, F., *Phys. Rev. Lett.* **30**, 1343–1346 (1973).
7. Politzer, H., *Phys. Rev. Lett.* **30**, 1346–1349 (1973).
8. Georgi, H., Quinn, H. & Weinberg, S., *Phys. Rev. Lett.* **33**, 451–454 (1974).
9. Gell-Mann, M., Ramond, P. & Slansky, R., in *Supergravity* (eds. van Nieuwenhuizen, F. & Freedman, D.) 315–321 (North-Holland, Amsterdam, 1979).
10. Yanagida, T., in *Workshop on the Unified Theory and Baryon Number in the Universe* (eds. Sawada, O. & Sugamoto, A.), 95–98 (KEK, Tsukuba, 1979).
11. Dimopoulos, S., Raby, S. & Wilczek, F., *Phys. Rev.* **D24**, 1681–1683 (1981).
12. Weinberg, S., *Phys. Rev.* **D26**, 287–302 (1982).
13. Sakai, N. & Yanagida, T., *Nucl. Phys.* **B197**, 533–542 (1982).

Masses and Molasses

Some time in the next few years a great discovery will be unveiled, with appropriate fanfare. The headlines will read "ORIGIN OF MASS DISCOVERED". Many readers will be blown away; many will be cynical. Some will scratch their heads and wonder, what do these words actually mean? One doesn't normally think of mass as something with an origin. But a wise and happy few will be prepared. They will leaf through their treasured back issues of *New Scientist*, fish out the right one, and refresh their memories. Welcome back!

What will have been discovered is a new kind of heavy, highly unstable particle, the so-called Higgs particle. And we might see it in just a few months, at one of the two high-energy accelerators: the Large Electron-Positron collider (LEP) at CERN near Geneva or the Tevatron at Batavia, Illinois.

The Higgs is more than just another expensive, highly unstable particle: it embodies the mechanism that gives other fundamental particles mass. But isn't mass just a fact of life? Not necessarily. In fact, ours would be a much simpler world if particles didn't have mass. For one thing, mass messes up the theory of the weak nuclear force. The weak force, as befits its name, is much weaker than the strong force which holds atomic nuclei together and the electromagnetic force that holds atoms together. But it does things that no other interaction can: it causes the slow decay of various otherwise stable particles, and it is the only interaction aware of neutrinos. So what's the problem? Well, the existence of mass means that particles feeling the weak force don't all spin in the same way (see "Messy mass"). It would be neater if they did.

That is merely untidiness; but there is another, more disturbing problem with the particles that carry the weak force. All forces in Nature work by the action of such carrier particles; photons carry the electromagnetic force, for example. And in 1954, Chen Ning Yang and Robert Mills hypothesised the existence of particles called vector mesons, generalized versions of the photon, which looked like good candidates to carry the weak nuclear force. Then in 1961 Sheldon Glashow used them in a theory that unified weak and electromagnetic forces. According to this theory, vector mesons are massless, like the photon. But unlike electromagnetism, the weak force is short-ranged, a sign that its carrier particles must have mass. To fix this, Glashow fudged the equations by just sticking in a mass, without understanding where it came from.

Cosmic Molasses

It would be easier, then, to understand an imaginary world with only massless particles, forever whizzing around at the speed of light. But we know that in our world particles do have mass. So to get from that ideal world to ours, we need some kind of cosmic molasses that fills all space and slows down these massless speed demons. But if this molasses is everywhere, why can't we see it?

To understand, imagine you're living in a bar magnet. An ordinary magnet is really an extraordinary thing. For whereas the laws of physics do not have a preferred direction, the magnet does: its pole. Where does this direction come from? Each electron in any material acts as a small magnet, pointing in the direction of its spin axis. An isolated electron would be equally happy with its spin in any direction, an indifference that we call rotational symmetry. But in some materials, such as iron, neighboring electrons prefer to point in the same direction. Like insecure teenagers, they don't care what they are doing, as long as they are all doing the same thing. So to make all the electrons happy or, in more dignified language, to obtain the configuration of minimum energy, all the spins have to pick up a common direction — it doesn't matter which. That direction defines the magnetic pole.

The rotational symmetry of an isolated spin is gone, but not forgotten. For if we heat an iron magnet above 870°C, the spins get enough energy to break free from their neighbors and point in random directions again — the material loses its magnetism. If the iron is later cooled, it will once again become magnetic. But the new pole will usually point in a different direction from the old.

And rotational symmetry can reappear in another, subtler way. Give the spins just a little energy, and you can make the preferred spin direction (the local magnetic North) change slowly as a function of location. Configurations in which the preferred direction varies periodically are called spin waves. And just as quantum mechanics parcels up light waves into photons, it parcels these spin waves into particles known as magnons.

Particle Swarm

Intelligent creatures living inside a magnet would be used to seeing magnons, but they would have trouble figuring out why magnons exist. Evolution would adapt their senses to ignore the unchanging aspects of their environment. So what we think of as the material of the magnet, they would commonly regard as empty space. And it would seem obvious that there was a preferred direction to space, because everything the creatures experienced would be colored by the pervasive magnetism of their world. Eventually, though, some visionary might imagine the true situation: an underlying set of laws with full rotation symmetry, a symmetry hidden by the spontaneous alignment of spins in the pervading medium. Our visionary would have deduced that the "vacuum" is really a structured medium, explained the existence of magnons, and so become a hero of physics.

This is just what happened on Earth. We have known since the 1930s that our vacuum is really a swarm of short-lived "virtual" particles, appearing and disappearing at random. But where is the organised structure in this melee? The visionaries who first saw it were Yoichiro Nambu and Jeffrey Goldstone. In the early 1960s they noticed a symmetry by which the laws of physics stay the same if certain particles are substituted for others. (It would take an article several times the length of this one to attach proper names and identifiable faces to these particles, and unless you are a very unusual person you would not stay awake to the end. Trust me.) But, just as in the magnet, at low temperature the symmetry is broken: from the symmetrical swarm of virtual particles, one kind condenses out in large numbers. So a preference is formed among the otherwise interchangeable types of particles. Instead of a preferred direction like the magnet, our space has a preferred particle composition.

The Higgs particle cookbook

Because Higgs particles interact most strongly with other high-mass particles, it is hard to make them directly in the collisions of lightweights like electrons. Instead, we reach the Higgs particle indirectly, through virtual *Z* or *W* bosons or pairs of top quarks, which then decay emitting Higgs.

In diagrams like these, only the particles with free ends extending backwards exist for a noticeable time in the past, and only the particles with free ends extending forwards exist for a noticeable time in the future. The lines with no free ends have only a very fleeting existence and cannot be observed — they are said to be virtual particles.

In part (A) of the diagram, we see how an electron and a positron create a virtual *Z* boson, which then emits a Higgs boson and becomes real. This is the process LEP experimenters hope to see. At the Tevatron, instead of electrons, experimenters will use quarks and antiquarks (B) found within their colliding protons and antiprotons, and produce *W* boson in place of *Z*s.

(C) Alternatively at the Tevatron, and especially at the LHC, gluons — found within colliding protons — should make pairs of virtual top quarks that will annihilate to form Higgs particles. This process is my own contribution to the Higgs particle cookbook.

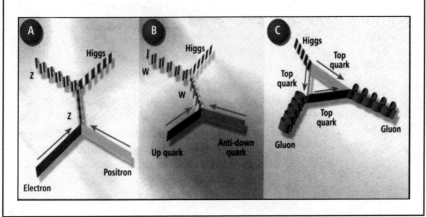

And this is where the cosmic molasses oozes into our story. In 1966 Peter Higgs of the University of Edinburgh, and independently Robert Brout and François Englert of the Free University in Brussels added this idea to the theory of vector mesons. They discovered that when the symmetry breaks, producing a condensate of virtual Higgs particles, the vector mesons become massive.

> **Messy mass**
>
> Once, we thought that the fundamental laws of physics made no distinction between left and right — for any behavior you can observe in the real world, its mirror image can also happen. So if you filmed the real world and its reflected image, someone watching the movies later wouldn't be able to tell which was which. This is called parity symmetry.
>
> Then in 1956 Tsao-Dai Lee at Columbia University, New York, and Chen Ning Yang at the Institute for Advanced Study, Princeton, suggested that the weak interaction breaks parity symmetry. They turned out to be right. For example, neutrons decay through the weak interaction into protons, electrons and electron antineutrinos. The electrons emitted in this decay are moving at nearly the speed of light, and they are also spinning. About 98 per cent of them are left-handed, meaning that if you pointed the thumb of your left hand in the direction of its motion your fingers would curl in the direction the electron spins. This bias violates parity symmetry, because it distinguishes left from right.
>
> So the weak interaction likes left-handed particles (electrons, muons, quarks and so on) and right-handed antiparticles. But, irritatingly, this seems to be no more than a rule of thumb. Weak processes involving right-handed particles or left-handed antiparticles are rare, but not absent.
>
> Here's why. According to the theory of relativity, the laws of physics should look the same to a moving observer. But consider an observer moving in the same direction as a left-handed electron emitted in neutron decay, but faster. They will see the electron going backwards, but spinning the usual way — it will seem right-handed.
>
> But what if the electron had zero mass? Then it would, like a photon, always move at the speed of light. No observer could overtake the electron, and the problem would no longer arise. In a world without mass, the "rule of thumb" that only left-handed particles and right-handed antiparticles participate in the weak interaction could be an exact principle.

Better still, interactions with the condensate could generate the masses of all the other elementary particles, the quarks and leptons. Nambu and Goldstone had constructed a form of cosmic molasses using particles already known to exist. But this isn't quite enough, because it exerts too little drag on the vector mesons, and none at all on the leptons. In 1967, however, Steven Weinberg (and later Abdus Salam) postulated an additional stickier form, and showed how it could give an improved, fudge-free version of Glashow's weak interaction model. This stickier stuff is what physicists usually mean when they talk about the Higgs condensate.

How can we test this extraordinary conception? We could try to heat up the vacuum, by concentrating a lot of energy in a small space, and watch to see if its symmetry is restored as the condensate evaporates. All particles in this region would become massless. Unfortunately, that will only happen at temperatures approaching 10^{16} Kelvin. Although such temperatures were universal in the early stages of the Big Bang, they are out of reach on Earth for the foreseeable future. The Relativistic Heavy Ion Collider at Brookhaven, New York, due to turn on this summer, will peak at only 10^{13} Kelvin.

Stir It Up

A much more modest project is feasible, however. Rather than restore symmetry completely, we can stir up the Higgs condensate a bit. This being a quantum world, we can only stir it up in discrete units. The minimal excitation — a ripple in the cosmic molasses — is the Higgs particle.

How hard will it be to make this particle? Who gets to taste the joy of discovery depends on the value of the Higgs mass, as does the nature of particle physics. We can already narrow down the range.

If the Higgs particle were lighter than 95 gigaelectronvolts (GeV), about 100 times the mass of a proton, LEP would already have seen it. If it were heavier than 600 GeV, virtual Higgs particles would affect many particle reactions in a way that experiments have already ruled out. And the promising theoretical idea of supersymmetry — an extension of the Standard Model that proposes a host of extra fundamental particles, partners of the familiar bunch — predicts masses well below 200 GeV for the Higgs particle; probably between 100 and 130 GeV.

That is why so much excitement surrounds the upcoming explorations (see "The Higgs particle cookbook"). Scientists at LEP will drive their machine to the limits of its energy and luminosity, pushing the mass window up to 105 GeV or so within two years. Meanwhile, scientists at the Tevatron hope to explore all the way up to 160 GeV. If they fail, then a final effort will be made at the Large Hadron Collider (LHC) being constructed in Geneva due to open around 2005. Its reach extends beyond 600 GeV. If that fails, we theoretical physicists will be exceedingly embarrassed, and I hesitate to predict what we'll do.

The Standard Model requires just one Higgs particle. But theories with more symmetry imply several new particles — Higgs galore. The theory of supersymmetry predicts at least five Higgs-type particles. In the most popular version, the lightest member of the Higgs family has the properties we just described. There is no consensus on the masses of the others, although they should not be much heavier than 1000 GeV, and might be much lighter. The masses of these particles will tell us how the supersymmetric partners of ordinary particles hide themselves from us. At present it is a big mystery, and wild concepts are in the air, including their infection by otherwise inaccessible "dark" matter, or exotic condensates living only in extra dimensions of space. The LHC should shed light on this mystery.

More ambitious models that unify the strong and electroweak forces predict a bizarre tribe of very much heavier Higgs particles. We probably won't be able to make them directly anytime soon, but we might sense the effect of their exchange as virtual particles. Some of them can make protons decay, at rates close to current experimental limits.

I hope I've conveyed why we physicists find cosmic molasses to our taste, and look forward to sampling it soon, perhaps in several varieties.

Note added, January 2006: LEP finished by establishing a lower limit of 114 GeV for the Higgs particle mass. LHC is now scheduled to begin operation in summer 2007. In short, we're still waiting …

In Search of Symmetry Lost

Powerful symmetry principles have guided the quest for Nature's fundamental laws. The successful gauge theory of electroweak interactions postulates a more extensive symmetry for its equations than are manifest in the world. We ascribe this discrepancy to a pervasive symmetry-breaking field, which fills all space uniformly, rendering the Universe a sort of exotic superconductor. Soon the theory will undergo a critical test depending on whether the quanta of this symmetry-breaking field, the so-called Higgs particles, are produced at the Large Hadron Collider (due to begin operation in 2007).

It has been almost four decades since our current, wonderfully successful theory of the electroweak interaction was formulated.[1-4] Central to that theory is the concept of spontaneously broken gauge symmetry.[5,6] According to this concept, the fundamental equations of physics have more symmetry than the actual physical world does. Although its specific use in electroweak theory involves exotic hypothetical substances and some sophisticated mathematics, the underlying theme of broken symmetry is quite old. It goes back at least to the dawn of modern physics, when Newton postulated that the basic laws of mechanics exhibit full symmetry in three dimensions of space — despite the fact that everyday experience clearly distinguishes 'up and down' from 'sideways' directions in our local environment. Newton, of course, traced this asymmetry to the influence of Earth's gravity. In the framework of electroweak theory, modern physicists similarly postulate that the physical world is described by a solution wherein all space, throughout the currently observed

Universe, is permeated by one or more (quantum) fields that spoil the full symmetry of the primary equations. Thus, modern physicists hypothesize that what we perceive as empty space is actually a highly structured medium. In fact, as I will elaborate below, we strongly suspect that the world is a multilayered, multicolored, cosmic superconductor.

Fortunately this hypothesis, which might at first hearing sound quite extravagant, has testable implications. The symmetry-breaking fields, when suitably excited, must bring forth characteristic particles: their quanta. Using the most economical implementation of the required symmetry breaking, one predicts the existence of a remarkable new particle, the so-called Higgs particle. More ambitious speculations suggest that there should be not just a single Higgs particle, but rather a complex of related particles. The very popular and attractive idea of low-energy supersymmetry,[7,8] to be discussed further below, requires at least five 'Higgs particles'.

The primary goal of fundamental physics is to discover profound concepts that illuminate our understanding of Nature. Discovering new particles, as such, is secondary. In recent times, however, physicists have often found that their most profound concepts, when implemented with rigorous logic, are reflected in the existence of new particles. This happens because both quantum mechanics and special relativity are important in the regime of short distances and high energies, where high-energy physics explores fundamental laws. It is difficult to combine quantum mechanics and special relativity in a consistent way. The only way we know how to do it is by using quantum field theory, and the basic objects of quantum field theory are space-filling entities (quantum fields) whose excitations are what we perceive, concretely, as particles.[9,10] So, when our concepts are made consistent with quantum mechanics and relativity, they tend to be reflected rather directly in predictions about particles.

The W and Z bosons, carriers of the weak nuclear force, and gluons, carriers of the strong nuclear force, are outstanding examples of ideas embodied as particles. These so-called gauge particles are physical embodiments of the symmetry of physical law (gauge invariance)[11–13] — not merely metaphorically but in a very precise sense. Indeed, as a fact of history, the existence of these particles and their detailed behavior was predicted before their experimental observation, starting from the concept of gauge symmetry. Harmony between mind and matter, in the form of mathematical abstractions conjuring up sensuous reality, has long figured in the dreams of mystics and the inspiration of visionaries —

the 'music of the spheres'. Here it is realized in a form that is genuine, reproducible and precise.

Now we are faced with the opportunity for another synthesis. Ironically, the concept whose embodiment we now seek is a special, structured sort of symmetry breaking. This concept is a necessary complement to what has come before; for our symmetry-based understanding of the W and Z bosons — that is, of the electroweak interaction — relies on postulating symmetries that are broken in a very specific way. They are supposed to be spoiled by a form of cosmic superconductivity, with newly hypothesized fields having the role performed by electrons in ordinary superconductors. It is these new quantum fields that are the progenitors of Higgs particles.

So far, no Higgs particle has been observed. As yet, this failure does not represent a crisis. Detection of Higgs particles that are sufficiently heavy — specifically, those whose mass exceeds 114 GeV, which is the current lower bound[14] — will have to await more powerful accelerators than we have now. But theory tells us that this evasion cannot be maintained indefinitely. If the Higgs particle, or an appropriate complex of Higgs particles, does not turn up at the Large Hadron Collider (LHC), a major revision of our thinking will be required. The LHC is now under construction at CERN (European Centre for Particle Physics) near Geneva. It is due to begin operation in late 2007.

There are already indirect but significant indications that at least one Higgs particle with a mass below 250 GeV does exist.[15] If there is such a particle, it will certainly be observed at the LHC. That observation, if and when it occurs, will bring a glorious chapter in physics to a glorious conclusion. It will also provide a key to unlock new volumes that are currently sealed; for the circle of ideas around symmetry breaking and the Higgs particle includes, quite close to its elegant central core, some of the darkest and most forbidding zones of ignorance in the existing landscape of fundamental physics. That is exciting, because it means we will have an opportunity to learn. Big ideas and speculations about the unification of forces, and the cosmology of the early Universe, as well as supersymmetry, are very much in play. Thus the Higgs sector is not only a destination, but also a portal.

There is a vast technical literature on most of the topics discussed here[16,17] and quite a few popular and semipopular presentations. My goal here is to present a brief but substantial and critical review of the main concepts and prospects, one that is accessible to scientifically sophisticated non-experts, yet reflects the essence of present-day thinking.

The Three Standard Systems

It has become conventional to say that our knowledge of fundamental physical law is summarized in a 'standard model'. But this convention lumps together three quite different conceptual structures. Instead, it is more accurate and informative to say that our current, working description of fundamental physics is made from three distinct parts: three 'standard systems'. These are the gauge system, the gravity system and the Higgs system.

Each of these systems concerns the interactions of a specific kind of particle: gauge bosons, gravitons and Higgs particles, respectively. It is remarkable that everything we know, or reliably infer, about the fundamental laws of Nature can be interpreted as a statement about how one or another of these particles interacts with other forms of matter. To be precise, every known departure from trivial 'free' propagation — every nonlinear coupling of the quantum fields that describe matter in all its forms — involves interaction with a gauge particle, a graviton or a Higgs particle. Two of these three systems, the gauge and gravity systems, are governed by principled theories founded on deep, powerful concepts. Because of this they are tight, both conceptually and algorithmically.

The gauge system is constructed as the embodiment of extensive symmetries involving transformations among different kinds of 'color' degrees of freedom. Color used in this sense has nothing to do with optical phenomena; rather, it is a generalization of the concept of electromagnetic charge. Quantum chromodynamics (QCD) — our theory of the strong interaction[18,19] — works with three color charges. The weak interaction invokes two other color charges and ordinary electromagnetism introduces yet another charge. The precise symmetry is expressed in the language of group theory, as $SU(3) \times SU(2) \times U(1)$.

Gauge symmetries, when combined with the principles of quantum mechanics and special relativity, are extremely powerful. Gauge symmetries require the existence of appropriate gauge bosons, and vice versa. Through this connection between mathematics and physics — concept and reality — we arrive at a beautiful and tightly integrated theory of gauge bosons and their interactions with other forms of matter. A profound reflection of this is that the physics of the gauge system is almost fully determined from considerations of inner consistency. In other words, it contains few freely adjustable parameters. There are basically just three such parameters, one each for $SU(3)$, $SU(2)$ and $U(1)$. Given these three numbers, no further 'fudge factors' remain available. The gauge system provides precise predictions for many phenomena; predictions that are in excellent agreement with numerous accurate experiments. In fact no significant deviation has been found so far. Each term in the foundational equations underlying a complete description of strong, weak and electromagnetic interactions has been checked out repeatedly in precise experiments.

The gravity system is essentially Einstein's general theory of relativity. It is sometimes claimed that general relativity is impossible to reconcile with quantum mechanics, that there is a terrible crisis in quantum gravity, and so on. On hearing this, one might be puzzled about how physicists and astrophysicists manage to get on with their work. The truth is that there is a very concrete and precise working theory of gravity that fully conforms to the principles of quantum mechanics and which so far has proved adequate to describe all physical and astrophysical observations. It is simply the quantum field theory version of general relativity (for experts: the Einstein–Hilbert action for gravity itself, extended to matter using the minimal coupling procedure). This theory fails to give predictions for processes involving ultra-high-energy particles, but the energies at which it fails are much larger than those that are accessible to observation.

Like the gauge system, the gravity system is constructed as the embodiment of a powerful symmetry principle, in this case Einstein's general covariance. General covariance both requires the existence of the graviton, and tightly constrains its properties. It thereby generates a unique, principled theory of gravity. General relativity is fully specified in terms of just two freely adjustable parameters: one is Newton's gravitational constant G_N; the other is the so-called cosmological term, which parameterizes the energy density of empty space. Until very recently there were only upper bounds on the value of the cosmological term, but observations of the acceleration of distant galaxies, together with indirect inference from measured cosmic microwave background anisotropies, seem to require a positive value.[20–22] All other gravitational phenomena are predicted using only G_N. General relativity has scored many triumphs, both qualitative ones, such as providing foundations for Big Bang cosmology and black hole physics, and quantitative ones, such as accurately predicting the precession of Mercury and the time variation of binary pulsar frequencies.

The gauge system and the gravity system can both be written in appealing geometric forms. More precisely, quantum fields that describe different forms of curvature produce both gravitons and gauge bosons. For gravitons, the relevant curvature is that of space-time; for gauge bosons, it is curvature in so-called internal spaces, which are defined by the variables that describe configurations of color charges. The coupling of these particles to other fields is also defined geometrically, technically, by the promotion of ordinary derivatives into covariant ones.

The third system, where the Higgs particle and its couplings to other forms of matter reside, is another story entirely. We know of no deep principle, comparable to gauge symmetry or general covariance, which constrains the values of these couplings at all tightly. In the Higgs system, the number of freely adjustable parameters mushrooms into dozens. Nor do the values of the couplings we now infer from the masses and mixing of quarks and leptons conform to any easily discernible pattern (see the section '... and a nest of kluges' below).

Clearly the Higgs system of fundamental physics is its least satisfactory part. Whether measured in terms of the large number of independent parameters it requires, or in terms of the small number of powerful ideas it contains, our theory of this sector falls far short of the high level we have achieved elsewhere. In particular, despite the phrase's connotation, no 'theory of everything' hitherto proposed has in practice materially improved our theory of the Higgs system. Having placed the Higgs system in context, let us now scrutinize it more closely.

Symmetry Breaking in Superconductivity

The most fundamental phenomenon of superconductivity is the Meissner effect, according to which magnetic fields are expelled from the bulk of a superconductor. The Meissner effect implies the possibility of persistent currents.[23] Indeed, if a superconducting sample is subjected to an external magnetic field, currents of this sort must arise near the surface of a sample to generate a cancelling field.

An unusual but valid way of speaking about the phenomenon of superconductivity is to say that within a superconductor the photon acquires a mass. The Meissner effect follows from this. Indeed, to say that the photon acquires a mass is to say that the electromagnetic field becomes a massive field. Because the energetic cost of supporting massive fields over an extended volume is prohibitive, a superconducting material finds ways to expel magnetic fields.

Bardeen, Cooper and Schrieffer (BCS) developed a satisfactory microscopic theory of superconductivity in metals.[24] BCS theory traces superconductivity to the existence of a special sort of long-range correlation among electrons. This effect is purely quantum-mechanical. A classical phenomenon that is only very roughly analogous, but much simpler to visualize, is the occurrence of ferromagnetism owing to long-range correlations among electron spins (that is, their mutual alignment in a single direction). The sort of correlations responsible for superconductivity are of a much less familiar sort, as they involve not the spins of the electrons, but rather the phases of their quantum-mechanical wavefunctions. One cannot do full justice to this concept without some elaborate mathematics. But as it is the leading idea guiding our construction of the Higgs system, I think it is appropriate to sketch an intermediate picture that is more accurate than the magnet analogy and suggestive of the generalization required in the Higgs system.

First we imagine that the electrons organize themselves into pairs, the so-called Cooper pairs. The wavefunction of a Cooper pair is a complex-valued function; it has both an amplitude and a phase. If we have a uniform density of Cooper pairs, then the amplitude is constant, but the phase can vary in space and time. We can represent the different possible values of the phase by points on a circle. So in representing the quantum dynamics of the Cooper pairs at each point of space-time, we have an overlying circular 'internal space' — an extra dimension if you like — and the position of the wavefunction in this extra dimension must be specified (Fig. 1a).

A fundamental principle of electrodynamics is its gauge symmetry. A gauge transformation is a mathematical transformation of electromagnetic potentials and the wavefunctions of charged particles. When we say electrodynamics obeys gauge symmetry, we mean that while the gauge transformation changes the variables that appear in the equations of electrodynamics, and consequently rearranges those equations, it nevertheless leaves the overall physical content of the equations unchanged (Fig. 1b). Weyl and London discovered the gauge symmetry of quantum electrodynamics in the 1920s (refs. 11, 12). Yang and Mills proposed a more general concept of gauge symmetry[13] that supports a much wider class of transformations (nonabelian gauge theory). At first this generalization seemed

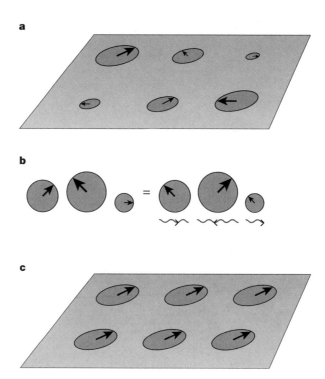

FIGURE 1. Visual metaphors for gauge symmetry and superconductivity. a, The quantum-mechanical wavefunction of an electrically charged field is a position-dependent complex number, which can be depicted as a collection of arrows with different sizes and orientations. b, Gauge symmetry states that the same physical situation can be described using different orientations of the arrows (phases of the wavefunction) with a compensating gauge field (electromagnetic potential, represented by wavy arrows). This is somewhat analogous to Einstein's equivalence principle, by which relatively accelerated reference frames become equivalent when a compensating gravity field is included. c, In the superconducting state, the charged-particle wavefunction is correlated over long distances and times — whereas in the non-superconducting state (a could be regarded as a typical, disordered slice of this state), the wavefunctions fluctuate randomly in space and time. The ordered structure of c is disrupted by gauge transformations, so now the electromagnetic potentials have an energetic cost — the photon has acquired a mass.

to be a mathematical curiosity, but as physics developed it has come to be more and more central. We have come to recognize both that gauge symmetry is necessary for the consistency of the theory, and conversely, that the equations of electrodynamics can be derived from gauge symmetry, assuming the validity of quantum mechanics and special relativity. One aspect of this is that gauge symmetry enforces zero mass for the photon.

For present purposes, what is crucial is that gauge transformations rotate the wavefunction in the extra dimension, through an angle that can vary depending on location in space-time. Superconductivity occurs when the phases of the Cooper pairs all align in the same direction; that is, when they all have the same position within their extradimensional circles (Fig. 1c). Of course, gauge transformations that act differently at different space-time points will spoil this alignment. Thus, although the basic equations of electrodynamics are unchanged by gauge transformations, the state of a superconductor does change. To describe this situation, we say that in a superconductor gauge symmetry is spontaneously broken.

The phase alignment of the Cooper pairs gives them a kind of rigidity. Electromagnetic fields, which would tend to disturb this alignment, are rejected. This is the microscopic explanation of the Meissner effect, or in other words, the mass of photons in superconductors.

Electroweak Symmetry Breaking: Some Elegant Details ...

Several basic properties of the $W^\pm$ and Z particles, which are responsible for mediating the weak interaction, are quite similar to properties of photons. They are all spin-1 particles that couple with universal strength to appropriate charges and currents. This resemblance, together with consistency requirements of the sort mentioned above, suggests that the equations of the weak interactions must have an appropriate gauge symmetry, distinct from, but of the same general nature as, the gauge symmetry of electromagnetism.

Elaboration of this line of thought leads to our modern theory of the electroweak interactions, which is firmly based on the postulate of gauge symmetry. The gauge symmetry involved is more elaborate than that of electromagnetism. It involves a more intricate internal space (Fig. 2a). Instead of just a circle (orientation) and a ray (magnitude), we have a three-dimensional sphere (orientation) and a ray (magnitude) over each point of space-time, so four 'internal dimensions'. Gauge transformations rotate the spheres, although not all types of rotation are allowed. Mathematically, the allowed transformations define the group $SU(2) \times U(1)_Y$.

There is an essential supplement to this generalization, which provides our primary motivation for postulating the existence of the Higgs system. Unlike the photon, which acquires mass only inside superconductors, $W^\pm$ and Z are massive particles even in empty space. The equations of gauge symmetry, in their pristine form, predict the existence of these particles, but require that their masses vanish. To account for the masses of $W^{\pm}$ and Z, we must suppose that what we perceive as empty space is, in reality, a new form of superconductor, not for electromagnetism, but for its near-relation gauge interactions.

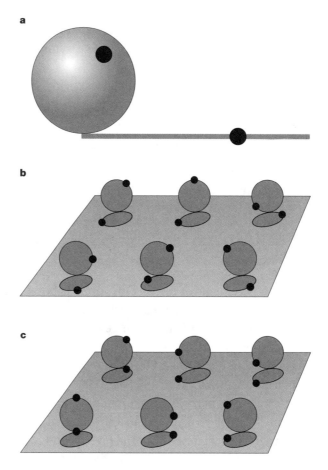

FIGURE 2. Gauge symmetry and the Higgs condensate. **a**, The minimal Higgs doublet takes values in an internal space that is the product of a three-dimensional sphere and a ray (indicated by the maroon points). **b**, To bring out the essential point, it is useful to simplify this geometry to two circles. Gauge transformations can rotate orientations within the vertical (*A*) or the horizontal (*B*) circles independently. **c**, This situation is analogous to Fig. 1c: it represents the nature of the condensation responsible for this more intricate, universal superconductivity. This condensation allows the free variation of an overall orientation (*A* + *B*), but locks the relative orientation within each pair of circles.

With this interpretation, what we perceive as empty space is not so empty. But what is it that plays the role — for this new universal superconductivity — of the Cooper pairs whose alignment is responsible for conventional superconductivity? No form of matter identified so far provides a suitable candidate. We are therefore led to postulate that there is a new form of matter doing the job. Accordingly, what we perceive as empty space is in fact filled with an exotic, suitably aligned substrate: the Higgs condensate.

'Observing' this condensate only through its effect on W and Z bosons gives us very limited insight into its nature. The minimal hypothesis, within the conventional framework of contemporary physics, is to postulate the existence of a quantum field with just enough structure to support the desired pattern of symmetry breaking, imparting mass to some, but not all, of the gauge fields.

Let me spell this out with reference to the slightly simplified model of Fig. 2b. The pattern we want in reality is $SU(2) \times U(1)_Y \to U(1)_\gamma$; the simplified model exhibits $U(1)_A \times U(1)_B \to U(1)_{A+B}$, but the essential points are the same.

We begin with two independent rotation symmetries, but only a combination survives, which will represent ordinary electromagnetism. To break the symmetry in the appropriate way, we require the condensate to have a component in both circles (Fig. 2c). Furthermore, its orientations within the circles must be equal to one another, although their common angle can vary. Similarly to what was discussed for the Cooper-pair condensate, the Higgs condensate has a form of rigidity that imposes an energetic price on fields that attempt to disrupt this favorable pattern. But the field that rotates both circles by the same amount preserves the pattern, so it remains massless. It can be identified with the electromagnetic field, which produces the photon γ. Note that the $A+B$ gauge symmetry of electromagnetism ($U(1)_\gamma$) is not the same as either 'pure' rotation A or B of the primary circles.

In the realistic $SU(2) \times U(1)_Y \to U(1)_\gamma$ case, a similar entwining of weak and electromagnetic gauge symmetry is mandatory, simply because the $W^\pm$ bosons are electrically charged. Thus, these interactions must be treated together, and we speak of 'electroweak' theory. It is often said that the $SU(2) \times U(1)_Y \to U(1)_\gamma$ theory of broken gauge symmetry is a unified theory of weak and electromagnetic interactions, but that seems to me an overstatement. The theory still involves two quite separate primary symmetries, which are related only by the fact that the same

Higgs condensate spoils them both. The simplest way of implementing the symmetry-breaking pattern $SU(2) \times U(1)_Y \to U(1)_\gamma$ makes specific predictions for the masses and couplings of the W and Z bosons, which seem to be quite accurate. More complicated implementations are conceivable, but they seem neither desirable nor necessary.

Together with its orientations in the internal sphere, the quantum field associated with the Higgs condensate must have another degree of freedom, representing its overall magnitude. The degrees of freedom corresponding to variations in orientation are associated with the (broken-symmetry) gauge transformations. They get absorbed into the massive gauge fields; their quanta are longitudinally polarized W and Z bosons. The degree of freedom associated with changes in overall magnitude, however, has independent meaning. Its elementary excitation, or quantum, is what is usually referred to as the Higgs particle. The inescapable minimal consequence of our postulate of a Higgs condensate is a new, electrically neutral spin-0 particle.

One embellishment of the minimal scheme is especially well motivated, because it arises in connection with the important concept of low-energy supersymmetry.[25] The generalization simply involves imagining that two independent fields of the same general character (both representing positions on internal spheres and circles, and undergoing the same sorts of gauge transformations) contribute to the Higgs condensate. A striking consequence is that we then expect to have not one but five new Higgs particles, because we have introduced a field with four additional degrees of freedom (specified by its location within its four-dimensional internal space). Of these, three are electrically neutral; the other two are positively and negatively charged, each being the other's antiparticle.

... and a Nest of Kluges

It is not only the masses of W and Z bosons that are inconsistent with pristine $SU(2) \times U(1)$ gauge symmetry and that get tied up with the Higgs condensate. The masses of quarks and leptons present a similar difficulty. In the end, these masses too can be ascribed to interaction with the Higgs condensate, but the details are quite different and frankly (in my view) seem quite ugly and clearly provisional.

The proximate source of the difficulty is the observed parity violation of the weak interaction.[26] In its most basic form, the observed phenomenon is that when fast-moving quarks or leptons are emitted in weak decays, their spin tends to be aligned opposite to their direction of motion; they are therefore said to be left-handed. To accommodate this in the framework of gauge symmetry, we would like to say that the symmetry acts only on left-handed particles. That formulation, however, is inconsistent with special relativity. Indeed, a sufficiently fast-moving observer could overtake the particle, and to such an observer, its direction of motion would appear reversed, and it would be right-handed. This difficulty would be avoided if the quarks and leptons had zero mass, for then they would move at the speed of light and could not be overtaken. So we can implement the interactions of W bosons with massless left-handed quarks and leptons in a consistent way using gauge symmetry; but non-zero masses for quarks and leptons must be tied up with gauge symmetry breaking.

The example of W and Z bosons might lead us to hope for nice geometrical pictures associated with the masses of quarks and leptons, and connections with profound concepts (gauge invariance, superconductivity). We are reduced to simply writing down the equations. For non-expert readers, the details are not vital, but I do want to convey a sense of what leads me to call them 'ugly' and 'provisional'.

At present there is no compelling theory that predicts the values of any quark or lepton masses. Yet there are striking facts to be explained that play a crucial part in the structure of the world as we know it. The mass of the electron, and of the up and down quarks are nearly a million times smaller than the 'natural' scale (250 GeV) set by the magnitude of the Higgs condensate, but the mass of the top quark is close to that scale; the CKM matrix (that describes the mixing of quark species) is measured to be almost, but not quite, the unit matrix. All this presumably indicates that principles — or perhaps accidents or even conspiracies — are lurking within the Higgs sector, which ensure that some, but not all, of the possible couplings are small. It is a great challenge for the future to discover these principles, or alternatively to understand why accident and conspiracy run rampant. So far, we have access only to the masses and the CKM matrix, which encode highly processed versions of the basic couplings. The discovery of Higgs particles and measurements of their interactions will give us better access to fundamentals.

On the Origin(s) of Mass

As we have just seen, the masses of W and Z bosons, and of quarks and leptons, arise from the interaction of these particles with the pervasive Higgs condensate. This has inspired references to the Higgs particle as 'the origin of mass', or even 'the God particle'.[27] The real situation is interesting but rather different from what this hyperbole suggests. A few critical comments seem in order.

First, most of the mass of ordinary matter has no connection to the Higgs particle. This mass is contained in atomic nuclei, which are built up from nucleons (protons and neutrons), which in turn are built up from quarks (mainly up and down quarks) and color gluons. Color gluons are strictly massless, and the up and down quarks have tiny masses, compared with the mass of nucleons. Instead, most of the mass of nucleons (more than 90%) arises from the energy associated with the motion of the quarks and gluons that compose them[28] — according to the original form of Einstein's famous equation, $m = E/c^2$. This circle of ideas provides an extraordinarily beautiful, overwhelmingly positive answer to the question Einstein posed in the title of his original paper[29]: "Does the inertia of a body depend on its energy content?" And it has nothing to do with Higgs particles!

Second, as we have just seen, for quarks and leptons the Higgs mechanism appears more as an accommodation of mass than an explanation of its origin. We map observed values of masses and mixings through some distorting lenses into corresponding values of Higgs-field couplings, but only for the W and Z bosons do we have reliable, independent insight into what these values ought to be. And third, the Higgs field in no sense explains the origin of its own mass. A parameter directly equivalent to that mass must be introduced into the equations explicitly.

Finally, there is no necessary connection between mass and interaction with any particular Higgs field. As an important example, much of the mass of the Universe is observed to be in some exotic, 'dark' form. The dark matter has been observed only indirectly, through the influence of its gravity on surrounding ordinary matter. It has evaded more conventional observation using optical, radio or other telescopes, so evidently it has only a feeble coupling to photons. We do not yet know what this dark stuff is precisely, but the leading theoretical candidates (axions and weakly interacting massive particles, WIMPs) are massive particles that do not acquire their mass by interacting with the electroweak Higgs condensate. The general point is that many kinds of hypothetical particle can have masses that, unlike the masses of W and Z bosons and quarks and leptons, do not violate the electroweak $SU(2) \times U(1)$ symmetry, and such masses need have no relation to that symmetry's spoiler, the electroweak Higgs condensate. The Higgs particle's own mass is also of this kind.

The genuine meaning of the Higgs field — that it embodies the concept of symmetry breaking and makes tangible the vision of a universal cosmic superconductor — is deep, strange, glorious, experimentally consequential and very probably true. This meaning has no need for embellishment, and can only be diminished by dubious oversimplification. So blush for shame, ye purveyors of hyperbole! End of sermon.

Searching for Higgs

From the observed masses of the W and Z bosons we infer the magnitude of the Higgs condensate to be 250 GeV. This sets the scale for structure in the Higgs system. The mass of the lightest excitation of the condensate could only be significantly larger than this if a large parameter appeared in the theory to amplify the scale. More specifically, what is relevant to this question is the coefficient of the nonlinear self-coupling of the Higgs field. But large self-coupling seems unlikely on both theoretical and experimental grounds. Theoretically, it takes us into a regime where the Higgs field undergoes violent quantum fluctuations, and appears to develop inconsistencies.[30] On the experimental side, precision measurements of various quantities can be compared to theoretical predictions at a level that is sensitive to the contribution of 'virtual particles'; that is, quantum fluctuations, including contributions that arise from coupling to the Higgs system. All indications so far are that these contributions are small, consistent with weak coupling.[15] Thus, on very general grounds, we expect to find new particles of some kind, recognizable as quanta of the Higgs condensate, with mass most probably smaller than 250 GeV, and in any case not much larger.

To go further and be quantitative, we must consider more specific models. The minimal implementation of symmetry breaking, which assumes that only a single $SU(2) \times U(1)$ doublet exists (and thereby, as discussed above, predicts the existence of exactly one physical Higgs particle), is of course the canonical choice. Within this framework, which I shall adopt for the next few paragraphs, everything of interest is determined in terms of just one unknown parameter, conveniently taken to be the value of the Higgs particle's mass.

Now, with everything else pinned down, the aforementioned contributions (due to virtual particles, and often called radiative corrections) to

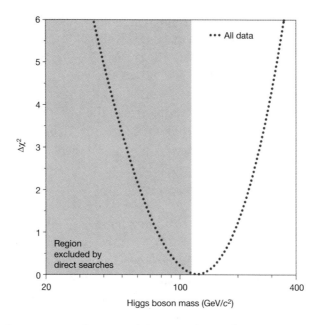

FIGURE 3. Constraints on the mass of the Higgs boson from experimental data. The dotted line shows, as a function of the Higgs mass, the χ^2 for a global fit to electroweak data (including such quantities as the mass of the W boson and the Z boson total width). Also included in the dataset is the current world-average value for the mass of the top quark.[48] A relatively light value for the Higgs mass is favored, less than 200 GeV. Shading indicates the region excluded by direct searches for the Higgs boson.[14]

a variety of measured quantities can be calculated precisely, as a function of the Higgs particle's mass. When this is done, one finds that the central value of the Higgs mass preferred by the experimental results is about 100 GeV, with a dispersion of about 50 GeV (Fig. 3).

The strength with which the Higgs particle couples to other fundamental particles is proportional to the mass of the particle involved (with a most important exception that I will come to in a moment). Thus, the dominant decay mode of the Higgs particle will be into the heaviest available particles. For a wide range of masses, including the most favored values, the dominant decay is therefore into a heavy quark pair — bottom/anti-bottom ($b\bar{b}$); channels with heavier particles are energetically forbidden. The rates of decay into other quark or lepton pairs — charm/anti-charm and tau/anti-tau — are lower but not insignificant.

The exceptional case is gluons.[31] The direct coupling of the Higgs particle to gluons, inferred from the classical form of the theory, vanishes —

as one might expect from the vanishing mass of the gluons. There is a purely quantum-mechanical process, however, in which the Higgs particle fluctuates into a virtual top/anti-top quark pair, which then materializes into a real gluon pair (Fig. 4a). As a fraction of all possible decays, this branch amounts to about 10%. A coupling to photons is similarly achieved by fluctuations through virtual W and Z bosons, as well as top quarks; it is smaller, owing to the weaker photon coupling.

Search strategies for the Higgs particle at electron-positron accelerators (the Large Electron-Positron collider, LEP, or the planned International Linear Collider, ILC) and at proton–proton accelerators (the Tevatron at Fermilab, United States, or CERN's LHC) are quite different. The coupling of electrons and positrons to the Higgs particle is extremely small. A favored strategy exploits processes in which the electron and positron annihilate into a Z boson, which then radiates the Higgs particle[32] (Fig. 4b). This has the further advantage that the final-state Z boson is very well characterized and can be reliably identified. Indeed, a general problem in Higgs-hunting is that one must distinguish the process of interest from possible 'background' processes that can lead, from the point of view of what is observable in practice, to effectively the same final state. This channel (with a final state of $Zb\bar{b}$) was the subject of very intense scrutiny at LEP, especially during the final stages of its operation in 2000. The result, however, was inconclusive. A lower bound of 114.1 GeV, at the

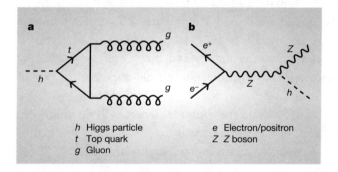

FIGURE 4. Higgs couplings. **a**, The dominant coupling of the Higgs particle to ordinary matter arises indirectly, through a virtual top/anti-top quark pair that annihilates into gluons. Direct couplings of the Higgs to up or down quarks, electrons, gluons or photons are very small, due to the tiny masses of these particles. **b**, Electrons can, however, give access to the Higgs through their coupling to the Z boson. The *Bremsstrahlung* process shown produces a distinctive Zh final state. Similar processes involving quark/antiquark annihilation produce either Zh or Wh final states.

95% confidence level, was put on the mass of the Higgs particle (assuming the minimal implementation). On the other hand, statistically weak (1.7σ) hints of excess events were observed at the upper limit of the machine's energy, consistent with what would be induced by a Higgs particle of mass 116 GeV.

At a hadron machine of sufficiently large energy, such as the LHC, Higgs particles are produced in abundance by the gluon-fusion process, which exploits the Higgs' indirect coupling to gluon pairs.[33] The difficulty here is not production, but the signal-to-noise ratio in detection. Final states at hadron machines typically contain much extraneous debris, because the initial projectiles — protons — are complex objects. Indeed, our gluon-fusion production process relies on the fact that protons, in addition to quarks, contain a substantial fraction of gluons. The dominant final result of Higgs decay, a $b\bar{b}$ pair, is also very easily produced by ordinary particle interactions that in no way involve the Higgs particle. These form, from an experimental point of view, an almost impenetrable background. For this reason, it might be best to focus on the rarer decay of the Higgs into a pair of photons. Another strategy is to look for 'tagged' events that contain a W or Z in the final state, accompanying the Higgs particle (as at electron-positron machines). These could result from the radiation of Higgs particles from the W or Z bosons produced by quark/antiquark annihilation.

Once we depart from the minimal implementation, many more possibilities for probing the Higgs sector open up, some much less demanding. For example, the additional Higgs particles predicted in the minimal supersymmetric model include electrically charged ones. These have very characteristic decay modes and they could be copiously produced at electron-positron machines once the energetic threshold for pair production is passed.

At this point I should mention that whereas in the minimal model the predicted value of the Higgs mass is only weakly constrained from above, the minimal supersymmetric model generically predicts the existence of at least one Higgs particle with mass below 140 GeV, whose properties resemble those of the (absolutely) minimal Higgs particle with the same mass.[17,34] The above-mentioned indications from radiative corrections, and perhaps from LEP, offer encouragement for the supersymmetric alternative.

This account of the practical strategy of Higgs searches is no more than a sketch of some of the simplest considerations involved. A vast, multifaceted and growing literature is devoted to the subject. There is a general consensus that discovery of one or more Higgs particles cannot elude experimenters at the LHC, but that an exploration of the Higgs system worthy of the opportunities it affords will require the clean environment and high energies available in electron-positron collisions at the ILC.

Extensions: Unification, Supersymmetry

If we merely count particles, then the minimal implementation of gauge symmetry breaking — using one doublet field and leading to one physical Higgs particle — recommends itself on grounds of economy. But considerations of logical coherence and structural integrity seem to lead us in another direction.

The structure of the gauge system gives powerful suggestions for its further fruitful development.[35] The product structure $SU(3) \times SU(2) \times U(1)$, the reducibility of the fermion representation (that is, the fact that the symmetry does not make connections linking all the fermions), and the peculiar values of the quantum number hypercharge assigned to the known particles all suggest a larger symmetry. The devil is in the details and it is not at all automatic that the observed, complex pattern of matter will fit neatly into a simple mathematical structure. But, to a remarkable extent, it does (Box 1).

This unification of quantum numbers, although attractive, remains purely formal until it is embedded in a physical model. This requires realizing the enhanced symmetry in a local gauge theory. The enhanced symmetry must be broken. The Higgs system of electroweak theory supplies a precedent for that. What we need is another condensate, with a vastly larger density. We are proposing that the world is a multi-colored, multi-layered superconductor.

There is an apparent difficulty with these ideas, which on closer scrutiny turns out to represent their greatest success. Non-abelian gauge symmetry requires that the relative strengths of the different couplings must be equal (universality), which is not what is observed. Fortunately, there is a compelling way to save the situation.[36] The higher symmetry is broken at a very large energy scale (equivalently, a small distance scale), but we observe interactions at much smaller energies (larger distances). The strength we observe differs from the intrinsic strength, because it is affected by the physics of vacuum polarization. A cloud of virtual particles surrounds the charge and can enhance or dilute its power. The resulting

Box 1
Unification

Within the standard gauge system the strong interactions are described by quantum chromodynamics (QCD), a theory based on the gauge symmetry group $SU(3)$ defined by transformations among three color charges; the weak interactions are described by an independent but mathematically similar gauge symmetry $SU(2)$ using two color charges, and finally an independent 'hypercharge' symmetry $U(1)$ based on a single type of charge. Ordinary electromagnetism involves a combination of $SU(2)$ and $U(1)$ symmetry that remains valid after these separate symmetries are broken by the Higgs condensate.

In part **a** of the figure, the relationships of the quarks and leptons under these transformations are shown. The strong gauge symmetries act horizontally, the weak gauge symmetries act vertically and the values of the hypercharges are indicated by subscripts. The 15 quarks and leptons shown here fall into five unrelated clans. (There is a threefold repetition of this entire structure, accommodating 45 quarks and leptons altogether.)

Theories of unified gauge symmetry propose that there is a more extensive symmetry that involves transformations among all these color charges. That symmetry must be spontaneously broken, to explain why we observe the consequences of different types of charge to be quite different (strong interactions really are strong, and weak interactions weak). Nevertheless, this idea is not without consequences: the quarks and leptons must furnish the material for building unified structures that remain coherent under the extended symmetry.

One particular unified symmetry passes this test with flying colors (and is shown in part **b** of the figure). Although the smallest simple group into which $SU(3) \times SU(2) \times U(1)$ could possibly fit is $SU(5)$ (ref. 49) — it fits all the fermions of a single family into two representations ($\mathbf{10} + \mathbf{\bar{5}}$) and the hypercharges click into place — a larger symmetry group, $SO(10)$ (ref. 50), fits these and one additional $SU(3) \times SU(2) \times U(1)$ singlet particle into a single representation (the spinor **16**). All 15 quarks and leptons appear on the same footing, and the additional particle, which has the quantum numbers of a right-handed neutrino, is quite welcome: it plays a crucial role in the attractive 'seesaw' model for neutrino masses.[43,44]

Where before we had a piecemeal accommodation of the observed particles, now we have a marvellous correspondence between reality and a unique, ideal mathematical object.

a

$\begin{pmatrix} u & u & u \\ d & d & d \end{pmatrix}_{1/6}$ $(u^c\ u^c\ u^c)_{-2/3}$

$(d^c\ d^c\ d^c)_{1/3}$

$\begin{pmatrix} \nu \\ e \end{pmatrix}_{-1/2}$ e^c_1

b

Quantum number / Particle	R	W	B	G	P
u	+	−	−	+	−
u	−	+	−	+	−
u	−	−	+	+	−
d	+	−	−	−	+
d	−	+	−	−	+
d	−	−	+	−	+
u^c	−	+	+	−	−
u^c	+	−	+	−	−
u^c	+	+	−	−	−
d^c	−	+	+	+	+
d^c	+	−	+	+	+
d^c	+	+	−	+	+
ν	+	+	+	+	−
e	+	+	+	−	+
e^c	−	−	−	+	+
N	−	−	−	−	−

Hypercharge $Y = -\frac{1}{6}(R + W + B) + \frac{1}{4}(G + P)$

'running' of couplings is an effect that can be calculated quite precisely, given a definite hypothesis about the particle spectrum. In this way we can test quantitatively the idea that the observed couplings originate from a single unified value at small distances.

The results of these calculations are quite remarkable and encouraging (Fig. 5). If we include vacuum polarization from the particles we know about in the minimal standard model, we find approximate unification. If we include vacuum polarization from the particles needed in expanding the standard model to include supersymmetry,[37,38] softly broken at the TeV (1,000 GeV) scale, we find accurate unification.[39,40] The unification occurs at a very large energy scale, of order 10^{16} GeV. This success is robust against small changes in the supersymmetry-breaking scale, and is not adversely affected by incorporation of additional particle multiplets, as long as they form complete representations of $SU(5)$ (ref. 37).

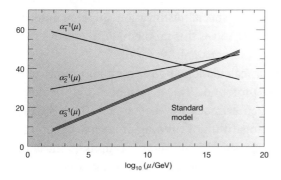

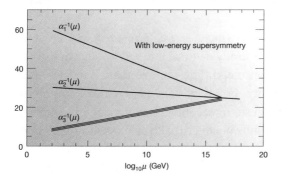

FIGURE 5. Unification of the forces. The strengths of the couplings of the weak, electromagnetic and strong forces are hugely disparate (represented here as α_1^{-1}, α_2^{-1} and α_3^{-1}). But their perceived strength changes with the energy scale of the process (μ), through corrections due to virtual particles. Assuming there are only the particles known to us in the standard model and extrapolating beyond the reach of experiment to very high energies, the couplings move towards each other but do not converge at a single point (top). If, however, the extra particles needed to implement low-energy supersymmetry are included in the calculation, the couplings meet neatly at an energy of about 10^{16} GeV (lower plot). Note that the energy scale is logarithmic (and the possible existence of other unknown particles is overlooked), so this calculation is a bold — perhaps reckless — extrapolation of the laws we know to apply to energies vastly larger than those at which these laws have been tested.

Low-energy supersymmetry is desirable on several other grounds as well. The most important has to do with the Higgs condensate. In the absence of supersymmetry, radiative corrections to the magnitude of this condensate diverge ('radiative corrections' is standard jargon for the effect of virtual particles or, alternatively, quantum fluctuations). One must fix the condensate's value, which sets the scale for electroweak symmetry breaking, by hand, as part of the definition of the theory. This 'renormalization' procedure leaves it utterly mysterious why the empirical value is so much smaller than unification scales. Moreover, enhanced unification symmetry requires that the Higgs doublet should come together with additional fields, to fill out a complete representation. Its partners, however, have the quantum numbers to mediate proton decay, so if they exist at all, their masses must be very large, at about the unification scale of 10^{16} GeV. This reinforces the idea that such a large mass is what is 'natural' for a scalar field.[41] The relatively small mass (of the order of 100 GeV) of the Higgs field that we need in our electroweak theory seems unnatural and requires some special justification. Supersymmetry, if it is not too badly broken, largely solves this problem, for it ensures that these unsavory radiative corrections are small.[42]

The fact that our unification calculations point to an enormous new mass scale for unification is profound. This enormous mass scale is inferred entirely from data taken at much lower energies. The disparity of scales arises from the slow (logarithmic) running of inverse couplings, which implies that modest differences in observed couplings must be made up by a long interval of running. The appearance of a very large mass scale is welcome on several grounds. I will mention three of the most important.

First, right-handed neutrinos, which as we have seen can enhance the symmetry of unification, naturally acquire masses of the order of the unification scale. Masses of that magnitude remove these particles from direct experimental accessibility, but they can have a most important indirect effect.[43,44] This is because, in second-order perturbation theory, the ordinary left-handed neutrinos make virtual transitions to their right-handed relatives and back. This exotic process generates non-zero masses for the ordinary neutrinos, but these are much smaller than the masses of other leptons and quarks. The magnitudes that arise in this way are broadly consistent with the tiny observed masses of neutrinos. No more than order-of-magnitude success can be claimed because many relevant details of the models are poorly determined.

Second, unification tends to obliterate the distinction between quarks and leptons, and hence to open up the possibility that protons decay (their building-block quarks turn into electrons or muons). Heroic experiments to observe this process have so far come up empty-handed, with limits on partial lifetimes approaching 10^{34} years for some channels. It is very difficult to ensure that these processes are sufficiently suppressed, unless the unification scale is very large. Even the high scale indicated by the running of couplings and neutrino masses is barely adequate. Spinning it positively, experiments to search for proton decay remain a most important and promising probe into the physics of unification. Similarly, it is difficult to avoid the idea that unification brings in new connections among the different families. There are significant experimental constraints on flavor-changing neutral currents, lepton number violation and other exotic processes that must be suppressed, and this makes a high mass scale for the virtual particles that mediate them most welcome.

Third, with the appearance of this large scale, unification of the strong and electroweak interactions with gravity becomes much more plausible. Newton's constant has dimensions of mass, so it runs even classically. Or, to put it another way, gravity responds to energy/momentum, so it gets stronger at large energy scales. Nevertheless, because gravity starts out extremely feeble compared to other interactions on laboratory scales, it becomes roughly equipotent with them only at enormously high scales, comparable to the Planck energy of 10^{18} GeV. By inverting this thought, we gain a deep insight into one of the main riddles about gravity: if gravity is a primary feature of Nature, reflecting the basic structure of space-time, why does it ordinarily appear so feeble? Elsewhere,[45] I have traced the answer down to the fact that, at the unification (Planck) scale, the strong coupling is about 1/2!

In view of all this, our accounting of the 'economy of ideas' is altered. For it seems that with five Higgs particles you can buy a lot more than with one.

Cosmological Implications

In the very early Universe, when temperatures were much higher, the Higgs condensate that now fills all space could not have maintained its alignment over extended distances. In a word, it melted.[46] Just as a superconductor heated beyond its critical temperature goes normal, or a magnet heated above its Curie temperature loses its magnetization, the

Universe would then have been in a different, more symmetric phase. In this phase, W and Z bosons — like photons, color gluons and gravitons — had zero mass, as did quarks and leptons. (Ironically, the Higgs particles themselves retained a finite mass.)

Thus, during the early evolution of the Universe there was a dramatic change in the properties of matter. The detailed physical nature of this change is at present unknown. It may have been a sharp phase transition in the thermodynamic sense, or a smooth crossover. Such a cosmic phase transition might have been accompanied by unusual or violent physical events that left lasting consequences. One possibility is that the current imbalance between the abundance of matter and antimatter might have been generated when the Higgs condensate froze in. Another is that the Higgs freeze-in catalysed an epoch of extremely rapid cosmic acceleration, akin to or even identical to the inflationary epoch, whose occurrence is widely conjectured in modern cosmology[47] but whose physical nature is highly uncertain. It is only by studying the Higgs system in detail that we can begin to assess these possibilities reliably.

The existence of a Higgs system with properties of the general sort I have discussed, notably including one or more accessible, recognizable Higgs particle, appears to be a compelling consequence of quantum field theory and the standard model of fundamental physics — a complex of ideas that has been tremendously successful. This would be a beautiful thing to observe, and extremely instructive.

And yet, our standard system of gravity — general relativity — incorporates the principle that all forms of energy exert gravitational influence. As a special case, the postulated Higgs condensate, which fills all space, should weigh something. In fact, it should weigh a lot: estimating the energy density of this condensate using straightforward dimensional reasoning gives a value much larger than is allowed by observations. It would show up as such a large contribution to the cosmological term that the size of the Universe would double every 10^{-38} seconds! Thus, either the Higgs condensate does not exist, or its energy density is cancelled out by some other, still more exotic, contribution, or there is a profound lacuna in our understanding of gravity. The second and third alternatives seem sufficiently improbable as to suggest that just maybe we will be dragged to the first. If so, then our search for Higgs particles as the quanta of gauge symmetry breaking might instead turn up something quite different. Theoretical physicists, roused from their dogmatic slumbers, would be forced back to the drawing board. Wouldn't that be fun?

References

1. Glashow, S. Partial symmetries of weak interactions. *Nucl. Phys.* **22**, 579–588 (1961).
2. Weinberg, S. A model of leptons. *Phys. Rev. Lett.* **19**, 1264–1266 (1967).
3. Salam, A. in *Elementary Particle Physics, Nobel Symp.* (ed. Svarthom, N.) No. 8, 367–377 (Almqvist & Wiksell, Stockholm, 1968).
4. 't Hooft, G. & Veltman, M. Regularization and renormalization of gauge fields. *Nucl. Phys.* **B44**, 189–213 (1972).
5. Englert, F. & Brout, R. Broken symmetry and the mass of gauge vector mesons. *Phys. Rev. Lett.* **13**, 321–323 (1964).
6. Higgs, P. Broken symmetries and the masses of gauge bosons. *Phys. Rev. Lett.* **13**, 508–509 (1964).
7. Wess, J. & Zumino, B. A lagrangian model invariant under supergauge transformations. *Phys. Lett.* **B49**, 52–54 (1974).
8. Ferrara, S. (ed.) *Supersymmetry* (North-Holland/World Scientific, 1987).
9. Weinberg, S. *The Quantum Theory of Fields 1: Foundations* (Cambridge Univ. Press, Cambridge, 1995).
10. Wilczek, F. Quantum field theory. *Rev. Mod. Phys.* **71**, S85–S95 (1999).
11. Weyl, H. Z. Elektron and gravitation I. *Z. Phys.* **56**, 330–352 (1929).
12. London, F. Quantenmechanische deutung der theorie von Weyl. *Z. Phys.* **42**, 375–389 (1927).
13. Yang, C. N. & Mills, R. Conservation of isotopic spin and isotopic gauge invariance. *Phys. Rev.* **96**, 191–195 (1954).
14. ALEPH Collaboration, DELPHI Collaboration, L3 Collaboration, OPAL Collaboration & The LEP Working Group for Higgs Boson. Search for the Standard Model Higgs boson at LEP. *Phys. Lett.* **B565**, 61–75 (2003).
15. LEP Electroweak Working Group [online] <http://lepewwg.web.cern.ch/LEPEWWG> (2004).
16. Einhorn, M. (ed.) *The Standard Model Higgs Boson* (North-Holland, Amsterdam, 1991).
17. Gunion, J., Haber, H., Kane, G., & Dawson, S. *The Higgs Hunter's Guide* (Addison-Wesley, New York, 1990).
18. Gross, D. & Wilczek, F. Ultraviolet behavior of non-Abelian gauge theories. *Phys. Rev. Lett.* **30**, 1343–1346 (1973).
19. Politzer, H. Reliable perturbative results for strong interactions? *Phys. Rev. Lett.* **30**, 1346–1349 (1973).
20. Riess, A. *et al.*, Observational evidence from supernovae for an accelerating universe and a cosmological constant. *Astron. J.* **116**, 1009–1038 (1998).
21. Permutter, S. et al. Measurements of Ω and Λ from 42 high-redshift supernovae. *Astrophys. J.* **517**, 565–586 (1999).

22. Spergel, D. et al., Determination of cosmological parameters. *Astrophys. J.* **148** (suppl.), 175–194 (2003).
23. London, F. *Superfluids 1: Macroscopic Theory of Superconductivity* (Wiley, New York, 1950).
24. Bardeen, J., Cooper, L. & Schrieffer, R. Theory of superconductivity. *Phys. Rev.* **108**, 1175–1204 (1957).
25. Weinberg, S. *Quantum Theory of Fields*, Vol. 3: *Supersymmetry* (Cambridge Univ. Press, Cambridge 2000).
26. Lee, T. D. & Yang, C. N. Question of parity conservation in weak interactions. *Phys. Rev.* **104**, 254–258 (1956).
27. Lederman, L. & Teresi, D. *The God Particle* (Bantam, New York, 1993).
28. Wilczek, F. The origin of mass. Physics@MIT **16**, 24–35 (2003).
29. Einstein, A. Does the inertia of a body depend upon its energy content? [in German] *Ann. Phys.* **18**, 639–641 (1905).
30. Kuti, J., Lin, L. & Shen, Y. Upper bound on the Higgs-boson mass in the Standard Model. *Phys. Rev. Lett.* **61**, 678–681 (1988).
31. Wilczek, F. Decays of heavy vector mesons into Higgs particles. *Phys. Rev. Lett.* **39**, 1304–1306 (1977).
32. Bjorken, J. in *Proc. SLAC Summer School* (SLAC publication 198, 1976).
33. Georgi, H., Glashow, S., Machacek, M. & Nanopoulos, D. Higgs bosons from two-gluon annihilation in proton–proton collisions. *Phys. Rev. Lett.* **40**, 692–694 (1978).
34. CMS collaboration. The compact muon solenoid [online] <http://www.lip.pt/~outreach/docs/cms2> (2004).
35. Pati, J. & Salam, A. Unified lepton-hadron symmetry and a gauge theory of the basic interactions. *Phys. Rev.* **D8**, 1240–1251 (1973).
36. Georgi, H., Quinn, H. & Weinberg, S. Hierarchy of interactions in unified gauge theories. *Phys. Rev. Lett.* **33**, 451–454 (1974).
37. Dimopoulos, S., Raby, S. & Wilczek, F. Supersymmetry and the scale of unification. *Phys. Rev.* **D24**, 1681–1683 (1981).
38. Dimopoulos, S. & Georgi, H. Softly broken supersymmetry and SU(5). *Nucl. Phys.* **193**, 150–162 (1981).
39. Amaldi, U., de Boer, W. & Fürstenau, H. Comparison of grand unified theories with electroweak and strong coupling constants measured at LEP. *Phys. Lett.* **B260**, 447–455 (1991).
40. Ellis, J., Kelly, S. & Nanopoulos, D. Precision LEP data, supersymmetric GUTs and string unification. *Phys. Lett.* **B249**, 441–448 (1990).
41. Gildener, E. & Weinberg, S. Symmetry breaking and scalar bosons. *Phys. Rev.* **D13**, 3333–3341 (1976).
42. Witten, E. Dynamical breaking of supersymmetry. *Nucl. Phys.* **B188**, 513–554 (1981).

43. Gell-Mann, M., Ramond, P. & Slansky, R. in *Supergravity: Proceedings of the Supergravity Workshop at Stony Brook*, September 27-29, 1979 (eds. van Nieuwenhuizen, P. & Freedman, D.) 315–321 (North-Holland, Amsterdam, 1979).
44. Yanagida, T. in *Workshop on Unified Theory and Baryon Number in the Universe* (eds. Sawada, O. & Sugamoto, A.) 95–98 (KEK, Tsukuba, 1979).
45. Wilczek, F. Scaling Mount Planck 3: Is that all there is? *Phys. Today* **55N8**, 10–11 (2002).
46. Kirzhnits, D. & Linde, A. Symmetry behaviour in gauge theories. *Ann. Phys.* **101**, 195–238 (1976).
47. Guth, A. Inflationary universe: a possible solution to the horizon and flatness problems. *Phys. Rev.* **D23**, 347–356 (1981).
48. A precision measurement of the mass of the top quark. *Nature* **429**, 638–642 (2004).
49. Georgi, H. & Glashow, S. Unity of all elementary-particle forces. *Phys. Rev. Lett.* **32**, 438–441 (1974).
50. Georgi, H. in *Particles and Fields 1974* (ed. Carlson, C.) 575–584 (AIP, New York, 1975).

From 'Not Wrong' to (Maybe) Right

Savas Dimopoulos is always enthusiastic about something, and in the spring of 1981 it was supersymmetry. He was visiting the new Institute for Theoretical Physics in Santa Barbara, which I had recently joined. We hit it off immediately — he was bursting with wild ideas, and I liked to stretch my mind by trying to take them seriously.

Supersymmetry was (and is) a beautiful mathematical idea. The problem with applying supersymmetry is that it is too good for this world. We simply do not find particles of the sort it predicts. We do not, for example, see particles with the same charge and mass as electrons, but a different amount of spin.

However, symmetry principles that might help to unify fundamental physics are hard to come by, so theoretical physicists will not give up on them easily. Based on previous experience with other forms of symmetry, we have developed a fallback strategy, called spontaneous symmetry breaking. In this approach, we postulate that the fundamental equations of physics have the symmetry, but the stable solutions of these equations do not. The classic example of this phenomenon occurs in an ordinary magnet. In the basic equations that describe the physics of a lump of iron direction is equivalent to any other, but the lump becomes a magnet with some definite north-seeking pole.

Understanding the possibilities for spontaneously broken supersymmetry requires model building — the creative activity of proposing candidate equations and analysing their consequences. Building models with spontaneously broken supersymmetry that are consistent with everything else we know about physics is a difficult business. Even if you manage to get the symmetry to break, the extra particles are still there (just heavier) and cause various mischief. I briefly tried my hand at model building when supersymmetry was first developed in the mid-1970s, but after some simple attempts failed miserably, I gave up.

Savas was a much more naturally gifted model-builder, in two crucial respects: he did not insist on simplicity, and he did not give up. When I identified a particular difficulty (let us call it A) that was not addressed in his model, he would say: "It's not a real problem, I'm sure I can solve it," and the next afternoon he would come in with a more elaborate model that solved difficulty A. But then we would discuss difficulty B, and he would solve that one with a completely different complicated model. To solve both A and B, you had to join the two models, and so on to difficulty C, and soon things got incredibly complicated. Working through the details, we would find some flaw. Then the next day Savas would come in, very excited and happy, with an even more complicated model that fixed yesterday's flaw. Eventually we eliminated all flaws, using the method of proof by exhaustion — anyone, including us, who tried to analyse the model would get exhausted before they understood it well enough to find the flaws.

When I tried to write up our work for publication, there was a certain feeling of unreality and embarrassment about the complexity and arbitrariness of what we had come up with. Savas was undaunted. He even maintained that some existing ideas about unification using gauge symmetry, which to me seemed genuinely fruitful, were not really so elegant if you tried to be completely realistic and work them out in detail. In fact, he had been talking to another colleague, Stuart Raby, about trying to improve those models by adding supersymmetry! I was extremely sceptical about this 'improvement', because I was certain that the added complexity of supersymmetry would spoil the existing success of gauge symmetry in explaining the relative values of the strong, electromagnetic and weak coupling constants. The three of us decided to do the calculation, to see how bad the situation was. To get oriented and make a definite calculation, we started by doing the crudest thing, which was to ignore the whole problem of breaking supersymmetry. This allowed us to use very simple (but manifestly unrealistic) models.

The result was amazing, at least to me. The supersymmetric versions of the gauge symmetry models, although they were vastly different from the originals, gave very nearly the same answer for the couplings.

That was the turning point. We put aside the 'not wrong' complicated models with spontaneous supersymmetry breaking, and wrote a short paper that, taken literally (with unbroken supersymmetry), was wrong. But it presented a result that was so straightforward and successful that it made the idea of putting gauge symmetry and supersymmetry unification together seem (maybe) right. We put off the problem of how supersymmetry gets broken. And even today, although there are some good ideas about it, there is still no generally accepted solution.

After our initial work, more precise measurements of the couplings made it possible to distinguish between the predictions of models with and without supersymmetry. The models with supersymmetry work much better. We all eagerly await operation of the Large Hadron Collider at CERN, the European particle physics laboratory, where, if these ideas are correct, the new particles of supersymmetry — or, you might say, the new dimensions of superspace — must make their appearance.

This little episode, it seems to me, is 179 degrees or so out of phase from Karl Popper's idea that we make progress by falsifying theories. Rather, in many cases, including some of the most important, we suddenly decide our theories might be true, by realizing that we should strategically ignore glaring problems. It was a similar turning point when David Gross and I decided to propose quantum chromodynamics (QCD) based on asymptotic freedom, putting off the problem of quark confinement. But that is another story ...

Unification of Couplings

by Savas Dimopoulos, Stuart A. Raby and Frank Wilczek

> Recent high-precision experimental results support the predictions of the minimal supersymmetric SU(5) model that unifies electromagnetism and the weak and strong interactions.

Ambitious attempts to obtain a unified description of all the interactions of Nature have so far been more notable for their ingenuity, beauty and *chutzpah* than for any help they have afforded toward understanding concrete facts about the physical world. In this article we wish to describe one shining exception: how ideas about the unification of the strong, weak and electromagnetic interactions lead to concrete, quantitative predictions about the relative strengths of these interactions.

The basic ideas in this subject are not new; they were all essentially in place ten years ago. For several reasons, however, the time seems right to call them back to mind. Most importantly, the accuracy with which the relevant parameters have been determined experimentally has improved markedly in the last few months, making a much more meaningful comparison between theory and observation possible. The results of this confrontation, as we shall see, are quite encouraging and suggestive.

Gauge Theories

It has been traditional to identify four fundamental interactions: strong, weak, electromagnetic and gravitational. In the 1960s and 1970s great progress was made in elucidating the principles underlying the first three of these interactions (see the articles by Howard Georgi and Sheldon Lee

Glashow in *Physics Today*, September 1980, page 30, and by David Gross, January 1987, page 39).[1] By comparison the elucidation of quantum gravity is at a comparatively primitive stage. Except for a few remarks toward the end, our discussion will be confined to the first three interactions — the traditional domain of high-energy physics.

To make a very long story short, it was found that a common mechanism underlies all three of these interactions: Each is mediated by the exchange of spin-1 particles, *gauge bosons*. The gauge bosons have different names in the three cases. They are called color gluons in the strong interaction, photons in the electromagnetic interaction, and W and Z bosons in the weak interaction. But despite the difference in names and some other superficial differences, all gauge bosons share a common mathematical description and deeply similar physical behaviors. Gauge bosons interact with quarks and leptons in several ways — mediating forces among them, being emitted as radiation when the quarks or leptons accelerate, and even changing one kind of quark or lepton into another.

The original and most familiar gauge theory is also the most basic. Quantum electrodynamics is properly understood, in modern terms, to be neither more nor less than the theory of a single gauge boson (namely, the photon) coupled to a single charge, or "color" (namely, electric charge). In mathematical language, it is the theory of the gauge group $U(1)$.

Chromatic terminology for charges is useful and evocative, but must not be taken too literally. Color charges are numerical quantities, which may be positive or negative integers (or zero). The charges associated with different colors are independent quantities. Thus a particle might carry blue charge +1 and yellow charge +1 but green charge 0.

The modern theory of the weak interaction is essentially the simplest nontrivial extension of this setup, to include *two* colors. An important new possibility for gauge boson physics first shows up with two colors: In addition to gauge bosons that, like the photon, respond to the color charges, there are also gauge bosons that change a unit of one charge into a unit of the other. In this fundamental process (see Figure 1), one kind of particle is changed into another carrying different color charge. Color charge is conserved overall because the difference in charge between the altered particles is carried by the gauge boson. The W bosons are of this identity-altering type, and their exchange is the mechanism underlying radioactive transmutations of atomic nuclei of one element into those of another. The Z boson, acting more like the photon, responds to but does not alter the weak color charges. In mathematical language, the modern

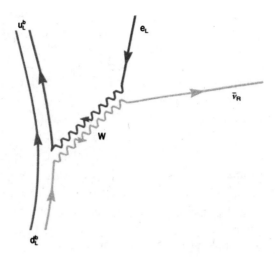

FIGURE 1. Exchange of color-changing gauge bosons can alter the identities of the quarks and leptons involved. Here, the elementary process underlying ordinary radioactivity is depicted as a process of weak color transmutation.

theory of the weak interaction is the theory of the gauge group SU(2) — the 2 here just indicates two colors.

Finally quantum chromodynamics, the modern theory of the strong interaction, is — you guessed it — the theory of three colors, based on the gauge group SU(3). It involves eight gauge bosons (color gluons), six that alter colors and two others that merely respond to them.

The color charges involved in the strong and weak interactions are completely distinct. It has become customary, at least in the US, to call the strong colors red, white and blue. The weak interaction gives us an opportunity to soften the chauvinism of this terminology to some extent, by adding two new colors: Call them yellow and green. It might seem that to complete the structure we would need a sixth color, for electromagnetic charge. But most remarkably, it appears that having identified the five strong and weak colors, we do *not* need to add a sixth, separate color for electromagnetism. Electric charge is not independent of the other charges. If we make the color assignments indicated in Figure 2 (whose true significance will emerge only below), then the electric charge Q of a particle is given in terms of its various color charges (R, W, B, Y and G) according to the simple formula

$$Q = -1/3(R+W+B)+G \qquad (1)$$

Unification: Triumphs and Challenges

The fact that all three major interactions of particle physics can be described using the concept of gauge bosons coupled to color charges hints at some deeper unity among them. So too, with more subtlety and power, does equation (1). The strong color gluons mediate all possible changes and responses among the red, white and blue colors, while the weak gauge bosons do the same between the yellow and green colors. What could be more natural than to postulate the existence of gauge bosons corresponding to all possible changes and responses among *all five* colors?[2] Such bosons would include the color gluons, weak bosons and photon, and also additional gauge bosons that would change (for example) red charge into yellow charge. Altogether 12 new gauge bosons must be added to the 12 known ones. The gauge theory for five colors is denoted SU(5). It includes the SU(3) × SU(2) × U(1) gauge theories of the strong, weak and electromagnetic interactions — and more.

This idea, on cursory examination, suggests two lovely qualitative successes and two quantitative disasters.

First, the successes. If we consider only the gauge bosons, the expansion of the theory appears as an appealing but quite speculative possibility. While it suggests the existence of new gauge bosons, it does not shed much light concerning the properties of the ones we already know to exist. However, if we widen our considerations to include quarks and leptons, a wonderful advantage of the larger theory comes into view. As indicated in Figure 2, the 15 quarks and leptons within a family can be grouped into two classes. One class consists of five particles, each carrying one unit of one of the five color charges. The other class consists of ten particles, each carrying a unit of each of two distinct color charges. Within either of these two classes, transformations carrying any given particle into any other one can be mediated by appropriate gauge bosons. In other words, the particles within either class are all related to one another by the gauge interaction. They are like different faces of a single die — inseparable, symmetrical pieces of a larger whole.

In mathematical terms, the particles fall into two irreducible representations of SU(5): a five-dimensional vector representation and a ten-dimensional antisymmetric tensor representation. By contrast, when we restrict ourselves to the transformations of SU(3) × SU(2) × U(1), the particles in a family fall apart into no less than five different classes. This striking gain in economy of description is one great qualitative success of the simplest SU(5) unification scheme.

a d_R^r	1	0	0	0	0	**b** $\bar{d}_L^r$	−1	0	0	0	0
d_R^w	0	1	0	0	0	$\bar{d}_L^w$	0	−1	0	0	0
d_R^b	0	0	1	0	0	$\bar{d}_L^b$	0	0	−1	0	0
$\bar{e}_R$	0	0	0	1	0	e_L	0	0	0	−1	0
$\bar{\nu}_R$	0	0	0	0	1	ν_L	0	0	0	0	−1
d_L^r	1	0	0	0	1	$\bar{d}_R^r$	−1	0	0	0	−1
d_L^w	0	1	0	0	1	$\bar{d}_R^w$	0	−1	0	0	−1
d_L^b	0	0	1	0	1	$\bar{d}_R^b$	0	0	−1	0	−1
u_L^r	1	0	0	1	0	$\bar{u}_R^r$	−1	0	0	−1	0
u_L^w	0	1	0	1	0	$\bar{u}_R^w$	0	−1	0	−1	0
u_L^b	0	0	1	1	0	$\bar{u}_R^b$	0	0	−1	−1	0
$\bar{u}_L^r$	0	1	1	0	0	u_R^r	0	−1	−1	0	0
$\bar{u}_L^w$	1	0	1	0	0	u_R^w	−1	0	−1	0	0
$\bar{u}_L^b$	1	1	0	0	0	u_R^b	−1	−1	0	0	0
$\bar{e}_L$	0	0	0	1	1	e_R	0	0	0	−1	−1

FIGURE 2. Quarks and leptons collect into two classes when assigned strong and weak colors inspired by SU(5) unification (**a**). We label the strong color red, white and blue and the weak colors green and yellow. Reading across each row gives the SU(5) color charge for each particle. Subscripts L (left) and R (right) indicate the chirality of the particles, while the superscripts r, w and b indicate the standard strong-color labels of the u (up) and d (down) quarks. Overbars indicate antiparticles (**b** explicitly displays the antiparticles of the particles in **a**). Notice that left-handed and right-handed versions of the same particle may be differently colored — this reflects the violation of parity in the weak interactions.

The other success concerns equation (1). This marvelous equation, in which the electromagnetic, strong and weak charges all come into play, was an encouraging hint toward unification. Within SU(5), its potential is brilliantly fulfilled. Although it is a little too complicated for us to derive here, it is not terribly difficult to show that equation (1) is an automatic consequence of unification in SU(5). The photon only fits within this symmetry group if it responds to precisely the combination of color charges that occurs in equation (1). Thus unification offers a framework in which the apparently chaotic spectrum of electric charges of quarks and leptons can be understood rationally.

In a more precise treatment we would actually have to worry about the spectrum of weak hypercharges, which is even worse. One of us (Wilczek) recalls that as a graduate student he considered the now standard $SU(2) \times U(1)$ model of electroweak interactions to be "obviously wrong" just because it requires such ugly hypercharge assignments. That was going too far, but it still seems fair to call the model "obviously incomplete" for this reason.

Now we must describe two daunting difficulties that such attempts at unification face. The first disaster is that the different gauge bosons, although they do similar things, do not do them with the same vigor. In other words, they couple to their respective color charges with different strengths. The strong interaction, as befits its name, really is much stronger than the weak interaction, which in turn is slightly stronger than the electromagnetic. Thus the perfect symmetry among colors required in a truly unified gauge theory doesn't seem to be in the cards. We shall return to this problem below.

The second disaster concerns the processes mediated by the extra gauge boson, particularly the ones that change strong into weak color charges. It is always at least slightly worrisome to postulate the existence of hitherto unobserved particles, but these fellows are especially objectionable, because their exchange mediates, through the mechanism indicated in Figure 3, processes capable of destabilizing protons. However, protons are rather reliably reported not to decay. Even in 1974, when unified theories of the type we are discussing were first proposed, the lifetime of the proton was known to be upwards of 10^{21} years. Since then, systematic experiments have raised the lower limit to over 10^{31} years (for most plausible decay modes).[3] Comparing this to the rates for comparable weak decays, which are measured in microseconds, we realize with a start what an enormity is being perpetrated — these new gauge bosons must be indeed very different from, and in some sense much less potent than, the old (that is, known) ones.

Nevertheless, given the qualitative successes of gauge theory unification, and its ineluctable beauty, one must not give it up without a fight. And indeed both difficulties can be overcome in triumphal style.

Let us take the second difficulty first. It is actually not so difficult to explain this problem away. To do so, we must now mention a very important aspect of gauge theories that for simplicity we have so far neglected: These theories may exist in different phases and exhibit properties at low energies that differ somewhat from their symmetrical

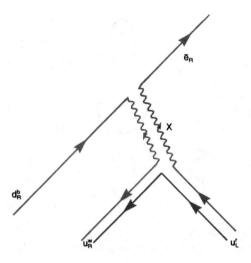

FIGURE 3. Proton decay would be caused by the exchange of some of the extra gauge bosons (X) needed for unification. These bosons would change strong into weak colors and would lead to processes wherein three quarks change into an antilepton.

high-energy behavior. For our purposes, the most important point is that gauge bosons may become massive, through the so-called Higgs mechanism: and the heavier the gauge bosons, the rarer are the processes mediated by their exchange. (The Higgs mechanism is in a very direct sense simply a relativistic version of Fritz and Heinz London's superconducting electrodynamics.)

This, by the way, is why the weak interactions are much less prominent than electromagnetism, even though the intrinsic strengths of the weak-vector-boson couplings are somewhat greater than those of photons. The weak vector bosons are massive, which not only makes them difficult to produce and unstable in isolation, but also makes the processes they mediate less vigorous. Clearly then, to exorcise the specter of the dangerous extra gauge bosons we need only suppose that they are very heavy.

What about the first difficulty? Though perhaps less dramatic, it is more profound. Its resolution involves another order of ideas, and is rich in consequences. To this, we now turn.

Running Coupling Constants

The crucial concept is that of running coupling constants — coupling strengths that vary with energy or distance. This is very similar to the

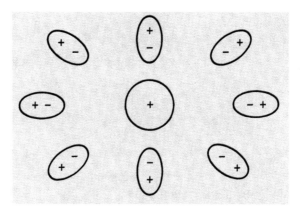

FIGURE 4. Dielectric screening occurs when a charge in a dielectric medium polarizes the molecules around it. This cloud of polarization partially hides, or screens, the central charge.

more familiar and intuitive notion of dielectric screening. In dielectric screening, a positively electrically charged particle within a material tends to pull negative charge toward it, for example, by distorting (polarizing) neutral molecules. This nearby enhancement of negative charge shields or screens the effect of the central positive charge, and so the electric field at large distances due to that charge is less than it would otherwise be (see Figure 4).

In modern quantum field theory, a similar effect happens even in empty space. This is because "empty space" is not a true void, but rather a dynamical medium full of virtual particle–antiparticle pairs that flicker briefly into existence and then reannihilate before traveling very far. (A less poetic but still visually appealing view of the same thing is afforded by the Feynman diagrams in Figure 5.) These denizens of the vacuum can be polarized, no less than molecules in a solid. As a result the charge and electric field distributions close to a nominal elementary "point particle" are in fact structured: The charge is partially screened. The vacuum is a dielectric.

Ordinary dielectric screening tends to make the effective charge smaller at large distances. Conversely, of course, if we work from the outside in, we see the effective charge gradually increasing from what we saw from far away. Virtual quarks and leptons also tend to screen any color charge they carry. This effect turns out to be very general: Spin-1/2, and spin-0 particles of any hypothetical type screen charge.

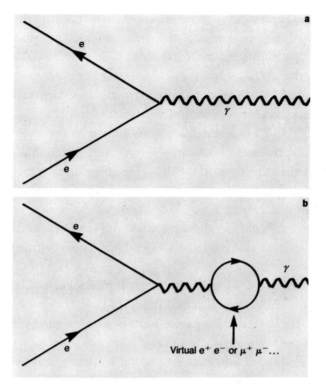

FIGURE 5. Empty space is a dielectric medium in quantum field theory and can screen charge. These Feynman diagrams represent the interaction between a bare charge and a photon (**a**) and the effect of virtual particles coming between the photon and the charge (**b**).

The discovery[5] in 1973 that spin-1 gauge bosons have the opposite effect was a wonderful surprise. It means that a charge that looks large and formidable at large distances can be traced to a weaker (and more manageable) source at short distances. The discovery of this dynamical effect — known as asymptotic freedom — led directly to the identification of SU(3) color gauge theory, or QCD, as the theory of the strong interaction.[6] This came about because the SLAC electroproduction experiments,[7] demonstrating the phenomenon of scaling, indicated that the strong interaction between quarks is much weaker at short distances than one would infer from afar. More precisely, what these experiments indicated is that rapidly accelerated quarks emit few gluons. In other words they behave when they are hit hard as if they are ideal structureless point particles: They recoil elastically; they have no "give." This behavior is in contrast to their appearance when hit softly. Then the more powerful, longer-range aspect of the strong interaction causes quarks to behave not like points but more like thick balls of virtual gluons, quarks and antiquarks.

The logic of the discovery of hard structureless particles — Richard Feynman's partons,[8] now identified with quarks and gluons — inside the proton in the SLAC experiments is quite similar to the logic underlying the classic experiment of Johannes Geiger and Ernest Marsden. Their observation that alpha particles impinging an gold foil may be violently deflected through large angles was interpreted by Ernest Rutherford as indicating the existence of hard, effectively point-like nuclei at the center of atoms. Replacing alpha particles with electrons, and nuclei with partons, we essentially map the Geiger-Marsden experiment onto the SLAC experiment.

Later experiments, as we shall discuss further below, have confirmed and sharpened the early indications from SLAC. When quarks are rapidly accelerated they usually propagate exactly as ideal structureless point particles, but occasionally radiate one or more color gluons instead. QCD gives a detailed quantitative account of these matters, and has been very successful in predicting the outcomes of experiments[9] (so much so, that experimentalists now rely on it to calculate their backgrounds).

Why do these bosons have the opposite effect from other particles? The mechanism of screening seems so clear and inevitable that its reverse seems implausible. However, it turns out, roughly speaking, that attractive magnetic dipole–dipole attractive forces between like-charge gauge gluons outweigh their electric repulsion, leading to an accumulation of the *same* charge — anti-screening![10]

In our present context, it is convenient to consider screening and asymptotic freedom as functions of energy rather than distance. In a sense that can be made precise, in quantum mechanics high energy or momentum corresponds to small distance. Roughly speaking, then, screening corresponds to the coupling's increasing with energy, while asymptotic freedom corresponds to its decreasing.

The coupling of SU(3) is more affected by asymptotic freedom than are the other couplings, simply because there are more strong color gauge bosons. It outweighs the effect of the quarks. For weak SU(2) the competition is more equal, while for electromagnetic U(1) there is no gauge boson contribution, and ordinary screening wins. As a result the strong

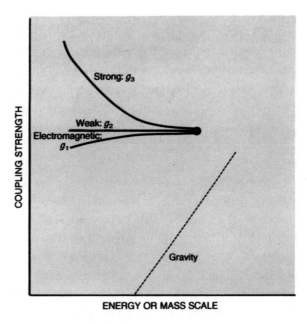

FIGURE 6. The running of the strong, weak and electromagnetic couplings, extrapolated to very high mass scales, can result in their meeting at a point. This occurs if these three interactions, as observed at low energy, all result from the spontaneous breakdown of a unified theory at a large mass scale. The gravitational coupling, shown schematically on this plot, starts out very much smaller than the other couplings but would join them if extrapolated to about 10^{19} GeV.

coupling decreases at large energies, while the weak stays nearly constant and the electromagnetic increases. But these are just the directions of change that can cause the couplings to merge!

This whole circle of ideas is beautifully summarized in the plot of effective couplings against energy or mass scale due to Howard Georgi, Helen Quinn and Steven Weinberg[11] (Figure 6). The energy scale for the running of the couplings is logarithmic, so it takes a big change in energy to see any change in the couplings. Thus the scale at which unification takes place will be very much larger than what we are accustomed to in accelerator physics.

The Logic of Prediction from Unification

A method for comparing unified theories and reality, using the observed strength of couplings, emerges from careful consideration of Figure 6.

On the left-hand side of the plot, we have three measurable parameters: the strong, weak and electromagnetic couplings. On the right-hand side, we have two unknown parameters: the mass scale for restoration of the full unified symmetry, and the strength of the coupling (there's just one!) when this occurs. Since the three measurable parameters are supposed to derive from two more fundamental (but *a priori* unknown) ones, they must obey a constraint. *The primary prediction from the logic of unification is a numerical relationship among the strong, weak and electromagnetic couplings.*

The form of this relationship is conveniently expressed in terms of the weak angle θ_W defined by the expression $\tan\theta_W = g_1/g_2$, where g_1 and g_2 are the coupling strengths of the U(1) and SU(2) gauge bosons. The coupling g_2 is directly the coupling strength of W bosons, analogous to the electromagnetic coupling e for photons. The physical photon is a mixture of the fundamental U(1) and SU(2) gauge bosons, and one finds $e = g_2 \sin\theta_W = g_1 g_2/(g_1^2 + g_2^2)^{1/2}$. Given the experimental values of the strong coupling g_3 and the electromagnetic coupling e, the logic of unification allows one to predict the value of $\sin^2\theta_W$, which is the quantity experimentalists generally report. The precise value predicted depends on the spectrum of virtual particles that enters into the calculation of the running of the couplings. We shall elaborate on the numerical aspect of these predictions and their comparison with experiment in a moment.

The predicted constraint on the observed couplings, however, does not exhaust the interest of this calculation. If the observed couplings do obey the constraint, we will also obtain definite predictions for the mass scale of restoration of symmetry and for the value of the coupling at this scale. Together these allow us to predict the mass of the dangerous gauge bosons whose exchange destabilizes protons and to obtain a rate for proton decay through this mechanism, which in principle (and perhaps in practice) can be compared with experiment. In the context of cosmology, these parameters determine the temperature at which a phase transition from unbroken to broken unified gauge symmetry occurred during the Big Bang.

Slightly more subtle but perhaps in the long run even more important for the future of physics is another aspect of the unified coupling and scale. A classic problem of physics for the past several decades has been the meaning of the numerical value of the fine-structure constant $\alpha = e^2/4\pi\hbar c$. This pure number largely controls the structure of the world (that is, all of chemistry and most of physics, as Dirac described the domain of quantum electrodynamics). Many attempts, ranging from crackpot numerology to serious efforts by leading physicists, have been made to calculate its value from deeper principles. None has succeeded. Unified theories radically alter the terms of this problem, but do not remove its substance. It becomes, if anything, grander. The fine-structure constant no longer appears as a simple primary ingredient of fundamental theory. Rather it, like the strong and weak couplings, derives from the primary unified coupling at short distances by processes of renormalization and symmetry breaking. The right problem, it seems, is not to try to calculate α but rather to explain why the unification coupling and scale are what they are.

Finally, the value of the scale of unification has important implications for the eventual reconciliation of gravity with the other, now unified interactions. We shall return to this point below.

Comparison with Experiment

But first let us return from these rarefied heights back down to earth, to discuss experimental measurements and their confrontation with models in a more concrete way.

The value of the strong interaction coupling (at some definite scale) can be measured in many ways. Perhaps the most intuitively appealing is to use electron-positron annihilation into hadrons. The fundamental process underlying the annihilation is production of a virtual photon, which converts to a quark-antiquark pair (Figure 7). Of course one does not see actual quarks in the laboratory, but only the hadron showers or jets they induce. At high energies the dressing process, whereby a bare quark is converted into physical hadrons, is soft. This means that the hadrons in the jet are all moving in very nearly the same direction as the underlying quark, and that the total energy and momenta of the particles in the jet add up very nearly to the underlying quark's energy and momentum (Figure 7a).

However, there is also a small probability — proportional to the strong coupling — for the quarks to radiate a hard gluon, that is, one with substantial energy and momentum of its own. In that case one should observe events with three jets, as shown in Figure 7b. Such events are indeed observed. That their angular distribution is observed to agree with the predictions of QCD provides splendid evidence for the existence of spin-1 gauge bosons coupled to color charge with the same space-time

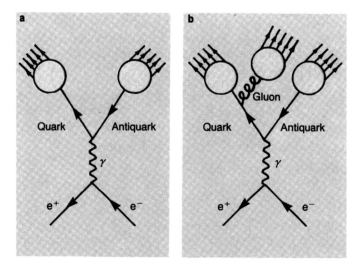

FIGURE 7. Electron-positron collisions at high energy can produce an elementary quark–antiquark pair, which materializes as two jets of particles moving in opposite directions (a). More rarely, the rapidly accelerated quarks rediate a color gluon, which produces a third jet (b).

structure as the coupling of the photon to ordinary charge. Most importantly for our purposes, the ratio of three-jet to two-jet events gives a direct quantitative measure of the strength of the strong coupling. Four-jet events are also observed at the expected (low) rate.

Conceptually, the most straightforward way to measure the weak coupling is simply to measure the mass of the W boson. Indeed, the rate of all weak processes at low energy, including, for example, the easily-measured rate of muon decay, is governed by the ratio of the weak coupling to the mass of the mediating W boson. (This follows from graphs like that shown in Figure 1, using the elementary Feynman rules.) In practice other, more complicated measurements, involving the mixing of the Z boson with the photon, are more easily made accurate.

The electromagnetic coupling, of course, has been known with extreme accuracy for a long time.

Though no true ambiguities arise if one uses the theory to calculate physically meaningful quantities, quantitative comparison with experiment requires great care. For instance it is not at all trivial to define the couplings properly and consistently, because they all run and in any physical process there are a variety of ways one might choose the nominal "scale." Once the experimental measurements are properly translated into values of the couplings, it becomes possible to confront them with the predictions of different unified models.

This involves at least two criteria. First, we must demand that the constraint on the observed couplings is satisfied. Second, we must demand that the predicted rate of proton decay through gauge boson exchange is not too large. Different models of unification will contain different numbers and kinds of virtual particles, which will cause the couplings to run differently. Therefore, in general, different models will lead to different constraints on the observed couplings and to different values for the unification scale and coupling.

It is good scientific strategy to check the simplest possibility first. The simplest unified model is the one based on SU(5) unification as described above. The minimal version of this model does not require any unobserved particles with mass significantly less than the unification scale beyond those already needed in $SU(3) \times SU(2) \times U(1)$ without unification (that is, the top quark and the Higgs boson). This minimal model does amazingly well by the first criterion. Until quite recently the measured values of the couplings did satisfy the constraint imposed by this minimal unified model, within experimental uncertainties. This is a truly remarkable result that encourages us to think that there is much truth captured within this circle of ideas. The minimal SU(5) unified model has difficulties, however, meeting the second criterion. It predicts too large a rate of proton decay or, equivalently, too small a unification scale. The predicted scale is roughly 10^{15} GeV, and the predicted lifetime is roughly 10^{29} years. This is not quite acceptable, as we mentioned before. On the other hand the predicted proton lifetime is not absurdly short — certainly it is a vast improvement on a microsecond! — and in fact special, heroic experimental efforts were required to rule it out.

Given these results, it seems wise merely to tinker with the basic ideas rather than simply junk them. Are there compelling alternatives to the minimal unified model? Do they manage to retain its successes while remedying its shortcomings?

Supersymmetry

Once we wander from the straight and narrow path of minimalism, infinitely many silly ways to go wrong lie open before us. In the absence of some additional idea, just adding unobserved particles at random to change the running of the couplings is almost sure to follow one of these. However, there are a few ideas that do motivate definite extensions of the minimal model and are sufficiently interesting that even their failure would be worth knowing about.

Surely supersymmetry[12] is in this class. In a well-defined sense, supersymmetry is the only possible way to unify the description of particles with different spins. Indeed, it is a symmetry whose basic operation is to transform particles or fields with one spin into other particles or fields whose spins differ by the minimal unit $\hbar/2$. In the process it transforms bosons into fermions, and vice versa. As yet there is no direct sign of supersymmetry in Nature (the developments reported below are probably the nearest thing so far), and if supersymmetry is relevant to the description of Nature it must be broken. However, a broken symmetry can still be rich in consequences if its breakdown occurs in a mild and orderly way.

Perhaps the most appealing idea in this direction is that the breakdown of supersymmetry is spontaneous.[13] This means that it remains a valid symmetry of the underlying *laws* of physics but is broken in the course of the evolution of the state of the universe. This process is similar to the way the alignment of spins in a ferromagnet spontaneously breaks rotational symmetry as the magnet is cooled through its Curie point: Rotational symmetry is still valid in a fundamental sense, even in a magnet, but the stable configurations of spins within the magnet do not respect it. The fundamental laws have more symmetry than any of their stable solutions.

Supersymmetry is a necessary ingredient in several other theoretical ideas. There are many hints that it may help to elucidate the gauge hierarchy problem — that is, the vast difference between the unification scale and the scale of electroweak symmetry breaking.[14,15] When one promotes supersymmetry to the status of a local gauge symmetry, one finds that Einstein's general relativity is a necessary consequence, thus finally bringing that theory within the circle of ideas used to describe the other interactions of particles.[16] And, of course, supersymmetry is a necessary ingredient in superstring theory, the most ambitious concrete approach to unifying the three interactions of particle physics with gravity currently known.

Pioneering attempts[17] to incorporate supersymmetry into realistic models of particle physics ran into various difficulties. Finally a consistent phenomenological model was found, using the idea of soft symmetry breaking.[18]

Ten years ago, we pointed out that extension of the minimal model to incorporate supersymmetry had important implications in the context of the ideas discussed above.[15] The most important change suggested by supersymmetry is that one should include many additional particles with masses of 10^4 GeV or less, whose properties are predicted with sufficient definiteness to allow a meaningful analysis of their effect on the running of couplings. Roughly speaking, one should expect a doubling of the spectrum of elementary particles at these energies. This doubling occurs because supersymmetry transforms particles into their superpartners, differing in spin by $\hbar/2$ but with closely related couplings, and none of the known quarks, leptons or gauge bosons can be identified with the superpartner of any other. The mass estimate for the superpartners is not quite firm, but if supersymmetry is to help address the hierarchy problem it seems necessary that its breaking (of which the mass difference between superpartners is a measure) not be too large.

What are the effects of adding these superpartners? The main effect is to raise the scale of unification *without much disturbing the successful SU(5) relation among couplings*. Indeed, the main reason superpartners tend to raise the scale of unification is that the gluinos, the spin-1/2 partners of the color gluons, partially cancel the asymptotic-freedom effect of the gluons themselves. Thus it takes a longer run in energy for the biggest difference between couplings — the anomalously large strength of the strong interaction — to get wiped away. On the other hand the group theoretic structure of the calculation, which controls the ratio of couplings, is not much affected by the new superpartners. This is because the new superpartners occur in the same symmetrical pattern as their known counterparts. (Indeed supersymmetry relates particles with different spins but the same gauge — that is, color — charges.) Thus, roughly speaking, the running of each coupling is slowed down by the same factor.

Of course, by raising the scale of unification, the supersymmetric unified models made it seem less certain that the proton would decay on schedule. As time went on and no decays were observed, it became clear that this might be just what the doctor ordered.

Until recently it was appropriate to emphasize that incorporating supersymmetry into simple unified models does not drastically change the relation that they predict among coupling constants, since this prediction was consistent with the available data. However, small deviations from the nonsupersymmetric predictions do exist, because it is not quite true that particles (and their superpartners) occur in a completely symmetrical pattern.[15,18,19] The "bad actors" are the scalar fields introduced to implement electroweak symmetry breaking — the Higgs fields. In constructing the standard electroweak $SU(2) \times U(1)$ theory one must introduce a complex doublet of Higgs fields carrying the weak color charges. However, on phenomenological grounds one must not introduce their counterparts carrying strong color charge. Indeed exchange of the strongly colored Higgs particles can destabilize protons, and it leads to catastrophic rates for proton decay unless the mass of these particles is extremely large. There is no compelling understanding of why the strong-color Higgs particles are so heavy compared to their weak-color counterparts: this is one aspect of the gauge hierarchy problem. (Actually what is puzzling is not so much the heaviness of the strong-color Higgs particles but the lightness of the ordinary ones.) At present it seems wise to be pragmatic and simply accept Nature's unequivocal indication that this is so. How does this mass difference affect the unification of couplings?

The normal, weak-color Higgs fields influence (to first order) only the running of the weak and, to a lesser extent, the electromagnetic couplings. They tend to make these couplings increase with energy. The inclusion of fermionic superpartners accentuates these effects. Furthermore, for technical reasons it turns out that in the minimal supersymmetric model one must introduce not just one hut two weak Higgs doublets. The contribution of all the Higgs fields and their superpartners to the running of couplings is quantitatively small compared with the contribution of quarks, leptons and gauge bosons, but recent accurate measurements (especially the beautiful results from LEP[20]) can resolve the small corrections the Higgs fields are predicted to make for the constraint on the couplings. Ugo Amaldi, Wim de Boer and Hermann Fürstenau[21] conclude that the minimal supersymmetric model gives an excellent fit to the data, whereas the minimal nonsupersymmetric model is definitely excluded by many standard deviations. Figures 8 and 9 show plots of effective coupling versus energy in the minimal nonsupersymmetric and supersymmetric SU(5) models, extrapolated from the latest data.

Together with the previous indications from proton decay, these new results provide highly suggestive, if circumstantial, evidence for virtual supersymmetry. They also greatly reinforce the case for color unification. A minimal supersymmetric model is certainly not the only way

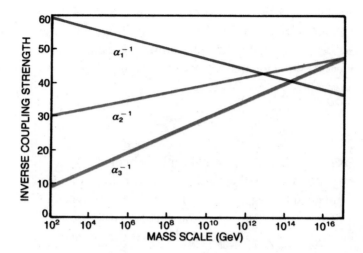

FIGURE 8. The most recent measurements of the low-energy couplings ($\alpha_i = g_i^2/4\pi\hbar c$) clearly fail to meet a point when they are extrapolated to high energies by computations incorporating the particle content of the minimal nonsupersymmetric SU(5) [or simply SU(3) × SU(2) × U(1)] model. This indicates that the observed low-energy couplings are not consistent with a unified model having just this particle content. The thickness of the lines indicates the experimental uncertainties. (Adapted from a figure provided by Ugo Amaldi, CERN.)

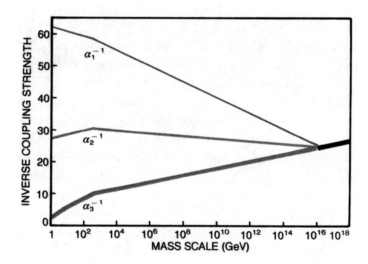

FIGURE 9. Minimal supersymmetric SU(5) model does cause the couplings to meet at a point. While there are other ways to accommodate the data, this straightforward, unforced fit is encouraging for the idea of supersymmetric grand unification. (Adapted from a figure provided by Amaldi.)

to reconcile the existing data with color unification. More complex unification schemes, typically involving new particles with exotic quantum numbers and more complicated symmetry-breaking patterns, are also contenders.[22] At the moment these other contenders seem less compelling than the minimal supersymmetric model.

Prospects

If we take these indications of unification of couplings and virtual supersymmetry at face value, they both brilliantly confirm old ideas in particle theory and augur a bright future for particle experimentation. Within the next year or so the electron–proton collider HERA should be gathering data that will both test QCD and refine the determination of the strong coupling constant, whose uncertainties are currently the most important factor limiting comparison of theory and experiment. If virtual supersymmetry is operative well below the unification scale, Nature would be perverse not to use it in addressing the gauge hierarchy problem. If Nature exhibits Her usual good taste, the masses of the superpartners cannot be too large. Real supersymmetry should not elude the next generation of acclerators (CERN's Large Hadron Collider or the SSC).

There is also good news regarding the search for proton decay. Taken at face value, the best fits of minimal superunified models to the couplings predict a unification scale of about 10^{16} GeV and a proton lifetime of about 10^{33} years through gauge boson exchange.[21] This lifetime is only slightly beyond the reach of existing experiments. Supersymmetric models offer additional mechanisms for proton decay — involving decay into virtual scalar quarks — that do not occur in nonsupersymmetric models.[23] The rate of decay through these modes depends on details of aspects of the models that are poorly understood, and so it cannot be predicted with precision. However, in a wide class of models these modes dominate the decay and lead to extremely unusual final states.[24] Thus if the proton is ever observed to decay, the nature of its decay modes may give strong clues as to the nature of the unified theory underlying its demise.

Finally, we would like to make some simple observations about how gravity might fit into this picture. The coupling of the graviton may also be considered to run, and much faster than the other couplings (see Figure 6). Because it is characterized by a dimensional coupling — Newton's constant — rather than the dimensionless couplings that characterize the other interactions, it increases (to a first approximation) linearly with energy, rather than logarithmically. However, it starts out so small that its extrapolation only meets the other couplings at approximately 10^{19} GeV, the Planck mass. This is comparable to, but definitely greater than, the unification scale. An important implication of this is that gravitational corrections do not drastically affect the running of couplings at or below the unification scale, on which the preceding discussion was based. The ratio of these scales — the Planck mass and the grand unification scale — is another fundamental dimensionless number, whose calculation presents an inspiring challenge to theoretical physics.

We wish to thank Ugo Amaldi, Wim de Boer, John Ellis, Hermann Füstenau and Luis Ibanez for valuable discussions.

References

1. Other good semipopular introductions to many of the main ideas of gauge theories include S. Weinberg, *Sci. Am.*, July 1974, p. 50; and G. 't Hooft, *Sci. Am.*, June 1980, p. 104. Standard textbooks on gauge theories include I. Aitchison, A. Hey, *Gauge Theories in Particle Physics*, Adam Hilger, Bristol, UK (1982); K. Gottfried, V. Weisskopf, *Concepts for Particle Physics I* and *II*, Clarendon, Oxford (1984); C. Quigg, *Gauge Theories for the Strong, Weak, and Electromagnetic Interactions*, Benjamin/Cummings, Reading, Mass. (1987); and T. P. Cheng, L.-F. Li, *Gauge Theories of Elementary Particle Physics*, Clarendon, Oxford (1984). See also the excellent annotated reprint collection *Gauge Invariance*, T. P. Cheng, L.-F. Li, eds., Am. Assoc. Phys. Teachers, College Park, Md. (1990).
2. An early attempt at unificatian of couplings is J. Pati, A. Salam, *Phys. Rev.* **D8**, 1240 (1973). Color unification along the lines discussed here was introduced in H. Georgi, S. L. Glashow, *Phys. Rev. Lett.* **32**, 438 (1974).
3. R. Becker-Szendy et al., *Phys. Rev.* **D42**, 2974 (1990). Particle Data Group, *Phys. Lett.* **B239**, 1 (1990).
4. For discussion of the Higgs phenomenon in quantum field theory, see Y. Nambu, *Phys. Rev.* **117**, 648 (1960); P. W. Higgs, *Phys. Rev. Lett.* **12**, 132 (1964); *ibid.*, **13**, 508 (1964); F. Englert, R. Brout, *Phys. Rev. Lett.* **13**, 321 (1964); G. S. Guralnik, C. R. Hagen, T. W. B. Kibble, *Phys. Rev. Lett.* **13**, 585 (1964). For the Higgs phenomenon in superconductivity, see F. London, H. London, *Proc. R. Soc. London, Ser.* **A149**, 71 (1935); V. L. Ginzburg, L. D. Landau, *Zh. Eksp. Teor. Fiz.* **20**, 1064 (1950); P. W. Anderson, *Phys. Rev.* **110**, 827 (1958); *ibid.*, **112**, 1900 (1958).
5. D. J. Gross, F. Wilczek, *Phys. Rev. Lett.* **30**, 1343, (1973). H. D. Politzer, *Phys. Rev. Lett.* **30**, 1346 (1973).

6. D. J. Gross, F. Wilczek, *Phys. Rev.* **D8**, 3633 (1973). S. Weinberg, *Phys. Rev. Lett.* **31**, 494 (1973). H. Fritzsch, M. Gell-Mann, H. Leutweyler, *Phys. Lett.* **B47**, 365 (1973).
7. G. Millee *et al.*, *Phys. Rev.* **D5**, 528 (1972). A. Bodek *et al.*, *Phys. Rev.* **D20** 1471 (1979).
8. R. P. Feynman, *Phys. Rev. Lett.* **23**, 1415 (1969). J. D. Bjorken, *Phys. Rev.* **178**, 1547 (1969).
9. For a recent review, see G. Altarelli, *Ann. Rev. Nucl. Part. Phys.* **39**, 357 (1984).
10. R. J. Hughes, *Phys. Lett. B* **97**, 246 (1980); *Nucl. Phys.* **B186**, 376 (1981). N. K. Nielsen, *Am. J. Phys.* **49**, 1171 (1981).
11. H. Georgi, H. Quinn, S. Weinberg, *Phys. Rev. Lett.* **33**, 451 (1974).
12. Yu. Gol'fond, E. Likhtman, *JETP Lett.* **13**, 323 (1971). D. Volkov, V. Akulov, *Phys. Lett.* **B46**, 109 (1973). J. Wess, B. Zumino, *Phys. Lett.* **B49**, 52 (1974). See also the very useful reprint collection *Supersymmetry* (2 vols.), S. Ferrara, ed., North-Holland/World Scientific (1987).
13. Y. Nambu, *Phys. Rev. Lett.* **4**, 380 (1960). Y. Nambu, G. Jona-Lasinio, *Phys. Rev.* **122**, 345 (1961); **124**, 264 (1961). J. Goldstone, *Nuovo Cimento* **18**, 154 (1961). P. Fayet, J. Iliopoulos, *Phys. Lett.* **B51**, 461 (1974). L. O'Raifeartaigh, *Nucl. Phys.* **B96**, 331 (1975).
14. Early papers mentioning the hierarchy problem include E. Gildener, *Phys. Rev.* **D14**, 1667 (1976); E. Gildener, S. Weinberg, *Phys. Rev.* **D15**, 3333 (1976); L. Susskind, *Phys. Rev.* **D20**, 2619 (1979). Early papers applying supersymmetry in attempts to ameliorate the hierarchy problem include M. Veltman, *Acta Phys. Pol.* **B12**, 437 (1981); S. Dimopoulos, S. Raby, *Nucl. Phys.* **B192**, 353 (1981); M. Dine, W. Fischler, M. Srednicki, *Nucl. Phys.* **B189**, 575 (1981); E. Witten, *Nucl. Phys.* **B188**, 513 (1981); S. Dimopoulos, F. Wilczek, in *Unity of the Fundamental Interactions*, A. Zichichi, ed., Plenum, New York (1983), p. 237.
15. S. Dimopoulos, S. Raby, F. Wilczek, *Phys. Rev.* **D24**, 1681 (1981).
16. D. Freedman, P. van Nieuwenhuizen, S. Ferrara, *Phys. Rev.* **D13**, 3214 (1976). S. Deser, B. Zumino, *Phys. Lett.* **B62**, 335 (1976). See also Ferrara, ref. 12.
17. P. Fayet, *Phys. Lett.* **B69**, 489 (1977); **84**, 416 (1979). S. Ferrara, L. Ghirardello, F. Palumbo, *Phys. Rev.* **D20**, 403 (1979).
18. S. Dimopoulos, H. Georgi, *Nucl. Phys.* **B193**, 150 (1981).
19. N. Sakai, *Z. Phys.* **C11**, 153 (1981). L. E. Ibanez, G. G. Ross, *Phys. Lett.* **B105**, 439 (1981). M. B. Einhorn, D. R. T. Jones, *Nucl. Phys.* **B196**, 475 (1982). W. J. Marciano, G. Senjanovic, *Phys. Rev.* **D25**, 3092 (1982).
20. These results from the DELPHI, ALEPH, L3 and OPAL collaborations are in a series of papers too numerous to list here; they mostly have appeared (and continue to appear) in *Phys. Lett. B*.

21. U. Amaldi, W. de Boer, H. Fürstenau, *Phys. Lett.* **B260**, 447 (1991). The possibility of such analysis was demonstrated earlier by J. Ellis, S. Kelly, D. Nanopoulos, *Phys. Lett.* **B249**, 441 (1990). A similar analysis was carried out independently by P. Langacker, M. Luo, U. Pennsylvania preprint (1991).
22. See, for example, H. Georgi, D. Nanopoulos, *Nucl. Phys.* **B155**, 52 (1979); L. E. Ibanez, *Nucl. Phys.* **B181** (1981); P. Frampton, S. L. Glashow, *Phys. Lett.* **B135**, 340 (1983); P. Frampton. B.-H. Lee, *Phys. Rev. Lett.* **64**, 619 (1990); A. Giveon, L. Hall, U. Sarid, U. Calif., Berkeley preprint (July 1991).
23. S. Weinberg, *Phys. Rev.* **D26**, 257 (1982). N. Sakai, T. Yanagida, *Nucl. Phys.* **B197**, 533 (1982).
24. S. Dimopoulus, S. Raby, F. Wilczek, *Phys. Lett.* **B112**, 133 (1982). J. Ellis, D. V. Nanopoulos, S. Rudaz, *Nucl. Phys.* **B202**, 43 (1982).

Methods of Our Madness

Why should we care? What's it good for? What's it really worth, in money and manpower? These questions, for the big science of high-energy physics, are not trivial — nor are worthy answers. Edmund Hilary's "Because it's there" won't do. Many "its" are "there," competing for attention and resources. For similar reasons, it won't do simply to rhapsodize about the beauty and profundity of what you're going to discover. That standard bluster doesn't deserve to work, and nowadays, with the prestige of the Manhattan Project and the fears of the Cold War receding, it doesn't. But there are genuinely good answers, I think. When I was asked to write an article on "The Social Benefit of High-Energy Physics" for the Macmillan Encyclopedia, I accepted immediately, because I was eager to get my own thoughts on these matters clear, and to set them out coherently. Of course, I'm aware that getting our leaders to listen and respond to ideas with subtlety and depth is a challenge of another order. I'm working on that one.

"When Words Fail" is, according to Betsy Devine, the best thing I ever wrote. Who am I to argue? But the best line in it comes from Hermann Weyl:

The objective world simply *is*, it does not *happen*.

Parmenides loves that, and so do I. (Unfortunately it appeared in the *Nature* version as "The objective world simply is, *it does not happen*." due to a typographer's error.)

"Why Are There Analogies between Condensed Matter and Particle Theory?" was my very first Reference Frame column for *Physics Today*. Gloria Lubkin had pestered me for years to do one, and I finally broke down. Gloria turned out to be the rare editor who's easy to work with and actually improves things; I enjoyed the opportunity to revive and embody my long-standing interests in (the good parts of) philosophy and history; and I got a lot of mostly positive feedback. Soon I was hooked. Now we're at eighteen and counting, with four more in the pipeline.

"Why Are There Analogies between Condensed Matter and Particle Theory?" begins to come to grips with the apparent "Unreasonable Success of Mathematics in the Natural Sciences" that Eugene Wigner identified. I hold this truth to be self-evident: If something is true, then it is reasonable. If an actual phenomenon, like the success of mathematical methods in physics, seems unreasonable, it's a sign that we haven't yet got to the bottom of it. To answer Wigner, I argue that specific, logically contingent features of physical laws, especially their locality and symmetry, are what make them mathematics-friendly (to the extent they are).

"The Persistence of Ether" takes on the vulgar misunderstanding that Einstein, with special relativity, eliminated the ether. That idea is ironic in many ways, but first of all because in its historical context the central innovation of Einstein's version of special relativity (missed by Lorentz and Poincaré) was to put the Lorentz symmetry of Maxwell's equations for Faraday's ether ahead of the traditional Galilean symmetry of Newton's ether-free mechanics. Ethers dominate our best models of the world, now more than ever.

"Reaching Bottom, Laying Foundations" introduced a collection of reprints of classic papers assembled by *Nature*, to celebrate the passing century. If human beings (or their analytic continuation) look back 10,000 years from now, they will point to the 20th century as the amazing time when mankind first properly understood what matter is. Good material to work with! And what do we do for an encore? It was great fun to write.

The Social Benefit of High-Energy Physics

Compared to most scientific endeavors, though not to space exploration or to some defense-related technology research, high-energy physics is an expensive enterprise. Modern accelerator facilities capable of expanding the high-energy frontier, such as Fermilab or the CERN Large Hadron Collider (LHC) project, are big science, involving the concerted efforts of thousands of people and costing several billions of dollars. High-energy physics has been supported almost entirely by government agencies, and thus ultimately by taxpayers. It is entirely appropriate that scientists who promote these expenditures should be expected to justify this investment by society as a whole, by explaining its benefits to society as a whole.

To address this challenge in an honest and meaningful way, we must begin by reviewing, in broad terms, the nature and goals of modern high-energy physics.

The primary aim of research in high-energy physics is easily stated. It is, simply, to produce a better understanding of fundamental physical law, by following a reductionist strategy. That is, we attempt to understand the behavior of matter in general by working up from profound understanding of the properties and interactions of its elementary constituents.

This strategy has proven remarkably fruitful and successful, especially over the course of the twentieth century. We have discovered that strange but precise and elegant mathematical laws, summarized in the so-called Standard Model, govern the laws of physics on subatomic scales. There is every reason to think that these laws, as presently formulated, are adequate to serve as the foundation for materials science, chemistry (including biochemistry), and most of astrophysics.

One must be careful in interpreting that sort of statement, which superficially might appear quite arrogant. Chemists in pursuit of their profession are rarely, if ever, concerned with the equations of QCD. They take the existence and basic properties of atomic nuclei as given. For most chemical purposes it is adequate to approximate nuclei as point-like concentrations of charge and mass. In a few chemical applications nuclear spin also plays a role, but rarely any other aspect of nuclear structure. So in saying that QCD provides part of the "foundation" for chemistry, one means no more (and no less) than that it provides equations which in principle should allow one to derive the existence of nuclei, and to calculate a few of their properties, from a few proven properties of their constituent quarks and gluons. It does not thereby directly solve, or even address, any properly chemical problems. In the same spirit, we might say that acoustics provides the foundation for music, or lexicography the foundation for literature.

As the inner frontier of the reductionist program has moved from explaining matter in terms of atoms, to explaining atoms in terms of electrons and nuclei, to explaining nuclei in terms of protons and neutrons, and these in turn using quarks and gluons, the models it creates have become ever more accurate and more broadly applicable. But with this progression, the domain of phenomena for which the new models provide qualitatively new insights, as opposed to better foundations, has grown increasingly remote from everyday life. Subatomic physics allowed us to understand and refine the basic principles of chemistry and to design materials with desired electric and magnetic properties; nuclear physics allowed us to understand the energy source of stars and the relative abundance of the elements; quark-gluon physics allowed us to understand the behavior of matter in the very early Universe. Future developments may help us to penetrate more deeply into the early moments of the Big Bang, or to recognize and understand yet undiscovered extreme astronomical environments, but apart from this it is hard to anticipate their direct application to the natural world. It would be quite disingenuous to hold out the promise of economically significant new technologies based on future discoveries in high-energy physics.

If we take a broader perspective, however, the picture looks quite different. Over recent history, again and again fundamental, curiosity-driven research has led to unexpected developments and spin-offs whose economic value far exceeds the cost of the investments that spawned them. Sometimes the payoff was delayed by many decades, and came from directions that no one remotely anticipated. The whole world of radio and wireless communication grew from Faraday's vision of empty space as a dynamical medium and the experiments it inspired. Lasers and digital cameras grew from the struggles of Planck and Einstein to understand the strange wave/particle dualism of light/photons. Modern microelectronics, with all its ramifications, grew out of Thomson's discovery of electrons and the revolutionary insights of Bohr, Heisenberg and Schrödinger in quantum theory.

Nor do we lack examples closer to the present, recognizably belonging to the modern era of high-energy physics.

The central tools of the field, accelerators, have become a ubiquitous medical device. Their simplest and most familiar incarnation, perhaps, are X-ray machines, but other particle beams are used in cancer therapy and for diagnosis. Who would have thought that reconciling quantum theory theoretically with special relativity would lead to important clinical technologies for medicine? Yet Dirac's theory predicted antimatter, and positron emission tomography (PET scan) has become a powerful tool for looking inside the brain. Another major application of accelerators is mass spectroscopy. This method of dating and analyzing the composition of materials has supported significant contributions to geology, archaeology, and art history.

At this moment, synchotron light sources are providing new, cutting-edge tools to structural biology and chemical dynamics. In high-energy physics the production of synchotron radiation, an inevitable accompaniment of charged particle acceleration, was initially regarded as a nuisance, since it drains energy from the particles we try to accelerate. But it turns out that this "waste-product" allows scientists to look at molecules with unprecedented resolution in space and time. So now special accelerators are designed specifically to be sources of synchotron radiation. They are used for medical diagnosis, drug design, and many other practical purposes.

Besides its direct impact, the development of high-energy accelerators has also spurred progress in a number of supporting technologies. To construct the accelerators, physicists had to design large powerful magnets, to guide the particles' orbits. Magnets of this sort have become the workhorse of magnetic resonance imaging (MRI), another major medical technology.

A completely unanticipated, quite recent spin-off of high-energy physics is in the process of becoming the most important of all. Modern high-energy physics experiments typically involve many tens or even hundreds of collaborators, who must share their data and their analyses. It was to facilitate that process of collaboration that Tim Berners-Lee, a software engineer working at CERN, developed the concept of the World Wide Web, and the first browser-editor, thus starting the Internet revolution. Many other innovations in high-speed electronics, less well-known but central to commercial computing and communication technology, were developed in response to the challenges of guiding vast numbers of particles moving at velocities very close to the speed of light, and interpreting the complicated results their collisions produce.

More difficult to identify specifically, but also important, are spin-offs from conceptual developments in high-energy physics. Quantum field theory was developed as the rigorous language of elementary processes, but also turns out to be the appropriate tool to understand superconductivity. The renormalization group, first developed as a technical tool within quantum field theory, turns out to be the key to understanding phase transitions, and is playing a dominant role in emerging theories of pattern-formation, chaos, and turbulence.

Why do such valuable surprises occur so regularly? I think there is a simple, yet profound explanation. In essence, it was put forward by William James, who spoke of "the moral equivalent of war." It is the fact that human beings can be inspired by difficult problems and challenges to work very hard and selflessly, and to find more in themselves than they knew existed. Especially in youth, they even seem driven to seek — or manufacture! — such problems. Perhaps evolution selected the ability to rise to challenges partly in response to the pressures of human conflict. In any case, we should exploit opportunities to direct that precious ability into constructive channels.

High-energy physics does not lack for tough challenges. Ultimate questions about the unification of fundamental forces and the origin of the Universe begin to seem accessible. On the experimental side the challenges are more tangible, and no less awesome. The next generation of accelerators will be engineering projects of grandeur, both in their size and in their precision. They will be our civilization's answer to the Pyramids of Egypt, but nobler, built to improve our understanding rather than to appease superstition and tyrannical theocracy.

We must learn how to handle the tremendous flow of data these accelerators will generate. The ATLAS experiment already planned for CERN's Large Hadron Collider is expected to collect 10^{15} bytes/year —

equivalent to a million human genomes. Amidst this torrent we must identify the fraction, probably a mere trickle, which does not fit the Standard Model. We will need to develop new ultrafast methods of communication and computation. It would be surprising if the effort of rising to these challenges failed to produce some spectacular by-products.

In short, the economic fruits of fundamental investigation, though unpredictable in detail, have arrived with wonderful reliability, and have been reliably wonderful. Investment in this area is ultimately an investment in people, specifically in the power of great problems to inspire great efforts.

The human effects of big scientific projects ramify far beyond their immediate research community. Construction of a modern high-energy accelerator, its detectors, and its information infrastructure brings engineers into intimate contact with exotic frontiers of technology and with problems of a quite different nature from those they would ordinarily encounter. Also, most of the young people going into these projects will not find permanent academic employment. They enter this life with open eyes, foregoing security for the opportunity to participate in something great. When these engineers and researchers return to the outside world, they bring with them unique skills and experience.

Finally, the visible commitment of society to high-profile scientific endeavors sends an important message to young people considering what career to enter, encouraging them in scientific and technological directions. That's important, because our society needs capable scientists and engineers, and they are always in great demand.

So much for spin-offs and indirect benefits. Now let us discuss the intrinsic worth of the prospective knowledge. There are several specific questions that seem ripe for progress:

❖ *Universal ether and the origin of mass* — Our theory of the weak and electromagnetic interactions postulates that what we ordinarily regard as empty space is in fact filled with a pervasive medium, or ether. It is only by interacting with this ether that many particles, notably including the W and Z bosons, which mediate the weak interaction, acquire their mass. Although the theory is extremely successful, this central aspect of it has not been tested directly. We hope to excite ether, which will either produce so-called Higgs particles, or reveal some more complex structure.

- *Unification of the Theory of Matter* — The Standard Model, containing the theory of the weak, electromagnetic and strong interactions, provides a remarkably complete theory of the behavior of matter. The different pieces of the Standard Model have related mathematical structures, embodying various symmetries. It is natural to speculate that there is a single master symmetry that includes them all, and goes beyond. We have some promising ideas about how this might occur, and some tantalizing hints that the basic idea is on the right track, but the decisive work has not yet been done.
- *Supersymmetry* — The logic of unification leads to another remarkable idea, supersymmetry. Supersymmetry postulates the addition of extra quantum-mechanical dimensions. Movement of particles into these dimensions will make them appear to be other kinds of particles that have quite different, but predictable properties. So far none of the new particles has been found, but according to theory they will begin showing up in higher-energy collisions, opening up a strange new world.
- *The Arrow of Time* — We have observed a few exceptional microscopic processes that exhibit a preferred direction in time (that look different when run backward). This phenomenon is central to our understanding of how the cosmic asymmetry between matter and antimatter arose. To understand it properly, we need to see more examples of how it works, especially at high energy.
- *Unification with Gravity* — The theory of gravity (Einstein's general relativity) is not deeply integrated into the theory of matter (the so-called Standard Model). But there are bold ideas for how a completely unified theory, that describes both matter and gravity, might be constructed. Some of those ideas lead to predictions of new particles, and of specific patterns among their masses. Discoveries here could open windows into the nature of quantum gravity, or extra curled-up spatial dimensions.

Continued pursuit of the ultimate foundations of physical behavior expresses our society's commitment to the deepest ideals of scientific culture: to pursue the truth wherever it leads; to ground our working picture of Nature in empirical realities; and to challenge it.

When Words Fail

Language is a social creation. It encodes the common experience of many people, past and present, and has been sculpted mainly to communicate our everyday needs. Ordinary language is most certainly not a product of the critical investigation of concepts. Yet scientists learn, think and communicate in it during much of their lives. Ordinary language is therefore an unavoidable scientists' tool — rich and powerful, but also quite imperfect.

One scientific imperfection of language, perhaps the most obvious, is its incompleteness. For example, there are no common words for several of the most central concepts of quantum theory, such as the linearity of state-space and the use of tensor products to describe composite systems. To be sure, we've developed some applicable jargon — 'superposition' and 'entanglement', respectively, are the words we use — but the words are unusual ones, not likely to convey much to outsiders, and their literal meaning is misleading to boot.

Although it creates cultural barriers and contributes to the balkanization of knowledge, such enrichment and slight abuse of language is not a serious problem. Much more insidious, and more fundamentally interesting, is the opposite case: when ordinary language is *too* complete. When something has a name, and the name is commonly employed in discourse, it is seductive to assume that it refers to a coherent concept, and an element of reality. But it need not. And the more pervasive the word, the more difficult it can be to evade its spell.

Few words are more pervasive than 'now'. According to his own account, the greatest difficulty Einstein encountered in reaching the special theory of relativity was the necessity to break free from the idea that there is an objective, universal 'now': "[A]ll attempts to clarify this paradox satisfactorily were condemned to failure as long as the axiom of the absolute character of times, *viz.*, of simultaneity, unrecognizedly was anchored in the unconscious. Clearly to recognize this axiom and its arbitrary character really implies already the solution of the problem."[1]

Einstein's original 1905 paper begins with a lengthy discussion, practically free of equations, of the physical operations involved in synchronizing clocks at distant points. He then shows that these same operations, implemented by a moving system of observers, lead to differing determinations of which events occur "at the same time".

As relativity undermines 'now', quantum theory undermines 'here'. Heisenberg had Einstein's analysis specifically in mind when, in the opening of his seminal paper on the new quantum mechanics in 1925, he advocated the formulation of physical laws using observable quantities only. But while classical theory has a naive conception of a particle's position, described by a single coordinate (a triple of numbers, for three-dimensional space), quantum theory requires this to be replaced by a much more abstract quantity. One aspect of the situation is that if you don't measure the position, you must not assume that it has a definite value. Many successful calculations of physical processes using quantum mechanics are based on performing a precise form of averaging over many different positions where a particle "might be found". These calculations would be ruined if you assumed that the particle was always at some definite place. You can choose to measure its position, but performing such a measurement involves disturbing the particle. It changes both the question and the answer.

Einstein himself was never reconciled to the loss of 'here'. In his greatest achievement, the general theory of relativity, Einstein relied heavily on the primitive notions of events in space-time and (proper) distance between nearby events. These notions rely on unambiguous association of times and spaces — 'nows' and 'heres' — to individual objects of reality (though not, of course, on the existence of a universal 'now'). Understandably impressed by the success of his theory, Einstein was loath to sacrifice its premises. He resisted modern quantum theory, and held aloof from its sweeping success in elucidating problem after great problem.

Ironically, the sacrifice he feared has not (yet) proved necessary. On the contrary, in the modern Theory of Matter, we retain 'nows' and 'heres' for the fundamental objects of reality. These primitives are no less important in the formulation of the subatomic laws of quantum theory than in general relativity. The new feature is that the fundamental objects of reality are one step removed from the directly observed: they are quantum fields, rather than physical events.

It is possible to avoid ordinary language and its snares. Within specific domains of mathematics, this is accomplished by constructing exact definitions and axioms. Purity of language is also forced on us when we interact with modern digital computers, since they do not tolerate ambiguity.

But the purity of artificial language comes at a great cost in scope, suppleness and flexibility. Perhaps computers will become truly intelligent when they learn to be tolerant of ordinary, sloppy language — and then to use it themselves! In any case, for us humans the practical and wise course will be to continue to use ordinary language, even for abstract scientific investigations, but to be very suspicious of it. Along these lines, Heisenberg's considered formulation, put forward in *The Physical Principles of Quantum Theory* in 1930, was: "[I]t is found advisable to introduce a great wealth of concepts into a physical theory, without attempting to justify them rigorously, and then to allow experiment to decide at what points a revision is necessary."

Looking to the future, after 'now' and 'here', what basic intuition will next acquire reformation? As the nature of mind comes into scientific focus, might it be 'I'? Perhaps the following remarks of Hermann Weyl, stimulated by deep reflection on the aspects of modern physics discussed here and stated in his *Philosophy of Mathematics and Natural Science* (1949), point in that direction: "The objective world simply *is*, it does not *happen*. Only to the gaze of my consciousness, crawling upward along the life line of my body, does a section of this world come to life as a fleeting image in space which continuously changes in time."

Reference

1. Einstein, A., "Autobiographical notes", in *Albert Einstein, Philosopher-Scientist*, ed. Schilpp, P. (Library of Living Philosophers, 1949).

Why Are There Analogies between Condensed Matter and Particle Theory?

The idea that the microcosm somehow reflects or embodies the macrocosm is deeply appealing to the human imagination, and is prominent in prescientific and mystical thinking. In fact, there once appeared to be an overwhelming argument for such a connection, often quoted in alchemical texts: One could not conceive how objects as complicated and structured as plants and animals could issue from tiny seeds, except by growth from miniature templates; and the homunculus would necessarily contain the seeds of future generations, even smaller … . This argument may strike us as naive, but let us remember that the elements of a true molecular explanation of genetic encoding, deciphering and development are only just now emerging, and they are no less amazing and inspiring! In any case, we can still readily sympathize with William Blake's longing "To see a World in a Grain of Sand / And a Heaven in a Wild Flower, / Hold Infinity in the palm of your hand / And Eternity in an hour."

In classical physics, it is a remarkable fact that the form of the laws for large and small bodies is essentially the same. Newton went to great pains, and according to legend delayed for many years publishing what became the central results of the *Principia*, to prove the theorem that the gravitational force exerted by a spherically symmetric body is the same as that due to an ideal point of equal total mass at the body's center. This theorem provides a rigorous and precise example of how macroscopic bodies can be replaced by microscopic ones, without altering the consequent behavior. More generally, we find that nowhere in the equations of classical mechanics is there any quantity that fixes a definite scale of distance. The same is true of classical, Maxwellian electrodynamics. In this sense, classical physics embodies a perfect match between the microscopic and the macroscopic.

For this very reason, however, classical physics cannot account for salient features of the actual world — specifically, the existence of atoms with definite sizes and properties.

The quantum revolution, as we know, changed all that. It is interesting that the reason for this change has often been misstated, or at least stated confusingly, starting with Max Planck himself. Planck was fascinated with the idea that, by combining his new constant h with the speed of light c and the gravitational constant G, one could form a definite length scale, $(Gh/c^3)^{1/2}$. This is indeed a remarkable length: the Planck length. It evaluates to about 10^{-35} m, and is thought to be the scale below which the effects of quantum gravity become significant. It has, however, nothing directly to do with the size of atoms, and thus far its roles in physics has been more inspirational than constructive. For practical purposes the crucial length is not the Planck length, but rather the Compton wavelength h/mc, which one can construct using the definite (quantized) value m of the electron mass. Also crucial is the quantized unit charge e, used to construct the dimensionless fine structure constant.

With the emergence of a fundamental length scale whose influence permeates every aspect of physical behavior, one might have anticipated that the theory of matter at larger scales (solid-state, or condensed matter, physics) and of matter at smaller scales (elementary particle, or high-energy, physics) — of macrocosm and microcosm — would irrevocably diverge. It is a profound, and at first sight astonishing, fact that this did not happen. One finds, instead, startling and far-reaching resemblances between phenomena at very different scales of time and distance, occurring in systems as different superficially as the electromagnetic ether and a crystal of diamond, or empty space and the inside of a metal, or the deep interior of a proton and a magnet near its Curie temperature.

Consider first the earliest history of quantum mechanics itself. Planck was led to discover his constant, which became supreme in the microworld, by analyzing an essentially macroscopic phenomenon: the behavior of the electromagnetic field at finite temperature (blackbody radiation). Planck's early use of his constant, however, was quite limited. He first introduced it as a parameter in an interpretation formula to fit the experimental results of Heinrich Rubens and Ferdinand Kurlbaum. He soon made a model for how their radiation spectrum could be achieved; in this model, the exchange of energy between atoms and radiation occurs only in discrete units proportional to h. Einstein, in work of almost supernatural genius, made analogies between Planck's formula and the corresponding formulas for gases of particles, and he insisted that the energy in radiation was not merely exchanged, but also propagated, in discrete units. In this way, the physical phenomenon underlying Planck's formula was stated in a universal fashion, independent of a detailed model of atoms: It was the existence of a new kind of elementary particle, the light-quantum, or photon. (Although this was the first step, a fully satisfactory derivation of Planck's formula required additional ideas, specifically stimulated emission and Bose statistics, and was not achieved until almost 20 years later.) Thus, Einstein was the first to predict the existence of a new elementary particle.

His next step was almost equally remarkable, and wonderfully illustrates my themes. Einstein applied Planck's formula, which we could say describes the vibrations of the electromagnetic ether at finite temperature, to the analogous problem of the vibrations of a crystal. He found that it fit data on the specific heat of diamond at low temperature very well. The underlying physical phenomenon, of course, is that the vibrations are created and transmitted in discrete units: phonons. It was the beginning of the quasiparticle concept that came to dominate much of condensed matter physics. For crystals, the immediate consequence was that one could not have high-frequency vibrations of very small amplitude. Their absence suppresses the vibrational specific heat of diamond at low temperatures, just as it removes the threatened ultraviolet catastrophe in the photon specific heat for blackbody radiation.

Another analogy between elementary particle and condensed matter phenomena straddled the birth of the new quantum mechanics. In 1923, Wolfgang Pauli, by analysis of spectroscopic data, was led to propose his exclusion principle: two electrons cannot occupy the same quantum state. He immediately applied this idea to explain the paramagnetism of metals. Subsequently, several physicists responded brilliantly to the challenge of working these ideas into the modern theory of solids, especially by developing the band concept.

So far, this progress reads like a standard reductionist triumph — macroscopic behaviors were "reduced" to microscopic laws. While it was occurring, however, there was a remarkable, unexpected reverberation of these ideas back toward microphysics. When Paul Dirac developed his relativistic wave equation for the electron, he found a host of unphysical, negative energy solutions. Inspired by the exclusion principle and its successful applications, he proposed that, in apparently empty space, the negative energy states were in fact occupied. Excitations above this state could make "holes" in the Dirac sea, similar to the electron deficits that were an important part of chemical valence theory and became the holes of band theory. Today, of course, positrons and other antiparticles are part of the bread and butter of elementary particle physics, and holes are central players in solid-state electronics.

Finally, a more recent example: Starting in the late 1960s, Kenneth Wilson developed conceptual and mathematical tools for describing the self-similar behavior that occurs near second-order phase transitions.

Superficially, it may seem that nothing could be further from the problems of elementary particle physics. Yet this toolkit, applied to quarks, led directly to the discovery of QCD, the modern theory of the strong interaction.

I hope that you find these examples of the flow of ideas across boundaries of scale and substrate impressive; others will appear in subsequent Reference Frame columns. Clearly, I have described instances of what Eugene Wigner called "the unreasonable success of mathematics." Why should such things occur?

If one is looking for a rational explanation, one must first recognize that it is certainly not logically necessary for there to be any deep resemblance between the laws of a macroworld and those of the microworld that produces it. For example, the rules governing "Super Mario World," or any computer game world involving magical transformations and non-Newtonian jumping abilities, have very little in common with the rules governing the microworld of semiconductor electronics (or ultimately elementary particles) that generates it.

To make such a flow of ideas possible, the laws must have some special properties. What are these properties? An important clue is that they must be upwardly heritable. (There does not seem to be a standard phrase for this important concept; it deserves one.) That is, we require microscopic laws that, when consistently applied to large bodies, retain their character. And indeed, the most basic conceptual principles governing physics as we know it — the principle of locality and the principle of symmetry — are upwardly heritable. If the influence of elementary units is limited in time and space, this will also be true of assemblies of such units; if there is symmetry in the action of elementary units, there will also be symmetry in the action of assemblies (provided, of course, that the assemblies are themselves put together symmetrically).

The fact that these upwardly heritable principles are so powerful goes a long way toward explaining, *a posteriori*, why a flow of ideas from the microworld to the macroworld is possible. An additional feature helps explain the reverse flow. In the modern theory of elementary particles, we learn that empty space — the vacuum — is in reality a richly structured, though highly symmetrical, medium. Dirac's sea was an early indication of this feature, which is deeply embedded in quantum field theory and the Standard Model. Because the vacuum is a complicated material governed by locality and symmetry, one can learn how to analyze it by studying other such materials — that is, condensed matter. I believe that the upwardly heritable principles of locality and symmetry, together with the quasimaterial nature of apparently empty space, together underlie most and possibly all of the remarkable modern analogies between our theories of microcosmos and macrocosmos.

The Persistence of Ether

Quite undeservedly, the ether has acquired a bad name. There is a myth, repeated in many popular presentations and textbooks, that Albert Einstein swept it into the dustbin of history. The real story is more complicated and interesting. I argue here that the truth is more nearly the opposite: Einstein first purified, and then enthroned, the ether concept. As the 20th century has progressed, its role in fundamental physics has only expanded. At present, renamed and thinly disguised, it dominates the accepted laws of physics. And yet, there is serious reason to suspect it may not be the last word.

As with most general ideas, the germs of the ether philosophy, and its main competitor, showed up in debates among the ancient Greeks. Aristotle taught that "Nature abhors a vacuum," while Democritus postulated "Atoms and the void." The modern history begins with the contest between the world system of René Descartes, who proposed to explain the motion of planets as caused by vortices that sweep them through in a universal medium, and the austere theory of Isaac Newton, who specified precise mathematical equations for the forces and motions, but "framed no hypotheses." Newton himself believed in a continuous medium filling all space and, in Query 21 of his *Optics*, speculated on how it could be responsible for a tremendous variety of physical phenomena. But his equations did not require any such medium, and his successors rapidly became more Newtonian than Newton. By the early 19th century the generally accepted ideal for fundamental physical theory was to discover mathematical equations for forces between indestructible atoms moving through empty space. In particular, it was in this form that leading mathematical physicists, including such giants as André Marie Ampère, Karl Friedrich Gauss, and Bernhard Riemann, tried to formulate the emerging laws of electrodynamics.

It was Michael Faraday, a self-taught and mathematically naive experimenter, who revived the idea that space was filled with a medium having physical effects in itself. His intuition led him to devise experiments looking for physical effects of magnetic flux lines in "empty" space, and of course, in his law of induction, he found them. Working to summarize Faraday's results, James Clerk Maxwell adapted and developed the mathematics used to describe fluids and elastic solids. To orient himself, and to understand Faraday's conceptions in terms of more familiar things, Maxwell postulated an elaborate mechanical model of electric and magnetic fields. In the end, though, his equations could stand by themselves.

The first sentence of Einstein's original paper on special relativity refers to "an asymmetry in the formulation of electrodynamics, which does not appear to inhere in the phenomena." His paper's achievement was to highlight and interpret the hidden symmetry of Maxwell's equations, not to change them. The Faraday–Maxwell concept of electric and magnetic fields, as media or ethers filling all space, was retained. What had to be sacrificed was only the false intuition that motion at a constant velocity would necessarily modify the equations of an ether.

Indeed, the argument can be turned around. One of the most basic results of special relativity, that the speed of light is a limiting velocity for the propagation of any physical influence, makes the field concept almost inevitable. For it implies that the influence of particle A on particle B depends not on the present position of A, but rather on where it was some time ago. This makes it very awkward to build up dynamical equations in terms of the position of particles.

The mathematics required to bring the equations of mechanics — that is, the motion of particles in response to given forces — into a form consistent with special relativity is not hard, though it required major conceptual readjustments. Einstein developed it swiftly and painlessly. The remaining foundational piece of classical physics, the theory of gravity, posed a much greater challenge. Although Newton's extremely economical, and extensively battle-tested, equations used forces depending on the present distance between particles, special relativity taught that the observers moving relative to one another would have different notions of distance, and that the speed of light bounded the transmission of all possible influences. Henri Poincaré formulated what is in retrospect the most straightforward response to these defects, modeling gravity as what we would now call a massless scalar field. (Of course, it was very far from straightforward in the contemporary state of the art!) But Einstein, influenced by the experimental results of Roland, Baron von Eötvös, and inspired by his own famous elevator thought-experiment, sought a formulation in which the equality of inertial and gravitational mass, and the universality of gravitational response, were organic features. As we know, he achieved this goal by identifying the gravitational interaction as the bending of space-time by matter.

Thus in 1917, following Einstein's revelations, the electromagnetic field remained essentially in the form bequeathed by Maxwell, satisfying his "ethereal" equations. Moreover space-time itself had become a dynamical medium — an ether, if ever there was one. For example, a major consequence of general relativity is that distortions of space-time can themselves produce further distortions, initiating gravitational waves.

To account for physical phenomena, one needs — apparently — more than the gravitational and electromagnetic fields. Electrons, for instance. By 1917, J. J. Thomson had discovered them, Hendrik Lorentz deduced many properties of matter from a theory in which they are prime players, and Niels Bohr had used them to make his brilliantly successful atomic model. In all these applications, the electrons were modeled as point particles. As such, they constituted an element of reality quite separate and distinct from any continuous ether.

Einstein was not satisfied with this dualism. He wanted to regard the fields, or ethers, as primary. In his later work, he tried to find a unified field theory, in which electrons (and of course protons, and all other particles) would emerge as solutions in which energy was especially concentrated, perhaps as singularities. But his efforts in this direction never succeeded.

The development of quantum theory changed the terms of the discussion. Paul Dirac showed that photons — Einstein's particles of light — emerged as a logical consequence of applying the rules of quantum mechanics to Maxwell's electromagnetic ether. The connection was soon generalized: Particles of any sort could be represented as the small-amplitude excitations of quantum fields. Electrons, for example, can be regarded as excitations of an electron field.

This formulation, which at first might sound extravagant, had a lot going for it right from the start. First, it answers one of the most basic and profound riddles about the physical world, which is otherwise quite mysterious: Why do electrons everywhere in the universe have precisely the same properties — the same mass, charge, magnetic moment? Answer: because they are all surface manifestations of a single more basic entity, the electron field, an ether that pervades all space and time uniformly.

Classic atomism sought to account for the world in terms of irreducible building blocks that could be rearranged, but neither created nor destroyed. This notion is incompatible with democratic treatment of the photon as a particle among others, since radiation and absorption of light are commonplace. In beta decay, a neutron is destroyed, and a proton, together with two particles of quite a different character, an electron and an antineutrino, are created. Evidently, neither protons nor neutrons nor electrons nor photons can be considered as abiding building materials. Instead, Enrico Fermi built a successful theory of beta decay in terms of excitation and de-excitation of the relevant fields. Particles come and go, but the ethers abide.

As hinted above, it is very much easier to incorporate the principles of locality and propagation of influence at finite speed when one deals with fields. Our current — extremely successful — theories of the strong, electromagnetic, and weak forces are formulated as relativistic quantum field theories, with local interactions. In fact, having told you that, I need only add a few more specifications to sum up pretty much everything we know about nongravitational fundamental interactions. The most ethereal of all theories, Einstein's general relativity, does the same for gravity.

Once I was fortunate enough to catch Richard Feynman alone and a little tired after a day of bravura performances. When I gently provoked him, he displayed a subdued, wistful side I never saw before, or again. He told me that he had been very disappointed when he realized that his theory of photons and electrons, the methods of calculating amplitudes by using Feynman graphs, was mathematically equivalent to the usual quantum electrodynamics. He had hoped that, by formulating his theory directly in terms of paths of particles in space-time, he would be avoiding the field concept, and constructing something essentially new and different.

Uniquely (as far as I know) among physicists of high stature, Feynman had hoped to remove the field–particle dualism by getting rid of the fields. For pure quantum electrodynamics, he came close. In retrospect, though, he was swimming against the tide for understanding the other interactions. Even in electrodynamics, his rules for dealing with virtual particles appear rather *ad hoc*, except when they are derived from standard quantum field theory. It gets much worse both in modern electroweak theory, which works smoothly only if we allow for a uniform excitation of the so-called Higgs field to fill space-time, and in quantum chromodynamics, where we operate with quark and gluon fields whose corresponding particles do not, properly speaking, exist at all.

How did I provoke Feynman? I asked him, "Doesn't it bother you that gravity seems to ignore all we have learned about the complications of the vacuum?" To which he immediately responded: "I once thought I had solved that one. I had a slogan: 'The vacuum is empty. It weighs nothing because *there's nothing there*.'" It was then he got wistful.

I was deeply impressed to realize that Feynman had been wrestling with the problem of the cosmological term already in the 1940s, long before it became a widespread obsession, and frustration. You have to admit that his slogan is catchy. So just maybe, despite everything I've said up to this point, eventually we really may have to do without ether.

Reaching Bottom, Laying Foundations

The twentieth century will be remembered as the century in which we reached bottom in our understanding of the physical world, grasping not only how things behave but also why they exist. And when, armed with this understanding, we were able to take a much more creative role in sculpting the material world, leading to new tools, new sensoria and ultimately new minds.

Earlier in history, the goal of physics — seldom made explicit but tacit in its practice — was to derive dynamical equations, so that given the configuration of a material system at one time, its future configuration could be predicted. The original, ideal realization of this goal was the description of the Solar System based on Newtonian celestial mechanics. Although this description accounts excellently for Kepler's laws, the tides, the precession of the equinoxes, and much else, it gives no *a priori* predictions for the number of planets, their relative sizes, or the number and sizes of their respective moons. Similarly, the great eighteenth- and nineteenth-century discoveries in electricity, magnetism and optics, culminating in Maxwell's dynamical equations of electromagnetism, provided a rich description of the behavior of given distributions of charges, currents and electric and magnetic fields, but gave no explanation of why there should be specific reproducible forms of matter. Thus, at the dawn of the twentieth century, physics offered no hint of why there should even be such a subject as chemistry, let alone why it is what it is.

Quantum mechanics changed that forever. Its early development featured two quite separate aspects, which converged only later. On the one hand, there was the primarily experimental discovery that the material world is constructed from vast numbers of identical copies of a relatively small number of building blocks — electrons, photons, protons and neutrons — that come in stereotyped units, having the same properties wherever and whenever they are observed. On the other hand, there was the primarily theoretical discovery of the strange but universal dynamics governing these building blocks. They are described by waves, which satisfy simple, definite equations. But the waves in question have the odd interpretation that their absolute square governs the probability for finding particles.

Together, these two ideas underlie the modern picture of atoms. An atom is a pattern of definite form, determined by minimizing the energy of electron waves bound to a small, positively charged nucleus (where protons and neutrons accumulate) by electrical attraction. A quantum-mechanical atom, unlike the Newtonian Solar System, has its size, shape and structure determined uniquely by the laws of physics.

This atomic model — or, more precisely, the mathematics that embodies it — provides the ultimate foundation for chemistry, materials science and physiology. But it begs the question of what holds the nuclei together. The pursuit of this question, beginning with heroic balloon flights and studies of cosmic rays, but soon using ever more powerful and sophisticated accelerators, revealed several new worlds of subnuclear phenomena. These included two new fundamental interactions, supplementing gravitation and electromagnetism, and a totally unexpected cornucopia of additional building blocks.

Truly satisfactory understanding came only after a long series of ingenious experiments and theoretical investigations, culminating in the so-called Standard Model. Here, I can only briefly mention a few key points.

It became clear early on that the proton and neutron are subject to additional forces, beyond gravity or electromagnetism, and more powerful than either, but acting only over short distances. Systematic study of the newly-named strong interaction suggested great complexity: the strong forces between protons and neutrons depend on the distances between the particles, their relative velocities, and even the relative orientation of their spins. Nevertheless, they could be measured and they were used to create a working quantum model of atomic nuclei that was comparable in power, if not in elegance, to the successful quantum model of atoms. Nor could complex details obscure the most fundamental point, that the rules of quantum dynamics still apply within the atomic nucleus, and that when applied to protons and neutrons subject to the strong interaction, they produce well-defined nuclei with unique structures.

Now we understand that the apparent complexity of the strong interaction occurs because protons and neutrons are composite objects. They are composed of quarks and gluons, whose mutual interactions are governed by beautifully simple (but hard to solve!) equations.

A second, new sort of interaction first revealed itself in processes that turn neutrons into protons (together with other particles), or vice versa. Such processes, when they occur within an atomic nucleus, change the chemical nature of the atom involved. Further investigation of comparatively feeble — that is, slow — processes whereby various otherwise-stable particles change into one another revealed many systematic patterns, and blossomed into the concept of a fourth fundamental interaction, the weak interaction.

The development of weak-interaction theory began a process whereby classic atomism, involving stable individual objects, has been replaced by a more sophisticated and accurate picture. In this new scenario, individual particles are not permanent. Rather they appear as excitations of universal quantum fields. These fields provide, in contemporary physics, the primary, permanent elements of reality.

It is only with the development of quantum-field theory that the first aspect of quantum mechanics became fully integrated into physical theory. Two electrons anywhere in the Universe, whatever their origin, have exactly the same properties. We understand this as a consequence of both being excitations of the same underlying feature, the electron field. The same logic, of course, applies to photons, quarks or gluons.

The Triumph of Symmetry

One might have expected that in achieving a detailed, accurate picture of Nature, taking into account increasingly disparate phenomena, our theories would necessarily become ever more complex. We find the opposite. Theoretical physics in the twentieth century has yielded bold syntheses and radical simplification of the foundations.

The first great moves in this direction were the two relativity theories. The special theory of relativity is, at its heart, the theory of the symmetry of Maxwell's equations. Be elevating that symmetry to a universal law of physics, the special theory of relativity allows this set of four separate partial-differential equations (two scalar and two vector) to appear as merely four different facets of a single inseparable, symmetric whole. Applied to mechanics, the new symmetry tied together the formerly separate conservation laws of mass, energy and momentum. The general theory of relativity exemplified how a postulate of symmetry — in this case, general covariance — could serve as a powerful guide for formulating new laws of physics.

Soon after the advent of modern quantum theory, theorists noted that they could derive the equations of quantum electrodynamics from postulates of symmetry — for quantum electrodynamics, gauge invariance and special relativity.

The discovery that the strong and weak interactions are also governed by equations of high symmetry, involving nonabelian generalizations of gauge invariance, marked a major advance in our understanding of Nature. The symmetries of these two interactions are, for different reasons, much harder to discern than in the cases of gravity or electromagnetism. In the case of strong interaction, the basic symmetry that dictates the equations of the theory (quantum chromodynamics) applies directly only to quarks and gluons, particles that are never observed directly in isolation. Thus its discovery and validation required several leaps of insight. In the case of weak interaction, the basic symmetry of the equations is broken by all their stable solutions. Thus, in particular, the symmetry is "spontaneously broken" in what we ordinarily regard as vacuum, and empty space must be regarded as a structured medium. In arriving at this conception, insights derived from the analysis of related phenomena in materials, particularly superfluidity and superconductivity, provided crucial (and, given the context, astonishing) inspiration and guidance.

The theories of the strong, electromagnetic and weak interactions together make up the Standard Model. Although correct and fruitful, the Standard Model is not fully satisfying in that the three theories are not tightly meshed, and gravity is left out altogether. Theoretical attempts to transcend these limitations have been dominated by the quest for yet higher degrees of better-hidden symmetry, involving either enlargement of gauge symmetry, or the ambitious but still poorly understood structures of string theory.

The fact that a few ideally simple, symmetrical equations suffice (in principle) to explain so much must also be read the other way. We have learned that even very simple equations can exhibit extremely rich and complex solutions. A celebrated example is how iteration of the one-dimensional logistic equation can lead, through a cascade of period doublings, to chaos. It is a great, continuing challenge to exploit our "in principle, more than adequate" knowledge of the basic equations to understand their solutions in useful detail.

New Productions

Our tentative, partial response to this challenge has raised our control of matter to new levels. The single image that epitomizes this control, unfortunately, is the mushroom cloud of a nuclear explosion. But there are other images, equally extraordinary and less problematic: the pure straight light of a laser beam, the double helix of DNA, and the X-ray diffraction patterns that encoded it. Perhaps of all these images, the purest monument to human curiosity is the valley-filling radio dish at Arecibo, with a maser amplifier at its focus, designed to distil every last bit of meaning from incredibly weak, garbled signals incident from the distant cosmos.

Arecibo is symbolic of the expansion in our cosmic awareness made possible by our detailed command over every portion of the electromagnetic spectrum. Our eyes have been opened to neutron stars (pulsars), to black holes with the mass of a billion suns (quasars), to the omnipresent lingering afterglow of the Big Bang (microwave background radiation), to the blasted corpses of exhausted stars (supernova remnants), to new solar systems, and more. Astronomy is being further enriched by its expansion beyond the electromagnetic spectrum, with more adequate monitoring of cosmic rays, development of neutrino observations, and deployment of gravitational wave detectors.

Beyond doubt, the most important series of productions to emerge from modern physics is digital microelectronics. Computer speed, power and miniaturization has institutionalized technological revolution, as exemplified by Moore's empirical law: the density of integrated circuits (transistors per unit area) doubles every 18 months. Although digital microelectronics has already utterly transformed communication and information processing, its ultimate potential has barely been scratched.

The Foreseeable Future

What's next? Surprises, for sure. But there are also several milestone advances that I think we can anticipate with some confidence over the next twenty years or so.

In fundamental physics, the triumph of the Standard Model will be crowned with direct observation of its last remaining ingredient, the so-called Higgs particle. The recent Super Kamiokande discovery of one non-zero neutrino mass will be fleshed out into a full picture of the properties of these elusive particles. Theoretical ideas that perfect the symmetry of the Standard Model, by embedding it in a large unified field theory, will come into their own. Such ideas have been greatly encouraged by the neutrino mass discovery, and will be consummated by observation of their other most spectacular consequence, the instability of protons. Supersymmetry, whose necessity is already indicated by quantitative aspects of unification, will be demonstrated explicitly by discovery of the parallel world of new elementary particles it requires.

The parameters of cosmology, including the lifetime of the Universe, its average density, the nature of the small fluctuations of density that led by means of gravitational instability to the formation of galaxies and other large-scale structures, and the nature of the "dark matter", will be determined. We will learn whether empty space weighs — that is, whether there is a non-zero cosmological term. Today, thinking about major questions in cosmology is dominated by metaphors and speculations inspired by high-energy physics, unified field theories and superstring theory. As our knowledge increases, the possibilities will become more narrowly defined and specific, and we will progress from metaphors towards rooted world-models.

As miniaturization approaches molecular scales, quantum mechanics in both its aspects — the discreteness of particles and the wave nature of their dynamics — will become increasingly crucial. Whether these turn out to be show-stoppers or, as I expect, opportunities, will depend on the creativity of a new generation of physicist-engineers. In recent issues of *Nature*, one finds considerable ferment around fascinating new ideas such as single-electron transistors, quantum dot arrays, optical switches, nanotube electronics and quantum computation. Many clever constructions will emerge, and some will also be useful.

And Beyond

Beyond that my crystal ball gets much cloudier. To respond to the challenge of sculpting matter at a molecular level, we will want to design molecular-scale tools. To marshal these tools effectively, we will want to have instructions available locally, on a molecular scale, which will drive us towards molecular tapes that store programs and data read out by molecular messengers. At this point, we will have reached a domain wherein physics, chemistry and biology are no longer clearly distinguishable.

As the limits of resolution are pushed down, the limits of overall scale and integration will be pushed up. At present, the useful size of closely integrated units (that is, chips) is limited by the necessity for perfect performance, and the difficulty of repairing flaws. Molecular monitoring and control will support new mechanisms of repair. Cheap redundancy, and innovations in software, will allow any remaining flaws to be worked around. Size constraints will effectively evaporate. Non-lithographic assembly will open up the third dimension.

Probably at present, and certainly within the next few years, the best chess players in the world will be non-human. In view of the above, it is difficult to avoid the idea that a hundred years from now the same will be true for physicists. Our future colleagues will work at a level and on scales we cannot at present meaningfully imagine. But they will, in more ways than one, be building upon the foundations we laid.

Inspired, Irritated, Inspired

I began reviewing books with considerable misgiving. I was concerned about the implied commitment to pay close attention, finish, and publicly reflect upon some book that might not be worth the effort. And of course, I'm mostly asked to review books written by significant people in the small world of physics. I know I'll be running into the authors, or friends of theirs, regularly. Tension between honesty and politeness looms ominously.

In short, it's like going on a blind date to play Russian roulette. Nevertheless I've done it, very selectively, and on the whole it's been an interesting and rewarding experience. Here I've gathered five reviews that are in some way memorable. Since the reviews themselves are short, I'll just telegraph some highlights, reflecting the cycles of inspiration and irritation that accompanied the work.

Niels Bohr's Times is a biography of Bohr by Abraham Pais. Bohr is an inspiring subject, both for what he did and for what he was. His interpretation of quantum mechanics is irritating, though. I used to think his pet idea, complementarity, was vapid, now I think it is deeply inspiring, but I'm not sure, maybe it's both, depending on how you measure it.

Great joke: Bohr, challenged as to why he kept a horseshoe over his door.

"Surely you don't believe in such stuff?"

"Of course not, but I've heard that it works even if you don't believe in it."

Dreams of a Final Theory by Steven Weinberg is a meditation on foundational physics — its current state and ultimate purposes. He has a lot of interesting things to say, and the quality of his prose is inspiring. (It bespeaks an inspiring quality of mind, of course.) Irritating (but maybe correct) viewpoint 1: The goal is to stop asking "Why?" Irritating (and silly) viewpoint 2: The payoff is to discredit religion. This book inspired me to have a think about reductionism, and thereby catalyzed my series "Analysis and Synthesis."

The Inflationary Universe is an exposition of that important cosmological theory, and some of its history, by its main discoverer, Alan Guth. The subject — getting something for nothing, on a grand scale — is inspiring, indeed awe-inspiring. It's irritating to me that I didn't think of it. It's inspiring to realize that the Universe really might have inflated — recent observations are encouraging.

Just Six Numbers and *The Nine Numbers of the Universe*, by Martin Rees and by Michael Rowan-Robinson, respectively, discuss how most of modern cosmology can be captured in a few numbers. It's profoundly inspiring that this is true. Unfortunately, they choose the wrong numbers (see "Analysis and Synthesis"). That's irritating. Rees' early advocacy of Multiverse and anthropic ideas inspires. Initially, in me, it inspired fear and loathing. Now it inspires my nuanced resignation.

Shadows of the Mind, by Roger Penrose, is really two books bound as one: exposition and thesis. The exposition of various grand theories, especially the circle of ideas around Gödel's theorem, is frequently inspired. But the original thesis, connecting the origin of consciousness to quantum gravity through microtubules, wave function collapse, and computation theory (How? — better not to ask), is extremely irritating. It inspired me to engage with, and to my own satisfaction utterly destroy, Lucas' philosophical chestnut, on which it heavily relies.

What Did Bohr Do?

Review of *Niels Bohr's Times* by Abraham Pais
(Oxford University Press, 1991)

This book by Abraham Pais is the third in a remarkable series on the history of modern science. The first two were *Subtle is the Lord*, a biography of Einstein, and *Inward Journey*, a history of particle physics. Pais is himself a physicist of great distinction. He writes with authority and, in addition, with unfailing grace and considerable charm.

Pais knew Bohr and his family well, and interacted with him scientifically over a span of sixteen years, often on a daily basis. Pais states at the outset that "... I loved Bohr. I have tried to exercise restraint in regard to these sentiments, which may or may not shine through ..." I believe they do. Some admirers of Pais' previous books may find themselves disappointed that the relevant science is generally treated in much less depth and detail in this one. In compensation, however, we receive a close-up portrait of a truly extraordinary, and extraordinarily appealing, personality. Niels Bohr would have been remarkable and fascinating even if he were not the author of truly monumental scientific achievements. But in Bohr's case the man and the science are inseparable.

Near the outset of the book Pais quotes three Nobel Prize-winning theoretical physicists of succeeding generations, regarding Bohr:

> Born (1923): "His influence on theoretical and experimental research of our time is greater than that of any other physicist."

Heisenberg (1963): "Bohr's influence on the physics and the physicists of our century was stronger than that of anyone else, even Einstein."

Anonymous modern physicist: "What did Bohr really do?"

I may add that also in my own experience, questions of this last sort are not rare in discussions among sophisticated modern physicists who have an active interest in the history of ideas.

An explanation, though certainly not a justification, for the change in perspective may follow from the peculiar nature of Bohr's major contributions.

First in any list of those contributions, and also first chronologically, must be his fruitful introduction, in 1913, of Planck's quantum of action into the dynamical description of Rutherford's new (1912) model of the atom. The boldness and depth of his ideas are perhaps belied by their mathematical simplicity and ultimately provisional character. Rutherford's model of electron orbiting a tiny charged nucleus, and held in place be electric attraction, was immediately suggested by experimental results from his lab. And yet that model contradicts basic principles of classical electrodynamics. For according to classical electrodynamics the orbiting electrons, being charged particles in accelerated motion, should continuously radiate electromagnetic waves, causing them to lose energy and spiral into the nucleus. Bohr simply postulated, that this classical picture was wrong, and that the electron could peaceably orbit in what he called stationary states. Transitions between these stationary states were supposed to occur only discontinuously, with the release of all the energy into light whose frequency obeys Planck's law (frequency equals energy divided by Planck's constant). Finally, with an eye toward a successful old piece of numerology describing the spectrum of hydrogen (Balmer's formula), Bohr postulated that the energy of the electron's orbit must, in stationary states, be related to the frequency of its classical motion in the almost the same way as for the photon — but with an additional numerical factor, 1/2 times a whole number. (The 1/2 is the licence of genius.)

From these postulates the Balmer formula could be "derived," with the important bonus that the numerical factor that appears in it could be related to fundamental quantities, that is, the charge and mass of the electron, and Planck's constant. Soon the fruitfulness of Bohr's line of thought proved itself in many new applications. Among the most remarkable of these was to the spectrum of ionized helium, where Bohr made the crucial observation that by going beyond his initial approximation of an infinitely heavy nucleus, subtle discrepancies between accurate hydrogen and ionized helium spectra could be understood (reduced mass correction).

This "theory", based on an uneasy mix of nearly contradictory concepts and cribbing from experimental data, was clearly meant to be used as a scaffold, fit to be discarded when a more finished structure could support itself. In fact it has been entirely superseded, by modern quantum theory. In a certain sense, then, Bohr's theory is no longer of direct scientific interest. But to neglect it for that reason would be to miss out on a circle of ideas having their own intrinsic beauty, whose appreciation requires historical understanding. Here is what a man with very different scientific taste and instincts, Einstein, had to say about it:

> That this insecure and contradictory foundation was sufficient to enable a man of Bohr's unique instinct and tact to discover the major laws of the spectral lines and of the electron shells of the atoms together with their significance for chemistry appeared to me like a miracle — and appears to me as a miracle even today. This is the highest form of musicality in the sphere of thought.

The deep structure of Bohr's work on the hydrogen atom exhibits his special style of thought, which remains visible throughout his major work. It is especially marked by three features: close attention to experimental reality; willingness to entertain ideas that are clearly provisional and logically incomplete; and, lurking in the background, the implication that *all* knowledge is provisional and incomplete. Here is Einstein again, giving what Pais calls the best characterization ever given of Bohr:

> He utters his opinions like one perpetually groping, and never like one who believes he is in possession of definite truth.

During the years 1913–1924 Bohr was the undisputed leader in developing the so-called old quantum theory. Roughly speaking, the old quantum theory was the extension of Bohrian modes of thought, and in particular the sort of ideas implicit in his atomic model, into ever wider domains. The *correspondence principle*, according to which the quantum laws must go over into the laws of classical physics in the limit of large quantum numbers, played a leading role in these developments. In skillful hands that typically Bohrian, apparently vague principle could be a tool of extraordinary power. For instance, it was used to understand and predict selection rules and the polarization laws for atomic radiation, and ultimately led to the exclusion principle and the prediction of electron spin to rationalize otherwise inexplicable features of atomic spectra.

Another astonishing Bohr contribution from those times was establishing fundamental understanding of the periodic table of elements, around 1920 — *prior* to the discovery of the exclusion principle, and of spin! We now know, of course, that these latter concepts are absolutely indispensable to understanding the periodic table. Nevertheless Bohr, with his deep knowledge of the phenomena and feeling for the possibilities of existing theoretical ideas, managed, in an incredibly complex and murky situation, to isolate concepts of lasting value (closed shells, building-up principle) and even to predict the existence and properties of a new element, hafnium, which was duly found by his Danish colleagues.

Starting in 1925 and 1926 with the discoveries of Heisenberg and Schrödinger, the new quantum mechanics replaced the old. The flavor of the subject changed: intuition and metaphorical reasoning from close acquaintance with the data was for the most part replaced by mathematico-deductive reasoning. Especially in atomic and molecular physics and in the foundations of condensed matter physics, where the semi-quantitative methods of the old quantum mechanics could be accommodated to the new, progress was extraordinarily rapid. Bohr's position evolved from that of intellectual leader into that of mentor for the new generation. For several years, his Copenhagen institute was the prime meeting-place and clearing-house for the new developments.

While he did not play a leading role in the technical development of the new quantum theory, Bohr was very much concerned with its logical and philosophical underpinnings. Perhaps today he is (alas) probably most familiar to the general public, and even to many practicing physicists, for his ideas concerning the interpretation and philosophical implications of quantum mechanics. I cannot begin to do justice to that subtle and controversial subject here. The spirit of the so-called Copenhagen interpretation of quantum mechanics, largely due to Bohr and his disciples and still widely regarded as the standard interpretation, is that the meaning of the formalism of quantum theory must always be referred to completely specified experimental situations, which in turn must be describable in classical terms. The Copenhagen interpretation is an interpretation of renunciation, as the following formulations of Bohr make clear:

> The unambiguous interpretation of any measurement must be essentially framed in terms of the classical physical theories, and we may say that in this sense the language of Newton and Maxwell will remain the language of physicists for all time.
>
> There is no quantum world. There is only an abstract physical description. It is wrong to think that the task of physics is to find out how Nature is. Physics concerns what we can say about Nature.

Many physicists, myself included, are not satisfied with these formulations. While most agree that quantum theory *can* be applied within the rigidly circumscribed domain Bohr assigned it, not all are content to leave it at that, nor to divide the world into intrinsically different "classical" versus "quantum" pieces. Also the world is more than a laboratory, and one really must describe its behavior even when there is clearly no well-defined experimental situation in Bohr's sense — a problem that becomes particularly severe in cosmological applications of quantum theory. In the famous Einstein–Bohr debates, Bohr defended quantum mechanics against Einstein's yearning for a more classical theory; but some of us are coming to feel in defending his valuable hard-won ground he compromised too much. Quantum mechanics should be pushed as hard as possible, to the point where it can describe within itself a recognizable caricature of the world as it is experienced, and thus begin to provide its own self-consistent interpretation — or else there should be some definite change in its equations. As yet this task has not been accomplished.

Bohr's renunciations were not made lightly. They were reactions to mistakes and contradictions that arise if we carry over intuitions about the behavior of physical objects from daily experience into the domain of quantum mechanics. A notorious example is our tendency to conceive entities like electrons or photons as particles or waves respectively (or *vice versa*), for either entity can behave "like" a particle or a wave under different conditions — in fact, of course, strictly speaking they behave like *neither*. Another is the tendency to visualize a particle as having both a position and a velocity, whereas Heisenberg's uncertainty principle tells us that in quantum mechanics it cannot have precisely defined values of both simultaneously. Bohr's main response to these and other related dualities was his concept of complementarity. According to this somewhat elusive but alluring notion, there may be alternative concepts or aspects of reality, each useful in itself, which cannot be used simultaneously. Thus any attempt to measure, or visualize a particle with a definite position precludes our ability to make some other measurement. Position and velocity are valid concepts separately, but not simultaneously — they are complementary.

In his later years Bohr made some tentative, provocative and perhaps playful attempts to apply the notion of complementarity well beyond physics. For example, he mooted the idea that to determine the state of a working brain sufficiently to predict its future behavior, you'd have to disturb the brain so much as to affect that behavior. Perhaps the flavor left by Bohr's more adventurous musings on quantum theory and complementarity is best conveyed by a joke he used to tell on himself:

> The first talk was brilliant, clear and simple. I understood every word. The second was even better, deep and subtle. I didn't understand much, but the rabbi understood all of it. The third was by far the finest, a great and unforgettable experience. I understood nothing and the rabbi didn't understand much either.

Beginning in the mid- and late 1930s, when he was close to fifty, Bohr had a sort of physicist's rebirth. He developed a new metaphor for atomic nuclei, based on an analogy to liquid drops, that has proved remarkably fruitful. Using ideas of this kind, he was able immediately to seize upon the discovery of nuclear fission by Hahn, Strassman, and Meitner in 1939, and to provide in very short order the foundations for a semi-quantitative understanding of its features, such as which nuclei were the most likely to fission, how much energy would be necessary to make fission likely, the likely decay products, and so forth. This work was epitomized in a truly remarkable paper written with John Wheeler, wherein several concepts (semiclassical quantization of extended objects, use of Morse theory in physics, instantons) that would be fully appreciated only decades later appear in germinal form.

The more dramatic immediate impact of this work, of course, was in the development of nuclear weapons and nuclear power. Bohr attempted, with complete lack of success, to influence the political fallout from these developments. That tragi-comic story that is recounted both by Pais here and, with different emphases, in Richard Rhode's *The Making of the Atomic Bomb*.

This recounting explains why in the ordinary course of their training most physicists (let alone others) may have a murky impression of Bohr's contribution. His most characteristic work was in provisional theories, often of a semi-phenomenological character, whose technical content has been largely superseded. Even in the area of interpretation of quantum mechanics, where his ideas are still very much alive, it seems unlikely that a doctrine of limitation and renunciation, however revolutionary and constructive in its time, can satisfy ambitious minds, or endure indefinitely. Like the rest it will be digested and transformed, and in its new form no longer bear Bohr's distinctive mark or his name. Yet, as his contemporaries realized, no one will have contributed more to the finished product. Pais' book, by telling the story as it happened, helps capture a rich and intrinsically interesting intellectual style and to preserve its achievements.

My discussion has emphasized the intellectual side of Bohr. It would be wrong not to mention the impression one gets, both from Pais' book and from the lovely collection of reminiscences *Niels Bohr: His Life and Work as Seen by His Friends and Colleagues* (ed. Rozental), of the rootedness and inner harmony of Bohr's life and personality. He was regarded with deep affection by all who knew him well. Pais' book contains many warm anecdotes and amusing stories, and some outright jokes, that help make it entertaining as well as edifying.

A fascinating man, Bohr; and a fascinating book, this, which should help do justice to his memory.

Dreams of a Final Theory

Review of ***Dreams of a Final Theory*** **by Steven Weinberg**
(Pantheon, New York, 1992)

The physics world is bound to have high expectations for a new book by Steven Weinberg, one of its most distinguished citizens. *Dreams of a Final Theory* does not disappoint.

Working physicists will want to read it for the insights and perspectives it offers on particular issues of physics, including the interpretation of quantum mechanics, the role of the anthropic principle and the problem of the cosmological constant. But broader questions of the place of physics in our culture and the place of "fundamental" (that is, for Weinberg, elementary particles) physics within physics as a whole are the main concern of the book. Although one of its chapters is entitled "Against Philosophy," the animating spirit of the book is clearly philosophy in the original Greek sense: love of wisdom.

Dreams of a Final Theory is a long meditation on what a final theory is, whether we are close to finding such a theory and what the implications of finding it would be.

What is meant by a final theory? It is the place where any series of whys comes to a satisfying end. Weinberg illustrates the concept by reference to a piece of chalk, showing how initially simple-minded questions about its color and composition lead ineluctably after a few whys into quantum electrodynamics, quantum chromodynamics and cosmic nucleosynthesis. He argues that no matter where you start in inquiring about the physical world, the arrows of explanation soon converge on a few basic underlying laws and principles that cannot be explained further. If these basic laws and principles were logically complete, aesthetically satisfying and difficult to modify or enlarge without spoiling their consistency, we would say that they constitute a final theory.

Are we close to a final theory? Although ancient Greek and later philosophers put forward encompassing world systems, their attempts did not reach the level of precision and detail that we aspire to today. The modern vision of a final theory emerged from Isaac Newton's triumphant mechanism. His fantastically successful mathematical astronomy inspired the program of explaining all of Nature in terms of forces between simple atoms. A decisive turn came in 1913 when Niels Bohr showed for the first time (in the only way that counts — by convincing example) that physicists could realistically expect to understand the behavior of matter on subatomic scales, although not along Newtonian lines. By 1929 Paul Dirac could declare that "the underlying physical laws necessary for the mathematical theory of a larger part of physics and all of chemistry are thus completely known."

During the mid-1970s what is now called the Standard Model of particle physics, based on gauge theories of the electroweak and strong interactions, took shape, with notable contributions from Weinberg himself. When supplemented with Einstein's theory of gravity — general relativity — the Standard Model gives what is widely believed to be an accurate and complete account of the behavior of ordinary matter under ordinary conditions, even with a definition of "ordinary" liberal enough to include neutron star interiors and conditions a billionth of a second after the Big Bang. The standard model is firmly based in quantum mechanics, and indeed it can be seen as a vast generalization of the quantum electrodynamics Dirac knew.

Weinberg devotes a chapter to a realistic description of the complex, messy sequence of events whereby three major physical theories — general relativity, the radiative corrections to quantum electrodynamics, and the electroweak gauge theory — were established. His emphasis on the empirical foundation of our knowledge is refreshing. Despite the Standard Model's uninterrupted success in confrontations with ever more precise experimental tests over the past 15 years, theoretical physicists cannot be entirely satisfied with it. The theory has too many loose ends. It fairly begs you to unite its disparate components, and recently concrete ideas for doing that have scored some impressive quantitative success. Perhaps the most profound problem is that the basic ideas of quantum mechanics and general relativity are not easily reconciled. Superstring theory makes the most ambitious attempt at a final theory. It seems to be rich enough to incorporate both the standard model and general relativity in a logically consistent way. Weinberg does not minimize the difficulties: the practical difficulty that the theory is very hard to work with and has not so far yielded specific testable predictions of any kind; and what may be (no one can calculate well enough to be quite sure) a fundamental difficulty in accounting for the fact that empty space doesn't weigh anything — the notorious cosmological constant problem.

We should recall that Newtonian gravity, thermodynamics and the electromagnetic wave theory of light were once regarded, with far better justification, as final theories in their respective domains. While the successful application of these classic theories to wide domains of experience survives, modern theories of gravity, statistical mechanics and light have completely different conceptual foundations.

What would it mean if a final theory were achieved? Weinberg compares it to the discovery of the North Pole or the source of the Nile: Heroic quests that finally came to a successful end. He mentions that it would put such "would-be sciences" as clairvoyance, telekinesis and creationism to rest. Perhaps, though I wouldn't underestimate the human capacity for self-delusion. (And anyway, why couldn't an advocate claim that these were emergent properties of complex systems, not immediately evident in the fundamental laws?)

An entire chapter, entitled "What about God?," is devoted to debunking hopes that a final theory will lead us closer to God or to any other idea one might call religious. One misses a sense of joy in the ultimate achievement; one senses a (premature) nostalgia for the quest. I felt let down by the discussion of this question. Although I agree with Weinberg that the difference between the existing Standard Model and a hypothetical final theory would be unlikely to make any practical impact in other fields of science (with the notable exception of early-universe cosmology) or provide anything of use to conventional religion, I can imagine it would instill new pride and confidence in the power of human reason and a sense of universal harmony, as did the triumphs of Newtonian physics during the Enlightenment. It is also just possible that the scientific virtues of cooperation and humility before the facts would gain prestige through striking success, and come to play, for a happy few, something of the moral role that religion has traditionally played.

In any case, Weinberg's subdued account of what it would mean to achieve a final theory leaves the burden of the motivations for expensive projects such as the Superconducting Super Collider (discussed in the final chapter) elsewhere. As he emphasizes, the most compelling motivations for building that machine are not to be found in hopes that it will give us a final theory or show us "the mind of God," but rather in the insight it will yield for particular pressing problems, such as the origin of gauge symmetry breaking, the role of supersymmetry and the missing matter puzzle.

I came away from a first reading of the book eager to re-read it again. A second reading left me mightily impressed with its subtlety, honesty and, with the exception just noted, vision. *Dreams of a Final Theory* deserves to be read and reread by thoughtful physicists, philosophers and just plain thoughtful people.

It would be wrong to conclude without mentioning one last vital fact about the book: It is a good read. The prose is limpid, and the texture is both learned and witty. Let me leave you smiling, with a representative quotation:

> Even where they do not attempt to formulate a science of war, military historians often write as if generals lose battles because they do not follow some well-established rules of military science. For instance, two generals of the Union Army in the Civil War that come in for pretty wide disparagement are George McClellan and Ambrose Burnside. McClellan is generally blamed for not being willing to come to grips with the enemy, Lee's army of northern Virginia. Burnside is blamed for squandering the lives of his troops in a headlong assault on a well-entrenched opponent at Fredericksburg. It will not escape your attention that McClellan is criticized for not acting like Burnside, and Burnside is criticized for not acting like McClellan.

Note added, January 2006: The Superconducting Super Collider never came to pass, in part because physicists did not persuade Congress that its mission deserved the substantial resources it required. As explained in "The Social Benefits of High Energy Physics," I think Congress erred.

Shadows of the Mind

Review of *Shadows of the Mind: A Search for the Missing Science of Consciousness* by **Roger Penrose**
(Oxford University Press, 1994)

In *Shadows of the Mind* celebrated Oxford physicist Roger Penrose continues, elaborates, and in some respects modifies arguments presented in his earlier book, *The Emperor's New Mind*. The claims put forward in these two extraordinary books are quite startling: that there are specific aspects of human experience, easily available to introspection, that cannot be explained, even in principle, within the framework of the known laws of physics; that in any case this framework is fundamentally flawed, in such a way that it fails in principle adequately to describe the behavior of macroscopic bodies; and that these two failings are related, so that only within a new physics, incorporating some yet-to-be-constructed quantum theory of gravity, will it be possible to understand the phenomenon of human consciousness.

Such claims, appearing as they do outside the regular scientific literature of the relevant disciplines, might not warrant serious consideration if they came from an anonymous source. But Penrose is hardly that, and his proposals deserve serious scrutiny. I have attempted such scrutiny, and — as will become obvious — have reached very negative conclusions. Before entering into the particulars, I should in fairness mention that large portions of this long book are devoted to exposition of background material, and that this exposition is generally sound and occasionally brilliant (here I have in mind particularly the mini-biography of Cardano on pp. 249–256). Thus discriminating readers might benefit from the book without buying into its original scientific claims. My focus here, however, will be on the main line of development, which attempts to justify the claims sketched above.

This main line proceeds as follows. First, Penrose argues that humans can perform mental feats that simply cannot be performed by a machine, however complicated, that follows a finite algorithm (that is, a Turing machine). He observes that the conventional laws of physics, as they operate on a finite material system, can be simulated by a Turing machine. Thus he concludes that the mental feats of humans cannot be explained within the conventional laws of physics. Second, he argues that the conventional laws of physics contain the seeds of their own destruction, in that the conventional rules of quantum mechanics are logically incoherent and, carried far enough, must give incorrect results. The breakdown is supposed to occur for small but "semi-macroscopic" bodies (more on this below). Finally, he argues that the required new laws of physics, as applied in the human brain, will explain our ability to transcend the Turing paradigm. Let us examine each of these steps, in turn.

The central — indeed essentially the only — exhibit in Penrose's case that humans do things Turing machines cannot is the supposed "obvious" ability of human mathematicians to transcend the limitations of Gödel's theorem. The core of Gödel's argument is his construction, in any sufficiently powerful formalized system (roughly speaking, in any system specified that is powerful enough to deal with arithmetic and whose procedures could be completely specified and mechanized *à la* Turing), of a proposition that can be interpreted as stating "I am not provable." If this proposition is true, then it cannot be proved, and if it is false the system is inconsistent. Thus any powerful, consistent formal system will allow statements that are true but not provable. Now, says Penrose, we can see that any attempt to capture the power of human mathematicians by a formal system must fail. For a human mathematician could understand the *meaning of* Gödel's argument, even as applied to a hypothetical formal description of herself, and thereby recognize the *truth* of the Gödelian proposition. This, according to Penrose, demonstrates that the human has methods of reaching truth that have the force of proof but that cannot be captured by the proof-process of any formalized system. Experts will recognize (and Penrose acknowledges) the descent of this argument from a famous 1961 article by Oxford philosopher John Lucas, which spawned a large, contentious literature and certainly has not won universal acceptance. Among the many counterarguments that have been offered, one that appears particularly clear and convincing to me is that the truth of the Gödel proposition only follows on the assumption of consistency — but consistency is, for a powerful system, not at all obvious. In fact, according to another closely related theorem of Gödel, consistency cannot be proved (if it is true). Thus the supposed royal road to truth involves a questionable shortcut, which arrives at something less than proof after all.

In any case, it seems quite strange to draw the battle line on this suspect terrain. Let us make a more modest request and ask for a demonstration that humans do something concrete that is not strictly impossible for Turing machines, but only difficult. And let us look where evolution suggests there might be something to find, in the perceptual processes that (unlike mathematical logic) are likely to be optimized by natural selection. Are there perceptual tasks that humans do much faster than any classical computer? (More precisely: Are there "holistic" perceptual tasks at the NP level of difficulty that humans can do in polynomial time?) Penrose seems to be edging toward this sort of question in his discussion of tiling problems, but he does not report on any attempts at systematic experimentation. Any convincing demonstration of such abilities, though giving only a much *weaker* result than Penrose claims, would have revolutionary implications.

Turning now to physics: Penrose perceives deep trouble in the foundations of quantum theory. He accepts that the physical interpretation of quantum theory *requires* something he calls the R-process, which is the moral equivalent of the "reduction of the state vector" invoked in some discussions of the Copenhagen interpretation. This is to be distinguished from ordinary deterministic dynamics described by the Schrödinger equation, which he calls the U-process. Some version of the R-process is, according to Penrose, necessary to make a bridge between the quantum rules for adding amplitudes and the classical rules for adding probabilities.

Again, this is suspect terrain. It is by no means the case that all informed physicists see the need for an R-process; indeed, the modern tendency (to which this reviewer is fully sympathetic) seems to be to see if we can get by with just the U-process. Though it would be very premature to declare final victory for this approach — that will require, I believe, construction of a recognizable caricature of an intelligent observer within the formalism of quantum theory, so that model experiences could be compared against our subjective experience as real observes — there have not been any decisive defeats, either, in some challenging battles.

What this means can to some extent be illustrated with reference to Penrose's tentative (and, it is fair to say, extremely vague) proposals. He claims that including the effects of gravity within the quantum-mechanical formalism will make a qualitative change in its predictions for the behavior of macroscopic bodies, in such a way that the laws of classical probability will apply for sufficiently large bodies. Well: what about the behavior of K-mesons, neutrons, and even photons, which are known by exquisite interferometric experiments to maintain quantum coherence after traversing what are by any reasonable measure macroscopic distances? What about superconductors, which conduct perfectly "through a mile of dirty wire" (Casimir), or similarly superfluid helium? What about the profound and beautiful work now being done on mesoscopic systems, which probes quantitatively how characteristic quantum mechanical behavior gradually becomes more subtle — but does not seem to break down or disappear — for pure, small, cold but definitely macroscopic systems (involving many thousands of electrons)?

The quantum theory of gravity is fraught with difficulties regarding its behavior at higher energies and short distances, difficulties that may or may not be resolved by superstring theory. However, quantization of the low-energy, long-wavelength part of the Einstein theory is not problematic, and effects of the sort Penrose proposes ought to be discussed within this theoretical framework, as a first step. My own conclusion is that the predicted effects are exceedingly small and are likely to be overwhelmed, under any remotely practical circumstances, by more mundane processes. In any case this framework is eminently *computable*, in the technical Turing-machine sense, so it is not sufficient for Penrose, who wants somehow to introduce a *noncomputable* R-process.

For the next step in Penrose's synthesis is to invoke the hypothetical noncomputable R-process to explain the supposed noncomputable abilities of human brains. Regarding this, I (like Penrose) will be brief. He claims, inspired in part by ideas of Hameroff, that microtubules perform crucial information-processing roles in the human brain and that they behave in essentially quantum-mechanical ways that allow them to transcend the limitations of Turing machines. A lot is known about information processing in the brain, particularly in the early stages of visual processing, and as far as I know it has never proved necessary to assign an important role to microtubules. Microtubules are not particularly characteristic of the human nervous system — indeed they are common in single-cell organisms — and on the face of it they appear to be versatile structural elements in many classes of cells. Moreover, the conditions of heterogeneity and temperature characteristic of biological activity hardly seem conducive to quantum coherence on a macroscopic scale. So speculations about a spectacular computational ability of microtubules based on quantum coherence and central to human consciousness appear quite bold at this time. They, at least, would seem to be open to experimental investigation in the near future.

It appears to me, in summary, that Penrose's argument, from formal logic and philosophy, that human beings perform noncomputable operations is simply mistaken; that his argument that quantum theory is incomplete is unconvincing and his proposed remedy implausible; that his conclusion that an essentially classical description of microtubule function must fail is premature to say the least; and that his discussion of this topic, and of neurobiology in general, does not do justice to a large important body of relevant empirical knowledge. Moreover, the whole grand structure of his arguments is exceedingly fragile, being at no point buttressed by specific reference to nontrivial experimental facts. Perhaps not since the heyday of the great rationalist metaphysicians — Descartes, Leibniz, Spinoza — has there been a comparable performance. Although there are several brilliant passages in the book and the distinguished author is evidently sincere in his convictions, in the end one can only agree with Francis Crick, who wrote (commenting on *The Emperor's New Mind*), "It will be remarkable if his main idea turns out to be true."

The Inflationary Universe

Review of *The Inflationary Universe: The Quest for a New Theory of Cosmic Origins* by Alan Guth
(Addison-Wesley, 1997)

The inflationary universe scenario postulates that at some time in its very early history of the universe underwent a period of extremely rapid (superluminal) expansion. Inflation is probably the single most important idea to arise in scientific cosmology since the early 1960s, when the foundations of modern Big Bang cosmology were laid. It potentially explains several observed features of the universe for which there is no other known explanation, as will be discussed below. The best ideas on why inflation might occur are inspired by exotic but established concepts of modern particle physics. Indeed, a period of inflation can be triggered by phase transitions of the sort that particle physics models predict under extreme conditions of temperature and density, such as occurred close to the Big Bang.

The intellectual origin of the inflationary universe scenario can be traced, with a precision unusual for modern science, to a specific date and author. On 6 December 1979 the author of the book under review, Alan Guth, realized that the models of particle physics he was analyzing for other purposes could, under reasonable assumptions, trigger an inflationary epoch. Equally important, he realized that the occurrence of such an epoch would answer some major cosmological riddles. There were partial anticipations before, and many refinements and applications were

"The inflationary solution to the flatness problem is illustrated by this sequence of perspective drawings of an inflating sphere. In each successive frame the sphere is inflated by a factor of three, while the number of grid lines is increased by the same factor. By the fourth frame, it is difficult to distinguish the image from that of a plane. In cosmology, a flat geometry corresponds to a universe with omega [the ratio of the actual mass density of the universe to the critical mass density] equal to one. 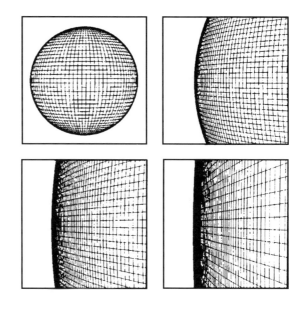 Therefore, as inflation drives the geometry of the universe toward flatness, the value of omega is driven toward one." [From *The Inflationary Universe*]

added later, but clearly Guth's contribution was the central one. Moreover, he is an exceptionally lucid and painstaking writer. It is uniquely fitting, therefore, that he should present this subject to the general public.

There are no equations in the text of the book, but much use of quantitative reasoning, graphic illustrations, and so forth. In other words, the book will be quite challenging for the innumerate, but otherwise generally accessible. It is an intellectually honest book that comes to grips with difficult concepts such as spontaneous symmetry breaking, blackbody radiation, and percolation, even when doing so slows down the narrative. There is a helpful glossary, and some slightly more technical material in appendixes at the end. It may be dangerous for a fellow "expert" to pronounce on the matter, but I think the author has succeeded remarkably well in conveying the profound concepts vital to his story in a simple, but never simplistic, way.

The Inflationary Universe falls naturally into three parts. The first part, constituting roughly half the book, sets the stage. After two brief, eclectic chapters that seem to belong elsewhere, we settle down to a very nice account of standard Big Bang cosmology, and a brief but well-thought-out and coherent account of relevant parts of particle physics. The material is presented in the context of a narrative of discovery that — regarded as history — is inevitably oversimplified, but always in good taste.

This part flows smoothly into the heart of the book, which is a personal account of how the author came to discover the inflationary universe scenario, and — together with others — to refine and apply it. Especially interesting is his very detailed account of his actual line of reasoning, which evolved from work with Henry Tye attempting to reconcile grand unified models of particle physics (and their specific prediction of magnetic monopole production) with standard Big Bang cosmology. Guth's work was also heavily influenced by an earlier, visionary seminar by the late Robert Dicke. The story is presented "warts and all" and does not stint in describing early misconceptions and effects of the pressure of competition. Guth also mentions, rather too schematically I think, the dramatic recent developments in measurements of microwave background anisotropies that might within a few years allow a much more meaningful check of very-early-universe cosmology, including inflation.

Finally, there is a distinct change of tone in the final 50 pages or so, where significantly more speculative topics are taken up. Here the connection to established physical laws becomes much more tenuous and the relevance to any practically conceivable observations remote. Although the speculations are clearly labeled as such, the length and style in which they are presented belie the warning. It might have been better instead to elaborate more deeply on the equally fascinating and much more realistic prospects for learning more about the early universe from better mapping of the microwave sky and for finding the elusive "dark matter" that appears to constitute most of the universe by weight.

It seems appropriate, in this journal, to say a few words about the status of inflation as a scientific theory. In its simplest form, this idea immediately explains one of the most striking and yet, from the point of view of general relativity, most puzzling facts about the present-day universe, namely that its spatial geometry is approximately Euclidean: that the universe is "flat." As emphasized by Dicke, this fact is quite peculiar, because deviations from flatness tend to grow with time and the universe is quite old. Since expansion will tend to make even a curved surface become flat, a period of drastic inflation can resolve Dicke's puzzle. Related to this, inflation can explain the approximate homogeneity of the observed universe, which — as attested in the microwave background — was accurate to a part in ten thousand in the early history of the universe. Yet the homogeneity is not perfect, as the recent observations have revealed; indeed, the early universe must have contained the seeds that later, by gravitational accretion, grew into galaxies and other large-scale structures. Inflationary models offer a very promising way of getting insight into these inhomogeneities, by tracing them to (inflated) quantum fluctuations.

On the other hand, one must admit that the phenomena so far explained by the inflationary scenario are few in number and not entirely characteristic: the flatness of the universe had been a working assumption of many cosmologists earlier, and the simplest (scale-invariant, or Harrison–Zeldovich) spectrum of inhomogeneities, which inflation models tend to give as a first approximation, was also hypothesized well before, by its namesakes. Also, the existing models of particle physics that give inflation are only metaphorically related to concrete worldmodels and contain some disturbingly "unnatural" features, in the form of unexplained small parameters. Most troubling of all, perhaps, is that the fundamental mechanism driving inflation, the negative pressure associated with the energy density of empty space, is closely related to perhaps the weakest point in current physical theory, the problem of the cosmological constant. Roughly speaking, the issue is as follows. We know by observation that space devoid of matter, in the present state of the universe, does not weigh very much. Strange though it sounds, this familiar fact is baffling to modern physicists, because according to our theories apparently empty space is actually highly intricate and structured. At any rate, we do not understand why empty space weighs so little now. In order to have inflation we must believe that the early universe was very different. Thus challenges remain to derive and test more distinctive consequences of inflation and to root it more firmly in physical theory.

All this is just to caution, as Guth himself does repeatedly, that there continue to be lively debates on the fundamentals of scientific cosmology. It is undeniable, and truly remarkable, that increasingly the terms of these debates are set by Guth's bold, yet coherent and believable extrapolation of known physical laws to produce an epoch of cosmic inflation. Anyone interested in ideas, or the history of ideas, should read this book.

Is the Sky Made from Pi?

Review of
Just Six Numbers: The Deep Forces that Shape the Universe by Martin Rees
(Weidenfeld & Nicolson, 1999)
and
The Nine Numbers of the Universe by Michael Rowan-Robinson
(Oxford University Press, 1999)

More than two millennia have passed since Pythagoras proclaimed, "All things are Number". Was his proclamation reverie or revelation? In the books reviewed here, two distinguished cosmologists, from profoundly different perspectives, weave popular accounts of their subject around that question.

Developments in physics and astronomy over the past century have brought the Pythagorean vision into sharp focus. Close observation has established homogeneity of the Universe, and the sameness (or 'universality') of the physical laws throughout. Also, we have a well-tested picture of evolution from a hot, dense, extremely homogeneous phase — the Big Bang. We can codify the physical laws in a remarkably simple, beautiful and mathematically precise set of equations — the so-called Standard Model, plus general relativity.

To specify cosmology fully requires at least the amplitude and slope of the spectrum of primeval fluctuations, the mass fraction in one or several kinds of dark matter, the baryon fraction, the Hubble parameter and the value of the cosmological term. Similarly, to fully specify the

Standard Model (including all the heavy quarks and leptons) requires many logically independent parameters.

Thus, our fundamental working models of nature, although containing an elegant and powerful Pythagorean core, begin to look complicated and unsatisfying when fully fleshed out. Moreover, we have learned that there are many questions we should not expect to be able to answer from first principles. Whereas it was reasonable for Johannes Kepler to try to fit the ratios of planetary orbits using ideal mathematical constructions (regular polyhedra), we now understand that these ratios are highly contingent facts, accidents of a unique history.

The title and introduction of Martin Rees's book led me to expect something quite different from what the rest of the book delivers. I had expected to find a celebration of the triumph of Pythagorism — an account of how, given the values of just a few numbers (say 6), you can account in detail for a wealth of physical phenomena and, on the way, do some pretty impressive engineering, by pure calculation. Instead, Rees delivers six case studies of how the Pythagorean programme very nearly goes catastrophically wrong. If any of six specific, chosen quantities had been significantly different, Rees argues, our world would be unrecognizable and intelligent life would most probably be impossible.

Taking a count: Pythagoras believed that "All things are Number."

Two examples will give a flavor of the argument. The binding energy of protons and neutrons into helium-4 is 0.007 of their rest mass. That number arises, according to modern elementary particle physics, from a rather intricate interplay between the masses of 'up' and 'down' quarks (relative to the basic quantum chromodynamic scale) and the fine-structure constant. From the point of view of fundamental theory the binding energy is a remote epiphenomenon, and without absurdity one can easily imagine worlds in which it had a slightly different value. Yet, Rees argues, if it were as small as 0.006, deuterium would be unbound and no elements other than hydrogen would be created; while if it were as large as 0.008, stellar evolution would be so fast and violent that life as we know it could not evolve.

The ratio of the strength of gravitational to electric forces, for protons, is about 10^{-36}. This extraordinary number arises, according to a rather more uncertain extrapolation of modern elementary particle physics, as a result of the slow running of the strong coupling constant. If it were significantly larger, gravity on bound systems such as planets would be much more oppressive, and large multicellular creatures would be crushed by their own weight (also stars would evolve faster). If it were significantly smaller, planets would not form in the first place.

In documenting these examples, Rees is moving towards a deeply subversive position. It is becoming increasingly difficult to believe, he argues, that the complex 'fine tunings' of physical parameters that seem to be necessary for the emergence of intelligent life are unique consequences of a simple fundamental theory. He favors instead the idea that we live not in a Universe but rather in a Multiverse, which contains regions with drastically different properties (for example, different values of the nuclear binding fraction, gravitational strength, or even space-time dimensionality). In this framework, the 'fine tunings' could be explained anthropically. They need not have occurred, indeed in most places they do not occur, but where they don't there's no one around to watch the botch!

Twenty years ago the Multiverse concept would have seemed utterly far-fetched, but now it shows signs of becoming conventional wisdom. The most serious objection to the Multiverse is, of course, simply the observational fact that extensive astronomical observations of distant regions of the cosmos have disclosed the basic uniformity, not diversity, of physical law. But the theory of cosmic inflation has made it plausible that the portion of the world we can currently observe might be just a small portion of the whole, an initially tiny (and hence uniform) patch inflated to gigantic proportions. And high-energy theorists working on supersymmetric unified theories and string/M theory find themselves confronted with a bewildering variety of apparently consistent solutions, with nothing to choose between them. Might each have its homeland?

If these ideas are correct, then the irreducible element of contingency we noted above extends much further than commonly allowed. Several of the seemingly 'fundamental' parameters of physics would, in fact, be features of our environment. They could never be calculated directly from fundamental theory without taking a serious detour through anthropism, because they would be consequences of our position in the Universe amidst the larger Multiverse, not of any universal truth.

Thus, Rees is the anti-Pythagoras. Whether or not they are convinced by his major thesis, readers will find Rees's short, well-written book an enjoyable and provocative intellectual adventure.

Michael Rowan-Robinson's book represents, by comparison, down-to-earth cosmology. He discusses the key observations — past, present and future — in much greater depth, including frequent and strenuous warnings about their uncertainties, yet he does so concisely. In the penultimate chapter he discusses his own involvement in investigating the starburst-galaxy phenomenon. Although perhaps not quite as grand as the other material, it is in compensation fresh and personal. The two books are in many respects complementary and, by looking at both, a reader could get an excellent, rounded view of the exciting state of contemporary physical cosmology.

Neither of these books, with their heavy focus on the wild-and-woolly frontiers of cosmology, begins to do justice to the truly remarkable triumphs of Pythagorism closer to home. Given five pure numbers — the electron, up- and down-quark masses, the quantum chromodynamic scale, and the fine-structure constant — one can accurately account for all the phenomena of chemistry, and the structure of ordinary matter. Add a couple more — Newton's gravitational and Fermi's weak-coupling constants — and essentially all of astrophysics and most of cosmology comes within the charmed circle of understanding. Small parts of this great scientific success story have been told, but it has yet to find its Milton.

Big Ideas

"Quantum Field Theory" was written as part of a comprehensive survey of modern physics to celebrate the 100th anniversary of the American Physical Society. The subject is vast. Quantum field theory dominated theoretical physics over most of the 20th century, has many brilliant applications, and still captures our deepest insights into how Nature works.

At first I struggled with the assignment, unable to get a grip on it. Desperate, I invited my friend and mentor, Sam Treiman to lunch, and put my problem before him. How could I say something fresh, meaningful, and crisp about such a gigantic and notoriously difficult subject? Sam suggested an approach I thought was brilliant — through a question:

> What do you learn from quantum field theory, that you don't already know from quantum mechanics and field theory separately?

We began to brainstorm this over lunch, and soon ideas began to flow easily. I regard the result as one of my best intellectual achievements. Thank you, Sam.

A sidelight, about the answer: The single most important thing you learn, without a doubt, is why the building blocks of Nature occur in a few stereotyped modules, i.e. why all electrons have exactly the same properties, all photons have exactly the same properties, and so forth.

This is not brought out in courses or textbooks of the subject, and so I was startled to realize it myself. As a game, for a couple of years I asked gatherings of theoretical physicists over lunch, at conferences, or during my travels: What is the most important thing you learn from quantum field theory, that you can't understand without it? I posed that question to hundreds of physicists, who thought about it and even discussed it, but didn't arrive at the right answer (at least not in my presence). Only one did. That was Freeman Dyson, who immediately responded

Why, that all electrons are the same, of course.

"Some Basic Aspects of Fractional Quantum Numbers" was written as part of a selection of Bob Schrieffer's greatest works. An introductory essay was solicited for each topic, and I was asked to introduce the group on fractional charge. (That origin explains some references to papers to follow, and so forth, which appear odd out of context.) It was a joy to remember working with Bob and learning from him about fractional quantum numbers, and condensed matter physics in general. I took the fractional charge assignment broadly, though I limited myself to one- and two-dimensional systems. Most of the basic ideas appear already in those contexts, and for the moment it's where the action is experimentally. I incorporated some basic insights I got from Sidney Coleman, which have not appeared elsewhere. I anticipate a brilliant future for this circle of ideas.

In these two pieces, I don't hold back. They are much more demanding than the rest of the book. Their intended audience was, and is, the physics research community. In them, I tried to do justice, as best I could, to two genuinely Big Ideas. While writing them I thought very hard about the central concepts of quantum field theory and charge fractionalization, and I think I came to understand the foundations of these subjects in new ways that are lucid and likely to be fruitful. So while these pieces aren't for the faint of heart, true seekers should attempt them.

Quantum Field Theory

I discuss the general principles underlying quantum field theory, and attempt to identify its most profound consequences. The deepest of these consequences result from the infinite number of degrees of freedom invoked to implement locality. I mention a few of its most striking successes, both achieved and prospective. Finally, I consider the possible limitations of quantum field theory, in the light of its history.

I. SURVEY

Quantum field theory is the framework used to formulate the regnant theories of the electroweak and strong interactions, which together form the Standard Model. Quantum electrodynamics (QED), besides providing a complete foundation for atomic physics and chemistry, has supported calculations of physical quantities with unparalleled precision. The experimentally measured value of the magnetic dipole moment of the muon,

$$(g_\mu - 2)_{\text{exp.}} = 233\ 184\ 600\ (1680) \times 10^{-11}, \quad (1)$$

for example, should be compared with the theoretical prediction

$$(g_\mu - 2)_{\text{theor.}} = 233\ 183\ 478\ (308) \times 10^{-11}. \quad (2)$$

In quantum chromodynamics (QCD) we cannot, for the foreseeable future, aspire to to comparable accuracy. Yet QCD provides different, and at least equally impressive, evidence for the validity of the basic principles of quantum field theory. Indeed, because in QCD the interactions are stronger, QCD manifests a wider variety of phenomena characteristic of quantum field theory. These include especially running of the effective coupling with distance or energy scale and the phenomenon of confinement. QCD has supported, and rewarded with experimental confirmation, both heroic calculations of multi-loop diagrams and massive numerical simulations of (a discretized version of) the complete theory.

Quantum field theory also provides powerful tools for condensed matter physics, especially in connection with the quantum many-body problem as it arises in the theory of metals, superconductivity, the low-temperature behavior of the quantum liquids He^3 and He^4, and the quantum Hall effect, among others. Although for reasons of space and focus I will not attempt to do justice to this aspect here, the continuing interchange of ideas between condensed matter and high energy theory, through the medium of quantum field theory, is remarkable in itself. A partial list of historically important examples includes global and local spontaneous symmetry breaking, the renormalization group, effective field theory, solitons, instantons, and fractional charge and statistics.

It is clear, from all these examples, that quantum field theory occupies a central position in our description of Nature. It provides both our best working description of fundamental physical laws, and a fine tool for investigating the behavior of complex systems. But listing examples, however triumphal, serves more to pose than to answer more basic questions: What are the essential features of quantum field theory? What does quantum field theory add to our understanding of the world, that was not already present in quantum mechanics and classical field theory separately?

The first question has no sharp answer. Theoretical physicists are very flexible in adapting their tools, and no axiomization can keep up with them. However I think it is fair to say that the characteristic, core ideas of quantum field theory are twofold. First, that the basic dynamical degrees of freedom are operator functions of space and time — quantum fields, obeying appropriate commutation relations. Second, that the interactions of these fields are local. Thus the equations of motion and commutation relations governing the evolution of a given quantum field at a given point in space-time should depend only on the behavior of fields and their derivatives at that point. One might find it convenient to use other variables, whose equations are not local, but in the spirit of quantum field theory there must always be some underlying fundamental, local variables. These ideas, combined with postulates of symmetry (e.g., in the context of the Standard Model, Lorentz and gauge invariance) turn out to be amazingly powerful.

The field concept came to dominate physics starting with the work of Faraday in the mid-nineteenth century. Its conceptual advantage over the earlier Newtonian program of physics, to formulate the fundamental laws in terms of forces among atomic particles, emerges when we take into account the circumstance, unknown to Newton (or, for that matter, Faraday) but fundamental in special relativity, that influences travel no faster than a finite limiting speed. For then the force on a given particle at a given time cannot be deduced from the positions of other particles at that time, but must be deduced in a complicated way from their previous positions. Faraday's intuition that the fundamental laws of electromagnetism could be expressed most simply in terms of fields filling space and time was of course brilliantly vindicated by Maxwell's mathematical theory.

The concept of locality, in the crude form that one can predict the behavior of nearby objects without reference to distant ones, is basic to scientific practice. Practical experimenters — if not astrologers — confidently expect, on the basis of much successful experience, that after reasonable (generally quite modest) precautions to isolate their experiments they will obtain reproducible results. Direct quantitative tests of locality, or rather of its close cousin causality, are afforded by dispersion relations.

The deep and ancient historic roots of the field and locality concepts provide no guarantee that these concepts remain relevant or valid when extrapolated far beyond their origins in experience, into the subatomic and quantum domain. This extrapolation must be judged by its fruits. That brings us, naturally, to our second question.

Undoubtedly the single most profound fact about Nature that quantum field theory uniquely explains is *the existence of different, yet indistinguishable, copies of elementary particles*. Two electrons anywhere in the Universe, whatever their origin or history, are observed to have exactly the same properties. We understand this as a consequence of the fact that both are excitations of the same underlying ur-stuff, the electron field. The electron field is thus the primary reality. The same logic, of course, applies to photons or quarks, or even to composite objects such as atomic nuclei, atoms, or molecules. The indistinguishability of particles is so familiar, and so fundamental to all of modern physical science, that we could easily take it for granted. Yet it is by no means obvious. For example, it directly contradicts one of the pillars of Leibniz' metaphysics, his "principle of the identity of indiscernables," according to which two objects cannot differ solely in number. And Maxwell thought the similarity of different molecules so remarkable that he devoted the last part of his *Encyclopedia Brittanica* entry on Atoms — well over a thousand words — to discussing it. He concluded that "the formation of a molecule is therefore an event not belonging to that order of nature in which we live ... it must be referred to the epoch, not of the formation of the earth or the solar system ... but of the establishment of the existing order of Nature ..."

The existence of classes of indistinguishable particles is the necessary logical prerequisite to a second profound insight from quantum field theory: *the assignment of unique quantum statistics* to each class. Given the indistinguishability of a class of elementary particles, and the invariance of their behavior under interchange, the general principles of quantum mechanics teach us that solutions forming any representation of the permutation symmetry group retain that property in time, but do not constrain which representations are realized. Quantum field theory not only explains the existence of indistinguishable particles and the invariance of their interactions under interchange, but also constrains the symmetry of the solutions. For bosons only the identity representation is physical (symmetric wave functions), for fermions only the one-dimensional odd representation is physical (antisymmetric wave functions). One also has the spin-statistics theorem, according to which objects with integer spin are bosons, whereas objects with half odd integer spin are fermions. Of course, these general predictions have been verified in many experiments. The fermion character of electrons, in particular, underlies the stability of matter and the structure of the periodic table.

A third profound general insight from quantum field theory is *the existence of antiparticles.* This was first inferred by Dirac on the basis of a brilliant but obsolete interpretation of his equation for the electron field, whose elucidation was a crucial step in the formulation of quantum field theory. In quantum field theory, we re-interpret the Dirac wave function as a position (and time) dependent operator. It can be expanded in terms of the solutions of the Dirac equation, with operator coefficients. The coefficients of positive-energy solutions are operators that destroy electrons, and the coefficients of the negative-energy solutions are operators that create positrons (with positive energy). With this interpretation, an improved version of Dirac's hole theory emerges in a straightforward way. (Unlike the original hole theory, it has a sensible generalization to bosons, and to processes where the number of electrons minus positrons changes.) A very general consequence of quantum field theory, valid in the presence of arbitrarily complicated interactions, is the CPT theorem. It states that the product of charge conjugation, parity, and time reversal is always a symmetry of the world, although each may be — and is! — violated separately. Antiparticles are strictly defined as the CPT conjugates of their corresponding particles.

The three outstanding facts we have discussed so far: the existence of indistinguishable particles, the phenomenon of quantum statistics, and the existence of antiparticles, are all essentially consequences of *free* quantum field theory. When one incorporates interactions into quantum field theory, two additional features of the world emerge.

The first of these is *the ubiquity of particle creation and destruction processes*. Local interactions involve products of field operators at a point. When the fields are expanded into creation and annihilation operators multiplying modes, we see that these interactions correspond to processes wherein particles can be created, annihilated, or changed into different kinds of particles. This possibility arises, of course, in the primeval quantum field theory, quantum electrodynamics, where the primary interaction arises from a product of the electron field, its Hermitean conjugate, and the photon field. Processes of radiation and absorption of photons by electrons (or positrons), as well as electron-positron pair creation, are encoded in this product. Just because the emission and absorption of light is such a common experience, and electrodynamics such a special and familiar classical field theory, this correspondence between formalism and reality did not initially make a big impression. The first conscious exploitation of the potential for quantum field theory to describe processes of transformation was Fermi's theory of beta decay. He turned the procedure around, inferring from the observed processes of particle transformation the nature of the underlying local interaction of fields. Fermi's theory involved creation and annihilation not of photons, but of atomic nuclei and electrons (as well as neutrinos) — the ingredients of 'matter'. It began the process whereby classic atomism, involving stable individual objects, was replaced by a more sophisticated and accurate picture. In this picture it is only the fields, and not the individual objects they create and destroy, that are permanent.

The second is *the association of forces and interactions with particle exchange.* When Maxwell completed the equations of electrodynamics, he found that they supported source-free electromagnetic waves. The classical electric and magnetic fields thus took on a life of their own. Electric and magnetic forces between charged particles are explained as due to one particle acting as a source for electric and magnetic fields, which then influence others. With the correspondence of fields and particles, as it arises in quantum field theory, Maxwell's discovery of electromagnetic waves corresponds to the existence of real photons, and the generation of forces by intermediary fields corresponds to the exchange of virtual photons. The association of forces (or, more generally, interactions) with exchange of particles is a general feature of quantum field theory. It was used by Yukawa to infer the existence and mass of pions from the range of nuclear forces, and more recently in electroweak theory to infer the existence, mass, and properties of W and Z bosons prior to their observation, and in QCD to infer the existence and properties of gluon jets prior to their observation.

The two additional outstanding facts we just discussed: the possibility of particle creation and destruction, and the association of particles with forces, are essentially consequences of classical field theory supplemented by the connection between particles and fields we learn from free field theory. Indeed, classical waves with nonlinear interactions will change form, scatter, and radiate, and these processes exactly mirror the transformation, interaction, and creation of particles. In quantum field theory, they are properties one sees already in *tree graphs*.

The foregoing major consequences of free quantum field theory, and of its formal extension to include nonlinear interactions, were all well appreciated by the late 1930s. The deeper properties of quantum field theory, which will form the subject of the remainder of this paper, arise from the need to introduce *infinitely many degrees of freedom*, and the possibility that all these degrees of freedom are excited as quantum-mechanical fluctuations. From a mathematical point of view, these deeper properties arise when we consider *loop graphs*.

From a physical point of view, the potential pitfalls associated with the existence of an infinite number of degrees of freedom first showed up in connection with the problem which led to the birth of quantum theory: the ultraviolet catastrophe of blackbody radiation theory. Somewhat ironically, in view of later history, the crucial role of quantum theory in blackbody theory was to remove the disastrous consequences of the infinite number of degrees of freedom possessed by classical electrodynamics. The classical electrodynamic field can be decomposed into independent oscillators with arbitrarily high values of the wavevector. According to the equipartition theorem of classical statistical mechanics, in thermal equilibrium at temperature T each of these oscillators should have average energy kT. Quantum mechanics alters this situation by insisting that the oscillators of frequency ω have energy quantized in units of $\hbar\omega$. Then the high-frequency modes are exponentially suppressed by the Boltzmann factor, and instead of kT receive $\dfrac{\hbar\omega e^{-\frac{\hbar\omega}{kT}}}{1-e^{-\frac{\hbar\omega}{kT}}}$, which is very small for large ω. The role of the quantum, then, is to prevent accumulation of energy in the form of very small amplitude excitations of arbitrarily high frequency modes. It is very effective in suppressing the *thermal* excitation of high-frequency modes.

But as it solves the blackbody catastrophe, quantum theory brings in a new problem. For it introduces the idea that the modes are always *intrinsically* excited to a small extent, proportional to $\hbar$. This so-called zero point motion is a consequence of the uncertainty principle. For a harmonic oscillator of frequency ω, the ground state energy is not zero, but $\frac{1}{2}\hbar\omega$. In the case of the electromagnetic field this leads, upon summing over its high-frequency modes, to a highly divergent total ground state energy. For most physical purposes the absolute normalization of energy is unimportant, and so this particular divergence does not necessarily render the theory useless.[a] It does, however, illustrate the dangerous character of the high-frequency modes, and its treatment gives a first indication of the leading theme of renormalization theory: we can only require — and generally will only obtain — sensible, finite answers when we ask questions that have direct, operational physical meaning.

Physicists first confronted an infinite number of degrees of freedom in Maxwell's specific theory of the electromagnetic field, but it is a general phenomenon, deeply connected with the requirement of locality in the interactions of fields. For in order to construct the local field $\psi(x)$ at a space-time point x, one must take a superposition

$$\psi(x) = \int \frac{d^4 k}{(2\pi)^4} e^{ikx} \tilde{\psi}(k) \tag{3}$$

that includes field components $\tilde{\psi}(k)$ extending to arbitrarily large momenta. Moreover in a generic interaction

$$\int \mathcal{L} = \int \psi(x)^3$$
$$= \int \frac{d^4 k_1}{(2\pi)^4} \frac{d^4 k_2}{(2\pi)^4} \frac{d^4 k_3}{(2\pi)^4} \tilde{\psi}(k_1) \tilde{\psi}(k_2) \tilde{\psi}(k_3) (2\pi)^4 \delta^4(k_1 + k_2 + k_3) \tag{4}$$

we see that a low momentum mode $k_1 \approx 0$ will couple without any suppression factor to high-momentum modes k_2 and $k_3 \approx -k_2$. Local couplings are "hard," in this sense. Because locality requires the existence of infinitely many degrees of freedom at large momenta, with hard interactions, ultraviolet divergences similar to the ones Planck cured, but driven by quantum rather than thermal fluctuations, are never far off stage. As mentioned previously, the deeper physical consequences of quantum field theory arise from this circumstance.

First of all, it is much more difficult to construct non-trivial examples of interacting relativistic quantum field theories than

---

[a]One would think that gravity should care about the absolute normalization of energy. The zero-point energy of the electromagnetic field, in that context, generates an infinite cosmological constant. This might be cancelled by similar negative contributions from fermion fields, as occurs in supersymmetric theories, or it might indicate the need for some other profound modification of physical theory.

purely formal considerations would suggest. One finds that *the consistent quantum field theories form a limited class, whose extent depends sensitively on the dimension of space-time and the spins of the particles involved.* Their construction is quite delicate, requiring limiting procedures whose logical implementation leads directly to renormalization theory, the running of couplings, and asymptotic freedom.

Secondly, *even those quantum theories that can be constructed display less symmetry than their formal properties would suggest.* Violations of naive scaling relations — that is, ordinary dimensional analysis — in QCD, and of baryon number conservation in the standard electroweak model are examples of this general phenomenon. The original example, unfortunately too complicated to explain fully here, involved the decay process $\pi^o \to \gamma\gamma$, for which chiral symmetry (treated classically) predicts much too small a rate. When the correction introduced by quantum field theory (the so-called 'anomaly') is retained, excellent agreement with experiment results.

These deeper consequences of quantum field theory, which might superficially appear rather technical, largely dictate the structure and behavior of the Standard Model — and, therefore, of the physical world. My goal in this preliminary survey has been to emphasize their profound origin. In the rest of the article I hope to convey their main implications, in as simple and direct a fashion as possible.

II. FORMULATION

The physical constants $\hbar$ and c are so deeply embedded in the formulation of relativistic quantum field theory that it is standard practice to declare them to be the units of action and velocity, respectively. In these units, of course, $\hbar = c = 1$. With this convention, all physical quantities of interest have units that are powers of mass. Thus the dimension of momentum is $(\text{mass})^1$ or simply 1, since mass $\times c$ is a momentum, and the dimension of length is $(\text{mass})^{-1}$ or simply -1, since $\hbar c/\text{mass}$ is a length. The usual way to construct quantum field theories is by applying the rules of quantization to a continuum field theory, following the canonical procedure of replacing Poisson brackets by commutators (or, for fermionic fields, anticommutators). The field theories that describe free spin 0 or free spin $\frac{1}{2}$ fields of mass m, μ respectively are based on the Lagrangian densities

$$\mathcal{L}_0(x) = \frac{1}{2}\partial_\alpha \phi(x)\partial^\alpha \phi(x) - \frac{m^2}{2}\phi(x)^2 \tag{5}$$

$$\mathcal{L}_{\frac{1}{2}}(x) = \bar{\psi}(x)(i\gamma^\alpha \partial_\alpha - \mu)\psi(x). \tag{6}$$

Since the action $\int d^4x \mathcal{L}$ has mass dimension 0, the mass dimension of a scalar field like ϕ is 1 and that of a spinor field like ψ is $\frac{3}{2}$. For free spin 1 fields the Lagrangian density is that of Maxwell,

$$\mathcal{L}_1(x) = -\frac{1}{4}(\partial_\alpha A_\beta(x) - \partial_\beta A_\alpha(x))(\partial^\alpha A^\beta(x) - \partial^\beta A^\alpha(x)), \tag{7}$$

so that the mass dimension of the vector field A is 1. The same result is true for non-abelian vector fields (Yang-Mills fields).

Thus far all our Lagrangian densities have been quadratic in the fields. Local interaction terms are obtained from Lagrangian densities involving products of fields and their derivatives at a point. The coefficient of such a term is a coupling constant, and must have the appropriate mass dimension so that the Lagrangian density has mass dimension 4. Thus the mass dimension of a Yukawa coupling y, which multiplies the product of two spinor fields and a scalar field, is zero. Gauge couplings g arising in the minimal coupling procedure $\partial_\alpha \to \partial_\alpha + igA_\alpha$ are also evidently of mass dimension zero.

The possibilities for couplings with non-negative mass dimension are very restricted. This fact is important, for the following reason. Consider the effect of treating a given interaction term as a perturbation. If the coupling κ associated to this interaction has negative mass dimension $-p$, then successive powers of it will occur in the form of powers of $\kappa \Lambda^p$, where Λ is some parameter with dimensions of mass. Because, as we have seen, the interactions in a local field theory are hard, we can anticipate that Λ will characterize the largest mass scale we allow to occur (the cutoff), and will diverge to infinity as the limit on this mass scale is removed. So we expect that it will be difficult to make sense of fundamental interactions having negative mass dimensions, at least in perturbation theory. Such interactions are said to be nonrenormalizable.

The standard model is formulated entirely using renormalizable interactions. Is this a fundamental fact about Nature? If non-renormalizable interactions occurred in the effective description of physical behavior below a certain mass scale, it would simply mean that the theory must change its nature — presumably by displaying new degrees of freedom — at some larger mass scale. From this point of view, the fact that the Standard Model contains only renormalizable operators means that it does not require modification up to arbitrarily high scales (at least on the grounds of divergences in perturbation theory). Whether or not we choose to call that a fundamental fact, it is certainly a profound one.

Moreover, all the renormalizable interactions consistent with the gauge symmetry and multiplet structure of the Standard Model do seem to occur — "what is not forbidden, is mandatory". There is a beautiful agreement between the symmetries of the Standard Model, allowing arbitrary renormalizable interactions, and the symmetries of the world. One understands why strangeness is violated, but baryon number is not. (The only discordant element is the so-called θ term of QCD, which is allowed by the symmetries of the Standard Model but is measured to be quite accurately zero. A plausible solution to this problem exists. It involves a characteristic very light *axion* field.)

Estimating divergences by power counting assumes that there are no special symmetries cancelling off the contribution of high energy modes. It does not work in supersymmetric theories, in which the contributions of boson and fermionic modes cancels, or in theories derived from supersymmetric theories by soft supersymmetry breaking. In the latter case the scale of supersymmetry breaking plays the role of the cutoff Λ.

The power counting rules, as discussed so far, are too crude to detect logarithmic divergences, of the form $\ln \Lambda^2$. Yet divergences of this form are pervasive and extremely significant, as we shall now discuss.

III. RUNNING COUPLINGS

The problem of calculating the energy associated with a constant magnetic field, in the more general context of an arbitrary nonabelian gauge theory coupled to spin 0 and spin $\frac{1}{2}$ charged particles, illustrates in a concrete way how the infinities of quantum field theory arise, and of how they are dealt with. It introduces the concept of running couplings and leads directly to qualitative and quantitative results of great significance for physics. The interactions of concern to us appear in the Lagrangian density

$$\mathcal{L} = -\frac{1}{4g^2} G^I_{\alpha\beta} G^{I\alpha\beta} + \bar{\psi}(i\gamma^\nu D_\nu)\psi + \phi^\dagger(-D_\nu D^\nu - m^2)\phi \quad (8)$$

where $G^I_{\alpha\beta} \equiv \partial_\alpha A^I_\beta - \partial_\beta A^I_\alpha - f^{IJK} A^J_\alpha A^K_\beta$ and $D_\nu \equiv \partial_\nu + iA^I_\nu T^I$ are the standard field strengths and covariant derivative, respectively. Here the f^{IJK} are the structure constants of the gauge group, and the T^I are the representation matrices appropriate to the field on which the covariant derivative acts. This Lagrangian differs from the usual one by a rescaling $gA \to A$, which serves to emphasize that the gauge coupling g occurs only as a prefactor in the first term. It parametrizes the energetic cost of non-trivial gauge curvature, or in other words the stiffness of the gauge fields. Small g corresponds to gauge fields that are difficult to excite.

From this Lagrangian it would appear that the energy required to set up a magnetic field B^I is just $\frac{1}{2g^2}(B^I)^2$. This is the classical energy, but in the quantum theory it is not the whole story. A more accurate calculation must take into account the effect of the imposed magnetic field on the zero-point energy of the charged fields. Earlier, we met and briefly discussed a formally infinite contribution to the energy of the ground state of a quantum field theory (specifically, the electromagnetic field) due to the irreducible quantum fluctuations of its modes, which mapped to an infinite number of independent harmonic oscillators. Insofar as only differences in energy are physically significant, we could ignore this infinity. But the change in the zero-point energy as one imposes a magnetic field cannot be ignored. It represents a genuine contribution to the physical energy of the quantum state induced by the imposed magnetic field. As we will soon see, the field-dependent part of the energy also diverges.

Postponing momentarily the derivation, let me anticipate the form of the answer, and discuss its interpretation. Without loss of generality, I will suppose that the magnetic field is aligned along a normalized, diagonal generator of the gauge group. This allows us to drop the index, and to use terminology and intuition from electrodynamics freely. If we restrict the sum to modes whose energy is less than a cutoff Λ, we find for the energy

$$\mathcal{E}(B) = \mathcal{E} + \delta\mathcal{E} = \frac{1}{2g^2(\Lambda^2)}B^2 - \frac{1}{2}\eta B^2(\ln(\Lambda^2/B) + \text{finite}) \qquad (9)$$

where

$$\eta = \frac{1}{96\pi^2}[-(T(R_o) - 2T(R_{\frac{1}{2}}) + 2T(R_1))]$$
$$+ \frac{1}{96\pi^2}[3(-2T(R_{\frac{1}{2}}) + 8T(R_1))], \qquad (10)$$

and the terms not displayed are finite as $\Lambda \to \infty$. The notation $g^2(\Lambda^2)$ has been introduced for later convenience. The factor $T(R_s)$ is the trace of the representation for spin s, and basically represents the sum of the squares of the charges for the particles of that spin. The denominator in the logarithm is fixed by dimensional analysis, assuming $B \gg \mu^2, m^2$.

The most striking, and at first sight disturbing, aspect of this calculation is that a cutoff is necessary in order to obtain a finite result. If we are not to introduce a new fundamental scale, and thereby (in view of our previous discussion) endanger the principle of locality, we must remove reference to the arbitrary cutoff Λ in our description of physically meaningful quantities. This is the sort of problem addressed by the renormalization program. Its guiding idea is that if we are working with experimental probes characterized by energy and momentum scales well below Λ, we should expect that our capacity to affect, or be sensitive to, the modes of much higher energy will be severely restricted. The cutoff Λ, which was introduced as a calculational device to remove such modes, can therefore be removed (taken to infinity). In our magnetic energy example, for instance, we see immediately that the difference in susceptibilities

$$\mathcal{E}(B_1)/B_1^2 - \mathcal{E}(B_0)/B_0^2 = \text{finite} \qquad (11)$$

is well-behaved — that is, independent of Λ as $\Lambda \to \infty$. Thus once we measure the susceptibility, or equivalently the coupling constant, at one reference value of B, the calculation gives sensible, unambiguous predictions for all other values of B.

This simple example illustrates a much more general result, the central result of the classic renormalization program. It goes as follows. A small number of quantities, corresponding to the couplings and masses in the original Lagrangian, that when calculated formally diverge or depend on the cutoff, are chosen to fit experiment. They define the physical, as opposed to the original, or bare, couplings. Thus, in our example, we can define the susceptibility to be $\frac{1}{2g^2(B_0)}$ at some reference field B_0. Then we have the physical or renormalized coupling

$$\frac{1}{g^2(B_0)} = \frac{1}{g^2(\Lambda^2)} - \eta \ln(\Lambda^2/B_0). \tag{12}$$

(For simplicity I have ignored the finite terms. These are relatively negligible for large B_0. Also, there are corrections of higher order in g^2.) This determines the 'bare' coupling to be

$$\frac{1}{g^2(\Lambda^2)} = \frac{1}{g^2(B_0)} + \eta \ln(\Lambda^2/B_0). \tag{13}$$

The central result of renormalization theory is that once bare couplings and masses are re-expressed in terms of their physical, renormalized counterparts, the coefficients in the perturbation expansion of any physical quantity approach finite limits, independent of the cutoff, as the cutoff is taken to infinity. (To be perfectly accurate, one must also perform wave-function renormalization. This is no different in principle; it amounts to expressing the bare coefficients of the kinetic terms in the Lagrangian in terms of renormalized values.)

Diagrammatic analysis leaves open the question whether this perturbation theory converges, or is some sort of asymptotic expansion of a soundly defined theory. This loophole is no mere technicality, as we will soon see.

Picking a scale B_0 at which the coupling is defined is analogous to choosing the origin of a coordinate system in geometry. One can describe the same physics using different choices of normalization scale, so long as one adjusts the coupling to fit. We capture this idea by introducing the concept of a running coupling defined, in accordance with eq. (12), to satisfy

$$\frac{d}{d\ln B} \frac{1}{g^2(B)} = \eta. \tag{14}$$

With this definition, the choice of a particular scale at which to define the coupling will not affect the final result.

It is profoundly important, however, that the running coupling does make a real distinction between the behavior at different mass scales, even if the original underlying theory was formally scale invariant (as is QCD with massless quarks), and even at mass scales much larger than the mass of any particle in the theory. Quantum zero-point motion of the high energy modes introduces a hard source of scale symmetry violation as an unavoidable residue of the hocus-pocus of renormalization.

The distinction among scales, in a formally scale-invariant theory, embodies the phenomenon of *dimensional transmutation*. Rather than a range of theories, parametrized by a dimensionless coupling, we have a range of theories differing only in the value of a dimensional parameter (say, for example, the value of B at which $1/g^2(B) = 1$).

Clearly, the qualitative behavior of solutions of eq. (14) depends on the sign of η. If $\eta > 0$, the coupling $g^2(B)$ will get smaller as B grows. In other words, as we treat more and more modes as dynamical, we approach closer to a smaller 'bare' charge. Those modes were enhancing, or *antiscreening* the bare charge. This is the case of *asymptotic freedom*.

In asymptotically free theories, we can complete the renormalization program in a convincing fashion. There is no barrier to including the effect of very large energy modes, and removing the cutoff. We can confidently expect, then, that the theory is well-defined, independent of perturbation theory. In particular, suppose the theory has been discretized on a space-time lattice. That amounts to excluding the modes of high energy and momentum. In an asymptotically free theory one can compensate for these modes by adjusting the coupling in a well-defined, controlled way as one shrinks the discretization scale. Very impressive nonperturbative calculations in QCD, involving massive computer simulations, have exploited this strategy. They demonstrate the complete consistency of the theory and its ability to predict the masses of hadrons.

In the opposite case, $\eta < 0$, the coupling formally grows, and even diverges as B increases. $1/g^2(B)$ goes through zero and changes sign. On the face of it, that would seem to indicate an instability of the theory, toward formation of a ferromagnetic vacuum at large field strength. This conclusion must be taken with a big grain of salt, because when g^2 is large the higher-order corrections to eq. (13) and eq. (14), on which the analysis was based, cannot be neglected; but at best, we're losing control of the calculation. In a non-asymptotically free theory the coupling does not become small, there is no simple foolproof way to compensate for the missing modes, and the existence of an underlying limiting theory becomes doubtful.

Now let us discuss how η can be calculated. The two terms in eq. (10) correspond to two distinct physical effects. The first is the convective, diamagnetic (screening) term. The overall constant is a little tricky to calculate, and I do not have space to do it here. Its general form, however, is transparent. The effect is independent of spin: simply counts the number of components (one for scalar particles, two for spin $\frac{1}{2}$ or massless spin 1, both with two helicities). It is screening for bosons, while for fermions there is a sign flip, because the fermionic oscillators have negative zero-point energy.

The second is the paramagetic spin susceptibility. For a massless particle with spin s and gyromagnetic ratio g_m the energies shift, giving rise to the altered zero-point energy

$$\Delta \mathcal{E} = \int_0^{E=\Lambda} \frac{d^3k}{(2\pi)^3} \frac{1}{2} (\sqrt{k^2 + g_m s B} + \sqrt{k^2 - g_m s B} + 2\sqrt{k^2}). \quad (15)$$

This is readily calculated as

$$\Delta \mathcal{E} = -B^2 (g_m s)^2 \frac{1}{32\pi^2} \ln\left(\frac{\Lambda^2}{B}\right). \quad (16)$$

With $g_m = 2$, $s = 1$ this is the spin-1 contribution, and with $g_m = 2$, $s = \frac{1}{2}$, after a sign flip, it is the spin-$\frac{1}{2}$ contribution. The preferred moment $g_m = 2$ is a direct consequence of the Yang-Mills and Dirac equations, respectively.

This elementary calculation gives us a nice heuristic understanding of the unusual antiscreening behavior of nonabelian gauge theories. It is due to the large paramagnetic response of charged vector fields. Magnetic interactions, which can be attractive for like charges (paramagnetism) are, for highly relativistic particles, in no way suppressed. Indeed, they are numerically dominant. Because we are interested in very high energy modes, the usual intuition that charge will be screened, which is based on the electric response of heavy particles, does not apply.

Though I have presented it in the very specific context of vacuum magnetic susceptibility, the concept of running coupling applies much more widely. The basic heuristic idea is that in analyzing processes whose characteristic energy-momentum scale (squared) is Q^2, it is appropriate to use the running coupling at Q^2, i.e. in our earlier notation $g^2(B = Q^2)$. For in this way we capture the dynamical effect of the virtual oscillators that can be appreciably excited, while avoiding the formal divergence encountered if we tried to include all of them (up to infinite mass scale). At a more formal level, use of the appropriate effective coupling lets us avoid large logarithms in the calculation of Feynman graphs, by normalizing the vertices close to where they need to be evaluated. There is a highly developed chapter of quantum field theory that justifies and refines this rough idea into a form where it makes detailed, quantitative predictions for concrete experiments. I will not be able to do proper justice to the difficult, often heroic labor that has been invested, on both the theoretical and the experimental sides, to yield Figure 1; but it is appropriate to remark that quantum field theory gets a real workout, as calculations of two- and even three-loop graphs with complicated interactions among the virtual particles are needed to do justice to the attainable experimental accuracy.

An interesting feature visible in Figure 1 is that the theoretical prediction for the coupling focuses at large Q^2, in the sense that a wide range of values at small Q^2 converge to a much narrower range at larger Q^2. Thus even crude estimates of what are the appropriate scales (e.g., one expects $g^2(Q^2)/4\pi \sim 1$ where the strong interaction is strong, say for 100 Mev $\lesssim \sqrt{Q^2} \lesssim 1$ GeV) make it possible to predict the value of $g^2(M_Z^2)$ with $\sim 10\%$ accuracy. The original idea of Pauli and others that calculating the fine structure constant was the next great item on the agenda of theoretical physics now seems misguided. We see this constant as just another running coupling, neither more nor less fundamental than many other parameters, and not likely to be the most accessible theoretically. But our essentially parameter-free approximate determination of the observable strong interaction analogue of the fine structure constant realizes their dream in a different form.

The electroweak interactions start with much smaller couplings at low mass scales, so the effects of their running are less dramatic (though they have been observed). Far more spectacular than the modest quantitative effects we can test directly, however, is the conceptual breakthrough that results from application of these ideas to unified models of the strong, electromagnetic, and weak interactions.

The different components of the Standard Model have a similar mathematical structure, all being gauge theories. Their common

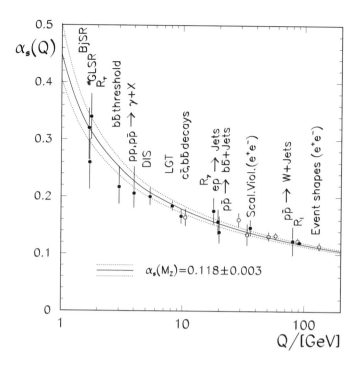

FIG. 1 Comparison of theory and experiment in QCD, illustrating the running of couplings. Several of the points on this curve represent hundreds of independent measurements, any one of which might have falsified the theory. Figure from M. Schmelling, hep-ex/9701002.

structure encourages the speculation that they are different facets of a more encompassing gauge symmetry, in which the different strong and weak color charges, as well as electromagnetic charge, would all appear on the same footing. The multiplet structure of the quarks and leptons in the Standard Model fits beautifully into small representations of unification groups such as $SU(5)$ or $SO(10)$. There is the apparent difficulty, however, that the coupling strengths of the different Standard Model interactions are widely different, whereas the symmetry required for unification requires that they share a common value. Running of couplings suggests an escape from this impasse. Since the strong, weak, and electromagnetic couplings run at different rates, their inequality at currently accessible scales need not reflect the ultimate state of affairs. We can imagine that spontaneous symmetry breaking — a soft effect — has hidden the full symmetry of the unified interaction. What is really required is that the fundamental, bare couplings be equal, or in more prosaic terms, that the running couplings of the different interactions should become equal beyond some large energy scale.

Using simple generalizations of the formulas derived and tested in QCD, we can calculate the running of couplings, to see whether this requirement is satisfied in reality. In doing so one must make some hypothesis about the spectrum of virtual particles. If there are additional massive particles (or, better, fields) that have not yet been observed, they will contribute significantly to the running of couplings once the scale exceeds their mass. Let us first consider the default assumption, that there are no new fields beyond those that occur in the Standard Model. The results of this calculation are displayed in Figure 2.

Despite the boldness of the extrapolation this calculation works remarkably well, but the accurate experimental data indicates unequivocally that something is wrong. There is one particularly attractive way to extend the Standard Model, by including super-

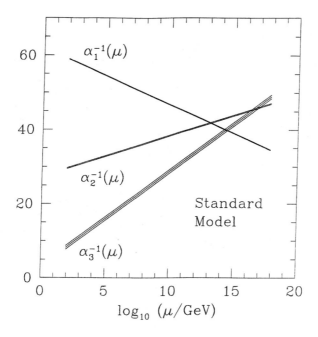

FIG. 2 Running of the couplings extrapolated toward very high scales, using just the fields of the Standard Model. The couplings do not quite meet. Experimental uncertainties in the extrapolation are indicated by the width of the lines. Figure courtesy of K. Dienes.

symmetry. Supersymmetry cannot be exact, but if it is only mildly broken (so that the superpartners have masses $\lesssim 1$ TeV) it can help explain why radiative corrections to the Higgs mass parameter, and thus to the scale of weak symmetry breaking, are not enormously large. In the absence of supersymmetry power counting would indicate a hard, quadratic dependence of this parameter on the cutoff. Supersymmetry removes the most divergent contribution, by cancelling boson against fermion loops. If the masses of the superpartners are not too heavy, the residual finite contributions due to supersymmetry breaking will not be too large.

The minimal supersymmetric extension of the Standard Model, then, makes semi-quantitative predictions for the spectrum of virtual particles starting at 1 TeV or so. Since the running of couplings is logarithmic, it is not extremely sensitive to the unknown details of the supersymmetric mass spectrum, and we can assess the impact of supersymmetry on the unification hypothesis quantitatively. The results, as shown in Figure 3, are quite encouraging.

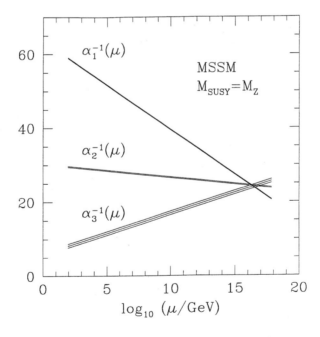

FIG. 3 Running of the couplings extrapolated to high scales, including the effects of supersymmetric particles starting at 1 TeV. Within experimental and theoretical uncertainties, the couplings do meet. Figure courtesy of K. Dienes.

With all its attractions, there is one general feature of supersymmetry that is especially challenging, and deserves mention here. We remarked earlier how the Standard Model, without supersymmetry, features a near-perfect match between the generic symmetries of its renormalizable interactions and the observed symmetries of the world. With supersymmetry, that feature is spoiled. The scalar superpartners of fermions are represented by fields of mass dimension one. This means that there are many more possibilities for low dimension (including renormalizable) interactions that violate flavor symmetries including lepton and baryon number. It seems that some additional principles, or special discrete symmetries, are required in order to suppress these interactions sufficiently.

A notable result of the unification of couplings calculation, especially in its supersymmetric form, is that the unification occurs at an energy scale which is enormously large by the standards of traditional particle physics, perhaps approaching 10^{16-17} GeV. From a phenomenological viewpoint, this is fortunate. The most compelling unification schemes merge quarks, antiquarks, leptons, and antileptons into common multiplets, and have gauge bosons mediating transitions among all these particle types. Baryon number violating processes almost inevitably result. Their rate is inversely proportional to the fourth power of the gauge boson masses, and thus to the fourth power of the unification scale. Only for such enormously large values of the energy scale is one safe from experimental limits on nucleon instability. From a theoretical point of view the large scale is fascinating because it brings us from the internal logic of the experimentally grounded domain of particle physics to the threshold of quantum gravity, as we shall now discuss.

IV. LIMITATIONS?

So much for the successes, achieved and anticipated, of quantum field theory. The fundamental limitations of quantum field theory, if any, are less clear. Its triumphant application to describe the other fundamental interactions has fallen short in the case of gravity.

All existing experimental results on gravitation are adequately described by a very beautiful, conceptually simple classical field theory — Einstein's general relativity. It is easy to incorporate this theory into our description of the world based on quantum field theory, by allowing a minimal coupling to the fields of the Standard Model — that is, by changing ordinary into covariant derivatives, multiplying with appropriate factors of $\sqrt{g}$, and adding an Einstein–Hilbert curvature term. The resulting theory — with the convention that we simply ignore quantum corrections involving virtual gravitons — is the foundation of our working description of the physical world. As a practical matter, it works very well indeed.

Philosophically, however, it might be disappointing if it were too straightforward to construct a quantum theory of gravity. We want it to be difficult, so that it will constrain our theoretical construction of the world. A great vision of natural philosophy, going back to Pythagoras, is that the properties of the world are uniquely determined by mathematical principles. A modern version of this vision was formulated by Planck, shortly after he introduced his quantum of action. By appropriately combining the physical constants c, $\hbar$ as units of velocity and action, respectively, and the Planck mass $M_{\text{Planck}} = \sqrt{\frac{\hbar c}{G}}$ as the unit of mass, one can construct any unit of measurement used in physics. Thus the unit of energy is $M_{\text{Planck}} c^2$, the unit of electric charge is $\sqrt{\hbar c}$, and so forth. On the other hand, one cannot form a pure number from these three physical constants. Thus one might hope that in a physical theory where $\hbar$, c, and G were all profoundly incorporated, all physical quantities could be expressed in natural units as pure numbers.

Within its domain, QCD achieves something very close to this vision — actually, in a *more* ambitious form! Indeed, let us idealize the world of the strong interaction slightly, by imagining that there were just two quark species with vanishing masses. Then from the two integers 3 (colors) and 2 (flavors), $\hbar$, and c — with *no* explicit mass parameter — a spectrum of hadrons, with mass ratios and other properties close to those observed in reality, emerges by calculation. The overall unit of mass is indeterminate, but this ambiguity has no significance within the theory itself.

The ideal Pythagorean/Planckian theory would not contain any pure numbers as parameters. (Pythagoras might have excused a few small integers.) Thus, for example, the value $m_e/M_{\text{Planck}} \sim 10^{-22}$ of the electron mass in Planck units would emerge from a dynamical calculation. That ideal might be overly ambitious, but it seems reasonable to hope that a quantum theory which consistently incorporates gravity will also produce constraints among other observables. Indeed, as we have already seen, one does find significant constraints among the parameters of the Standard Model by requiring that the strong, weak, and electromagnetic interactions emerge from a unified gauge symmetry; so there is precedent for results of this kind.

The unification of couplings calculation provides not only an inspiring model, but also direct encouragement for the Planck program, in two important respects. First, it points to a unification scale remarkably close to the Planck scale (though apparently smaller by $10^{-2} - 10^{-3}$). Thus there are pure numbers with much more 'reasonable' values than 10^{-22} to shoot for. Second, it shows quite concretely how very large scale factors can be controlled by modest ratios of coupling strength, due to the logarithmic nature of the running of couplings — so that 10^{-22} may not be so 'unreasonable' after all.

Perhaps it is fortunate, then, that the straightforward, minimal implementation of general relativity as a quantum field theory — which lacks the desired constraints — runs into problems. The problems are of two quite distinct kinds. First, the renormalization program fails, at the level of power-counting. The Einstein–Hilbert term in the action comes with a large prefactor $1/G$, reflecting the difficulty of curving space-time. If we expand the Einstein–Hilbert action around flat space in the form

$$g_{\alpha\beta} = \eta_{\alpha\beta} + \sqrt{G} h_{\alpha\beta} \qquad (17)$$

we find that the quadratic terms give a properly normalized spin-2 graviton field $h_{\alpha\beta}$ of mass dimension 1, as the powers of G cancel. But the higher-order terms, which represent interactions, will be accompanied by positive powers of G. Since G itself has mass dimension -2, these are non-renormalizable interactions. Similarly for the couplings of gravitons to matter. Thus we can expect that ever-increasing powers of $\Lambda/M_{\text{Planck}}$ will appear in multiple virtual graviton exchange, and it will be impossible to remove the cutoff.

Second, one of the main qualitative features of gravity — the weightlessness of empty space, or the vanishing of the cosmological constant — is left unexplained. Earlier we mentioned the divergent zero-point energy characteristic of generic quantum field theories. For purposes of non-gravitational physics only energy differences are meaningful, and we can sweep this problem under the rug. But gravity ought to see that energy. Our perplexity grows when we recall that according to the Standard Model, and even more so in its unified extensions, what we commonly regard as empty space is full of condensates, which again one would expect to weigh far more than observation allows.

The failure, so far, of quantum field theory to meet these challenges might reflect a basic failure of principle, or merely that the appropriate symmetry principles and degrees of freedom, in terms of which the theory should be formulated, have not yet been found.

Promising insights toward construction of a quantum theory including gravity are coming from investigations in string/M theory. Whether these investigations will converge toward an accurate description of Nature, and if so whether that description will take the form of a local field theory (perhaps formulated in many dimensions, and including many fields beyond those of the Standard Model), are questions not yet decided. It is interesting, in this regard, briefly to consider the rocky intellectual history of quantum field theory.

After the initial successes of the 1930s, already mentioned above, came a long period of disillusionment. Initial attempts to deal with the infinities that arose in calculations of loop graphs in electrodynamics, or in radiative corrections to beta decay, led only to confusion and failure. Similar infinities plagued Yukawa's pion theory, and it had the additional difficulty that the coupling required to fit experiment is large, so that tree graphs provide a manifestly poor approximation. Many of the founders of quantum theory, including Bohr, Heisenberg, Pauli, and (for different reasons) Einstein and Schrödinger, felt that further progress would require radically new innovations. These innovations would be a revolution of the order of quantum mechanics itself, and would introduce a new fundamental length.

Quantum electrodynamics was resurrected in the late 1940s, largely stimulated by developments in experimental technique. Those experimental developments made it possible to study atomic processes with such great precision, that the approximation afforded by keeping tree graphs alone could not do them justice. Methods to extract sensible finite answers to physical questions from the jumbled divergences were developed. Spectacular agreement with experiment resulted — all without changing electrodynamics itself, or departing from the principles of relativistic quantum field theory.

After this wave of success came another long period of disillusionment. The renormalization methods developed for electrodynamics did not seem to work for weak interaction theory. They did suffice to define a perturbative expansion of Yukawa's pion theory, but the large coupling observed in nature made that limited success academic (and soon it came to seem dubious that Yukawa's schematic theory could do justice to the wealth of newly discovered phenomena). In any case, as a practical matter, throughout the 1950s and 1960s a flood of experimental discoveries, including new classes of weak processes and a rich spectrum of hadronic resonances with complicated interactions, had to be absorbed and correlated. During this process of pattern recognition the elementary parts of quantum field theory were used extensively, as a framework, but deeper questions were put off. Many theorists came to feel that quantum field theory, in its deeper aspects, was simply wrong, and would need to be replaced by some S-matrix or bootstrap theory; perhaps most thought it was irrelevant, or that its use was premature, especially for the strong interaction.

As it became clear, through phenomenological work, that the weak interaction is governed by current × current interactions with universal strength, the possibility to ascribe it to exchange of vector gauge bosons became quite attractive. Models incorporating the idea of spontaneous symmetry breaking to give mass to the weak gauge bosons were constructed. It was conjectured, and later proved, that the high degree of symmetry in these theories allows one to isolate and control the infinities of perturbation theory. One can carry out a renormalization program similar in spirit, though considerably more complex in detail, to that of QED. It is crucial, here, that spontaneous symmetry breaking is a very soft operation. It does not significantly affect the symmetry of the theory at large momenta, where the potential divergences must be cancelled.

In parallel, phenomenological work on the strong interaction made it increasingly plausible that the observed strongly interacting particles — mesons and baryons — are composites of more basic objects. The evidence was of two different kinds: on the one hand, it was possible in this way to make crude but effective models for the observed spectrum with mesons as quark-antiquark, and baryons as quark-quark-quark, bound states; and on the other hand, experiments provided evidence for hard interactions of photons with hadrons, as would be expected if the components of hadrons were described by local fields. The search for a quantum field theory with appropriate properties led to a unique candidate, one that contained both objects that could be identified with quarks and an essentially new ingredient, color gluons.

These quantum field theories of the weak and strong interactions were dramatically confirmed by subsequent experiments, and have survived exceedingly rigorous testing over the past two decades. They make up the Standard Model. During this period the limitations, as well as the very considerable virtues, of the Standard Model have become evident. Whether the next big step will require a sharp break from the principles of quantum field theory or, like the previous ones, a better appreciation of its potentialities, remains to be seen.

Acknowledgments

I wish to thank S. Treiman for extremely helpful guidance, and M. Alford, K. Babu, C. Kolda, and J. March-Russell for reviewing the manuscript. F.W. is supported in part by DOE grant DE-FG02-90ER40542.

References

For further information about quantum field theory, the reader may wish to consult:

1. T.P. Cheng and L.F. Li, *Gauge Theory of Elementary Particle Physics* (Oxford, 1984).
2. M. Peskin and D. Schroeder, *Introduction to Quantum Field Theory* (Addison-Wesley, 1995).
3. S. Weinberg, *The Quantum Theory of Fields, I* (Cambridge, 1995) and *The Quantum Theory of Fields, II* (Cambridge, 1996).

Some Basic Aspects of Fractional Quantum Numbers

I review why and how physical states with fractional quantum numbers can occur, emphasizing basic mechanisms in simple contexts. The general mechanism of charge fractionalization is the passage from states created by local action of fields to states having a topological character, which permits mixing between local and topological charges. The primeval case of charge fractionalization for domain walls, in polyacetylene and more generally, can be demonstrated convincingly using Schrieffer's intuitive counting argument, and derived formally from analysis of zero modes and vacuum polarization. An important generalization realizes chiral fermions on high-dimensional domain walls, in particular for liquid He^3 in the A phase. In two spatial dimensions, fractionalization of angular momentum and quantum statistics occurs, for reasons that can be elucidated both abstractly, and specifically in the context of the quantum Hall effect.

Quantization of charge is a very basic feature of our picture of the physical world. The explanation of how matter can be built up from a few types of indivisible building-blocks, each occurring in vast numbers of identical copies, is a major triumph of local quantum field theory. In many ways, it forms the centerpiece of twentieth-century physics.

Therefore the discovery of physical circumstances in which the unit of charge can be divided, its quanta dequantized, came as a

---

*Commentary for *Selected Papers of J. Robert Schrieffer: In Celebration of His 70th Birthday*, eds. N. E. Bonesteel and L. P. Gor'kov, World Scientific, 2003, pp. 135–152. cond-mat/0206122 [MIT-CTP-3275].

shock to most physicists. It is remarkable that this fundamental discovery emerged neither from recondite theoretical speculation, nor from experiments at the high-energy frontier, but rather from analysis of very concrete, superficially mundane (even messy) polymers [1, 2]. In the process of coming to terms with charge fractionalization, we've been led to a deeper understanding of the logic of charge quantization itself. We have also been led to discover a whole world of related, previously unexpected phenomena. Exploration of this concept-world is far from complete, but already it has fed back significant value into the description of additional real world phenomena.

Bob Schrieffer's contributions in this field, partially represented in the papers that follow, started early and have run deep and wide. In this introduction I've tried to distill the core theoretical concepts to their simplest, most general meaningful form, and put them in a broader perspective. Due to limitations of time, space and (my) competence, serious analysis of particular materials and their experimental phenomenology, which figures very prominently in Schrieffer's papers, will not be featured here.

1. The Secret of Fractional Charge

To begin, let us take a rough definition of charge to be any discrete, additive, effectively conserved quantity, and let us accept the conventional story of charge quantization as background. A more discriminating discussion of different varieties of charge, and of the origin of quantization, follows shortly below.

The conventional story of charge quantization consists of three essential points: some deep theory gives us a universal unit for the charges to be associated with fields; observed particles are created by these fields, acting locally; the charges of particles as observed are related by universal renormalization factors to the charges of the fields that create them. The last two points are closely linked. Indeed, conservation of charge implies that the state produced by a local field excitation carries the charge of the field. Thus renormalization of charge reflects modification of the means to measure it, not properties of the carriers. This is the physical content of Ward's identity, leading to the relation $e_{\text{ren.}} = (Z_3)^{\frac{1}{2}} e_{\text{bare}}$ between renormalized and bare charge in electrodynamics, wherein only wave-function renormalization of the photon appears.

That reasoning, however, applies only to states that cannot be produced by local action of quantum fields. States that don't meet this criterion often occur. They can, for example, be associated with topologically non-trivial rearrangements of the conditions at infinity. Simple, important examples are domain walls between two degenerate phases in one-dimensional systems and flux tubes in two-dimensional systems. Such disruptive states are often associated additive quantum numbers, also called topological charges. For example, the amount of flux is an additive quantum number for flux tubes, given in terms of gauge potentials at spatial infinity by $\oint d\theta A_\theta$.

With two underlying charges, the general relation between renormalized and bare charge becomes

$$q_{\text{ren.}} = \epsilon_1 q_{\text{bare}}^{(1)} + \epsilon_2 q_{\text{bare}}^{(2)}. \tag{1}$$

Given this form, quantization of $q_{\text{bare}}^{(1)}$ and $q_{\text{bare}}^{(2)}$ in integers does not imply that renormalized charges are rationally related. In particular, suppose that the first charge is associated with local fields, while the second is topological. The ratio between renormalized charges for a general state, and the state of minimal charge produced by local operations is a conventionally normalized charge. In the absence of topology, it would be simply $q^{(1)}$, an integer. Here it becomes

$$q_{\text{normalized}} = q^{(1)} + \frac{\epsilon_2}{\epsilon_1} q^{(2)}. \tag{2}$$

The ratio $\frac{\epsilon_2}{\epsilon_1}$ is a dynamical quantity that, roughly speaking, measures the induced charge associated with a unit of topological structure. In general, it need not be integral or even rational. This is the general mechanism whereby fractional charges arise. It is the secret of fractional charge.

An important special case arises when the topological charge is discrete, associated with a finite additive group Z_n. Then the renormalized charge spectrum for $q^{(2)} = n$ must be the same as that for $q^{(2)} = 0$, since a topological charge n configuration, being topologically trivial, can be produced by local operations. So we have the restriction

$$\frac{\epsilon_2}{\epsilon_1} = \frac{p}{n} \tag{3}$$

with an integer p. The fractional parts of the normalized charges are always multiples of $\frac{1}{n}$.

The primeval case of 1-dimensional domain walls, which we are about to discuss in depth, requires a special comment. A domain wall of the type $A \to B$, going from the A to the B ground state, can only be followed (always reading left to right) by an anti-domain wall of the type $B \to A$, and *vice versa*; one cannot have two adjacent walls of the same type. So one does not have, for domain wall number, quite the usual physics of an additive quantum number, with free superposition of states. However, the underlying, spontaneously broken Z_2 symmetry that relates A to B also relates domain walls to anti-domain walls. Assuming that this symmetry commutes with the charge of interest, the charge spectra of the domain wall and the anti-domain wall must agree. (That assumption is valid in most cases of interest.) At the same time, the spectrum of total charge for domain wall plus anti-domain wall, a configuration that can be produced by local operations, must reduce to that for vacuum. So we have $2\frac{\epsilon_2}{\epsilon_1} = $ integer, and we find (at worst) half-integer normalized charges.

2. Polyacetylene and the Schrieffer Counting Argument

For our purposes, polyacetylene in its ground state can be idealized as an infinite chain molecule with alternating single and double bonds. That valence structure is reflected, physically, in the spacing of neighboring carbon nuclei: those linked by double bonds are held closer than those linked by single bonds. Choosing some particular nucleus to be the origin, and moving from there to the right, there are two alternative ground states of equal energy, schematically

$$\cdots 121212121212 \cdots \text{ (A)}$$
$$\cdots 212121212121 \cdots \text{ (B)} \tag{4}$$

Now consider the defect obtained by removing a bond at the fourth link, in the form

$$\cdots 121112121212 \cdots \tag{5}$$

By sliding bonds between the tenth and fifth links we arrive at

$$\cdots 121121212112 \cdots, \tag{6}$$

which displays the original defect as two more elementary ones. Indeed, the elementary defect

$$\cdots 12112121 \cdots , \tag{7}$$

if continued without further disruption of the order, is a minimal domain wall interpolating between ground state A on the left and ground state B on the right.

The fact that by removing *one* bond we produce *two* domain walls strongly suggests that each domain wall is half a bond short. If the bonds represented single electrons, then each wall would fractional charge $\frac{e}{2}$, and spin $\pm\frac{1}{4}$. In reality bonds represent pairs of electrons with opposite spin, and so we don't get fractional charge. But we still do find something quite unusual: a domain wall acquires charge e, with spin 0. Charge and spin, which normally occur together, have been separated!

This brilliant argument, both lucid and suggestive of generalizations, is known as the Schrieffer counting argument. In it, the secret of fractional charge is reduced to barest bones.

A simple generalization, which of course did not escape Schrieffer [3], is to consider more elaborate bonding patterns, for example

$$\cdots 112112112112 \cdots \tag{8}$$

Here removing a bond leads to

$$\cdots 111112112112 \cdots \tag{9}$$

which is re-arranged to

$$\cdots 111211121112 \cdots , \tag{10}$$

containing *three* elementary defects. True fractions, involving one-third integer normalized electric charges, are now unavoidable.

3. Field Theory Models of Fractional Charge

While the Schrieffer counting argument is correct and utterly convincing, it's important and fruitful to see how its results are realized formally, in quantum field theory.

First we must set up the field theory description of polyacetylene. Here I will very terse, since the accompanying paper of Jackiw and

Schrieffer sets out this problem in detail [4]. We consider a half-filled band in one dimension. With uniform lattice spacing a, the fermi surface consists of the two points $k_\pm = \pm\pi/2a$. We can parametrize the modes near the surface using a linear approximation to the energy-momentum dispersion relation; then near these two points we have respectively right- and left-movers with velocities $\pm\left|\frac{\partial \epsilon}{\partial k}\right|$. Measuring velocity in this unit, and restricting ourselves to these modes, we can write the free theory in pseudo-relativistic form. (But note that in these considerations, physical spin is regarded only as an inert, internal degree of freedom.) It is convenient here to use the Dirac matrices

$$\gamma^0 = \begin{pmatrix} 0 & 1 \\ 1 & 0 \end{pmatrix}$$

$$\gamma^1 = \begin{pmatrix} 0 & -1 \\ 1 & 0 \end{pmatrix}$$

$$\gamma^\chi = \begin{pmatrix} 1 & 0 \\ 0 & -1 \end{pmatrix}, \tag{11}$$

where $\gamma^\chi \equiv \gamma^0\gamma^1$ is used to construct the chirality projectors $\frac{1\pm\gamma^\chi}{2}$. In the kinetic energy

$$L_{\text{kinetic}} = \bar{\psi}(i\gamma\cdot\partial)\psi \tag{12}$$

the right- and left-movers $\frac{1\pm\gamma^\chi}{2}\psi$ do not communicate with one another. Nevertheless scattering on the optical phonon mode ϕ, with momentum π/a, allows electrons to switch from one side of the fermi surface to the other. This is represented by the local Yukawa interaction

$$\Delta L(x,t) = g\phi(x,t)\bar{\psi}(x,t)\psi(x,t). \tag{13}$$

One also has kinetic terms for ϕ and a potential $V(\phi)$ that begins at quadratic order. The wave velocity for ϕ of course need not match the fermi velocity, so there is a violation of our pseudo-relativistic symmetry, but that complication plays no role in the following.

This field theory description does not yet quite correspond to the picture of polyacetylene sketched in the previous section, because the breaking of translational symmetry (from $x \to x + a$ to $x \to x + 2a$) has not appeared. We need not change the equations, however, we need only draw out their implications. Since our optical phonon field ϕ moves neighboring nuclei in opposite directions, a condensation $\langle \phi \rangle = \pm v \neq 0$ breaks translational symmetry in the appropriate way. We might expect such a symmetry breaking phonon condensation to be favorable, at half filling, because it opens up a gap at the fermi surface, and lowers the energy of occupied modes near the top of the band. Since those modes have been retained in the effective field theory, that theory will contain the instability. Calculation bears out that expectation. The classical potential $V(\phi)$ is subject to quantum corrections, which alter it qualitatively. Upon calculation of a simple one-loop vacuum polarization graph, one finds a correction

$$\Delta V(\phi) = \frac{g^2}{\pi} \phi^2 \ln(\phi^2/\mu^2), \tag{14}$$

where μ is an ultraviolet cutoff. (This cutoff appears because the assumed Yukawa interaction $g\phi\bar{\psi}\psi$ is an appropriate description of physics only near the fermi surface. A more sophisticated treatment would use the language of the renormalization group here.) For small ϕ this correction always dominates the classical ϕ^2. So it is always advantageous for ϕ to condense, no matter how small is g. Indeed, one finds the classic "BCS type" dependence

$$\langle \phi \rangle^2 = \mu^2 e^{-\frac{m^2 \pi}{g^2}} \tag{15}$$

at weak coupling.

This elegant example of dynamical symmetry breaking was first discussed by Peierls [5], who used a rather different language. It was introduced into relativistic quantum field theory in the seminal paper of Coleman and E. Weinberg [6]. In four space-time dimensions the correction term goes as $\Delta V(\phi) \propto g^4 \phi^4 \ln \phi^2$, and it dominates at small ϕ only if the classical mass term ($\propto \phi^2$) is anomalously small.

3.1. Zero Modes

The symmetry breaking $\langle \phi \rangle = \pm v$ induces, through the Yukawa coupling, an effective mass term for the fermion ψ, which can be interpreted in the language of condensed matter physics as the opening of an energy gap. The choice of sign distinguishes between two degenerate ground states that have identical physical properties, since they can be related by the symmetry

$$\phi \to -\phi$$
$$\psi_L \to -\psi_L \qquad (16)$$

With this interpretation, we see that a domain wall interpolating between $\langle\phi(\pm\infty)\rangle = \pm v$, necessarily has a region where the mass vanishes. We might expect it to be favorable for fermions to bind there. What is remarkable, is that there is always a solution of zero energy — a mid-gap state — localized on the wall. Indeed, in the background $\phi(x) = f(x)$ the Dirac equation for zero energy is simply

$$i\partial_x \psi_1 = gf\psi_2$$
$$i\partial_x \psi_2 = gf\psi_1 \qquad (17)$$

with the normalizable solution

$$\psi_1(x) = \exp\left(-g\int_0^x dy f(y)\right)$$
$$\psi_2(x) = -i\psi_1(x). \qquad (18)$$

Note that the domain wall asymptotics for $\langle\phi(x)\rangle$ allows the exponential to die in both directions.

It is not difficult to show, using charge conjugation symmetry (which is not violated by the background field!), that half the spectral weight of this mode arises from modes that are above the gap, and half from modes that are below the gap, with respect to the homogeneous ground state.

When we quantize the fermion field, we must decide whether or not to occupy the zero-energy mode. If we occupy it, then we will have occupied half a mode that was unoccupied in the homogeneous ground state, and we will have a state of fermion number $\frac{1}{2}$. If we do not occupy it, we will have the charge conjugate state, with fermion number $-\frac{1}{2}$. It is wonderful how this delicate mechanism, discovered by Jackiw and Rebbi [1], harmonizes with the Schrieffer counting argument.

3.1.1. Zero Modes on Domain Walls

An abstract generalization of this set-up, with relativistic kinematics, is very simple, yet it has proved quite important. Consider massless, relativistic fermions in an odd number $2n+1$ of Euclidean dimensions, interacting with a scalar field ϕ according to $L_{\text{int.}} = g\phi\bar\psi\psi$ as before. Let $\langle\phi(z)\rangle = h(z)$ implement a domain wall, with $h(\pm\infty) = \pm v$. Thus away from the wall the fermion acquires mass^2 $m^2 = g^2v^2$. But, guided by previous experience, we might expect low-energy modes localized on the wall. Here we must look for solutions of the $2n+1$ dimensional Dirac equation that also satisfy the $2n$ dimensional Dirac equation. With the factorized form $\psi(x,z) = f(z)s(x)$, where f is a c-number and s a spinor satisfying the $2n$-dimensional Dirac equation, we must require

$$\gamma^{2n+1}\frac{\partial}{\partial z}f(z)s = -gh(z)s. \qquad (19)$$

For $s_\pm$ an eigenspinor of γ^{2n+1} with eigenvalue ± 1, this leads to

$$f_\pm(z) \propto e^{-\int_0^z dy(\pm gh(y))}. \qquad (20)$$

Only the upper sign produces a normalizable solution. Thus only a particular *chirality* of $2n$-dimensional spinor appears. This mechanism has been used to produce chiral quark fields for numerical work in QCD [7], avoiding the notorious doubling problem, and it has appeared in many speculations about the origin of chirality in Nature, as it appears in the Standard Model of particle physics.

A very much more intricate example of chiral zero modes on domain walls, in the context of superfluid He^3 in the A phase, is analyzed in the accompanying paper of Ho, Fulco, Schrieffer and Wilczek [8]. (Note the date!) A very beautiful spontaneous flow effect is predicted in that paper, deeply analogous (I believe) to the persistent flow of edge currents in the quantum Hall effect. I'm confident that we haven't yet heard the last words on this subject, neither theoretically nor experimentally.

3.2. Vacuum Polarization and Induced Currents

To round out the discussion, let us briefly consider a natural generalization of the previous model, to include two scalar fields ϕ_1, ϕ_2

and an interaction of the form

$$L_{\text{int.}} = g_1\phi_1\bar{\psi}\psi + g_2\phi_2\bar{\psi}\gamma^\chi\psi. \tag{21}$$

Gradients in the fields ϕ_1, ϕ_2 will induce non-trivial expectation values of the number current $j^\mu \equiv \bar{\psi}\gamma^\mu\psi$ in the local ground state. In the neighborhood of space-time points x where the local value of the effective mass^2, that is $g_1^2\phi_1^2 + g_2^2\phi_2^2$, does not vanish, one can expand the current in powers of the field gradients over the effective mass. To first order, one finds

$$\begin{aligned}\langle j_\mu \rangle &= \frac{1}{2\pi} \frac{g_1 g_2 (\phi_1 \partial_\mu \phi_2 - \phi_2 \partial_\mu \phi_1)}{g_1^2\phi_1^2 + g_2^2\phi_2^2} \\ &= \frac{1}{2\pi}\partial_\mu\theta \end{aligned} \tag{22}$$

where

$$\theta \equiv \arctan\frac{g_2\phi_2}{g_1\phi_1}. \tag{23}$$

We can imagine building up a topologically non-trivial field configuration adiabatically, by slow variation of the ϕs. As long as the effective mass does not vanish, by stretching out this evolution we can justify neglect of the higher-order terms. Flow of current at infinity is not forbidden. Indeed it is forced, for at the end of the process we find the accumulated charge

$$Q = \int j^0 = \frac{1}{2\pi}(\theta(\infty) - \theta(-\infty)) \tag{24}$$

on the soliton. This, of course, can be fractional, or even irrational. In appropriate models, it justifies Schrieffer's generalized counting argument [9].

Our previous model, leading to charge $\frac{1}{2}$, can be reached as a singular limit. One considers configurations where ϕ_1 changes sign with ϕ_2 fixed, and then takes $g^2 \to 0$. This gives $\Delta\theta = \pm\pi$, and hence $Q = \pm\frac{1}{2}$, depending on which side the limit is approached from.

4. Varieties of Charge

In physics, useful charges come in several varieties — and it seems that all of them figure prominently in the story of fractional charge. Having analyzed specific models of charge fractionalization, let us pause for a quick survey of the varieties of charge. This will both provide an opportunity to review foundational understanding of charge quantization, and set the stage for more intricate examples or fractionalization to come.

Deep understanding of the issues around charge quantization can only be achieved in the context of quantum field theory. Even the prior fact that there are many entities with rigorously identical properties, for example many identical electrons, can only be understood in a satisfactory way at that level.

4.1. Bookkeeping Charges

The simplest charges, conceptually, are based on counting. They encode strict, or approximate, conservation laws if the numbers thus calculated before and after all possible, or an appropriate class of, reactions are equal. Examples of useful charges based on counting are electric charge, baryon number, lepton number, and in chemistry 90+ laws expressing the separate conservation of number of atoms of each element.

Using operators ϕ_j to destroy, and their conjugates ϕ_j^+ to create, particles of type j with charge q_j, a strict conservation law is encoded in the statement that interaction terms

$$\Delta L_{\text{int.}} \sim \kappa \prod_m \phi_{j_m}^{k_{j_m}} \prod_n (\phi^+)_{j_n}^{k_{j_n}} \tag{25}$$

which fail to satisfy

$$\sum_m k_{j_m} q_{j_m} = \sum_n k_{j_n} q_{j_n} \tag{26}$$

do not occur. (In the first expression, already awkward enough, all derivatives and spin indices have been suppressed.) Alternatively, the Lagrangian is invariant under the abelian symmetry transformation

$$\phi_j \to e^{i\lambda q_j} \phi_j . \tag{27}$$

An approximate conservation law arises if the offending interactions occur only with small coefficients. One can also have discrete conservation laws, where the equality is replaced by congruence modulo some integer.

In all practical cases effective Lagrangians are polynomials of small degree in a finite number of fields. In that context, conservation laws of the above type, that forbid some subclass of terms, can always be formulated, without loss of generality, using integer values of the q_j. It will usually be simple and natural to do so. In a sense, then, quantization of charge is automatic. More precisely, it is a consequence of the applicability of local quantum field theory at weak coupling, which is what brought us to this class of effective Lagrangians.

Of course, the fact that we *can* always get away with integers does not mean that we *must* do so. For example, suppose we have a situation where there are two applicable conservation laws, with integer charges $q_j^{(1)}$, $q_j^{(2)}$ for particles of type j. If I define the master-charge

$$Q_j \equiv q_j^{(1)} + w q_j^{(2)}, \qquad (28)$$

with w irrational, then conservation of Q encodes both of the prior conservation laws simultaneously. This semi-trivial trick comes close to the heart of the fractional charge phenomenon, as exposed above.

4.1.1. Gauge Charges; Nonabelian Symmetry

Substantial physical issues, that are definitely not matters of convention, arise for conserved quantities that have independent dynamical significance. The prime example is electric charge, to which the electromagnetic field is sensitive.

Empirically, the electric charges of electrons and protons are known to be equal and opposite to within a part in 10^{-21}. Their cancellation occurs with fantastic accuracy despite the fact that the protons and electrons are very different types of particles, and in particular despite the fact that the proton is composite and is subject to the strong interaction. More generally, the accurate neutrality of all unionized atoms, not only hydrogen, can be tested with sensitive atomic beam experiments, and has never been found to fail.

Neither pure quantum electrodynamics nor its embedding into the Standard Model of matter explains why electrons and protons carry commensurate charges, though of course both theories can accommodate that fact. Specifically, either theory would retain its intellectual integrity if the photon coupled to a modified charge

$$\tilde{Q} = Q + \epsilon(B - L), \qquad (29)$$

where B is baryon number, L is lepton number, and ϵ a numerical coefficient, instead of the conventional charge Q. If ϵ is taken small enough, the modified theories will even continue to agree with experiment.

To produce a mandatory unit of charge, which cannot be varied by small amounts from particle to particle (or field to field), we must embed the abelian counting symmetry into a simple, nonabelian group. Unified gauge theories based on the gauge groups $SU(5)$ or $SO(10)$ accomplish this; moreover, they account nicely for the full spectrum of $SU(3) \times SU(2) \times U(1)$ quantum numbers for the particles observed in Nature [10]. This represents, at present, our best understanding of the origin of charge quantization. It indirectly incorporates Dirac's idea [11] that the existence of magnetic monopoles would force the quantization of charge, since these theories contain magnetic monopoles as regular solutions of the field equations [12, 13].

4.2. Topological Charges

Bookkeeping charges, as described above, reside directly in quantum fields, and from there come to characterize the small-amplitude excitations of these fields (that is, the corresponding particles). These particles are, at the level of the effective field theory, point-like. In addition to these objects, we can have collective excitations with a useful degree of stability, which then become significant, identifiable objects in their own right. These are usually associated with topological properties of the fields, and are generically called solitons. Of course, at a higher level of description, solitons themselves might be regarded as primary ingredients in an effective theory.

Solitons fall into two broad classes, boundary solitons and texture solitons. Boundary solitons are associated with non-trivial structure at spatial infinity. Domain walls in polyacetylene, as discussed above, provide a simple example. Texture solitons are associated with non-trivial mapping of space as a whole into the target field configuration space, with trivial structure at infinity. A simple example is a phason in 1 space dimension. To see it, specialize our two-field model by requiring $\phi_1^2 + \phi_2^2 = $ constant. Then the target space for the field θ is a circle, and a field configuration that starts with $\theta = 0$ on the far left and winds continuously to $\theta = 2\pi$ on the far right has non-trivial topology as a mapping over all space, though none at the boundaries. Skyrmions [14, 15] provide a higher-dimensional generalization of this type. Texture solitons can be produced by local operations, but generally not by means of a finite number of field operations (creation and destruction operators) so their topological quantum numbers can also appear in fractionalization formulae.

4.3. Space-Time and Identity Charges

Each of the charges we have discussed so far can be considered as a label for representations of some symmetry group. This is obvious for bookkeeping charges, which label representations of phase groups; it is also true for topological charges, which label representations of homotopy groups. There are also symmetry groups associated with space-time transformations, specifically rotations, and with interchange of identical particles. And there are corresponding quantum numbers. For rotations this is, of course, spin. For identity it is fermi versus bose character — an additive, Z_2 quantum number. These quantum numbers are quite familiar to all physicists. Less familiar, and perhaps unsettling on first hearing, is the idea that they can be dequantized. Let's focus on that now.

4.3.1. Space-Time Charges

In three space dimensions, rotations generate the nonabelian group $SO(3)$. The quantization of spin, in integer units, follows from that fact. Actually, not quite — we prove too much, because there are particles with half-integer spin. The point is that quantum mechanics only requires that symmetry groups are implemented "up to a phase," or, in the jargon, projectively. If the unitary transformation associated with a symmetry generator g is $U(g)$, then we need only have

$$U(g_1)U(g_2) = \eta(g_1, g_2)U(g_1 g_2), \qquad (30)$$

where $\eta(g_1, g_2)$ is a phase factor, since observables, based on inner products, will not depend on η. It turns out that projective representations of $SO(3)$ correspond to ordinary representations of $SU(2)$, so one still has quantization, but in half-integer units.

In two space dimensions the group is $SO(2)$. We can parametrize its elements, of course, in terms of an angle θ, and its irreducible representations by the assignments $U(\theta) = e^{is\theta}$, for $0 \leq \theta < 2\pi$. These are ordinary representations only if s is an integer; but they are perfectly good projective representations for any value of s. Thus in two space dimensions angular momentum is dequantized.

4.3.2. Identity Charges

Among all quantum-mechanical groups, perhaps the most profound is the symmetry group associated with interchange of identical particles. For the existence of this symmetry group, reflected in the existence of quantum statistics and associated exchange phenomena, lets us reduce the number of independent entities we need to describe matter decisively.

We teach undergraduates that quantum statistics supplies symmetry conditions on the wave function for several identical particles: the wave function for bosons must not change if we interchange the coordinates of two of the bosons, while the wave function for fermions must be multiplied by -1 if we exchange the coordinates of two of the fermions. If the interchange of two particles is to be accompanied by a fixed phase factor $e^{i\theta}$, it would seem that this factor had better be ± 1, since iterating the exchange must give back the original wave function.

Nevertheless we can make sense of the notion of fractional statistics; but to do so we must go back to basics [16].

In quantum mechanics we are required to compute the amplitude for one configuration to evolve into another over the course of time. Following Feynman, this is done by adding together the amplitudes for all possible trajectories (path integral). Of course the essential dynamical question is: how are we to weight the different paths? Usually, we take guidance from classical mechanics. To quantize a classical system with Lagrangian L we integrate over all trajectories weighted by their classical action $e^{i \int L dt}$. However, essentially new possibilities arise when the space of trajectories falls into disconnected pieces. Classical physics gives us no guidance as to how to assign relative weights to the different disconnected pieces of trajectory space. For the classical equations of motion are the result of comparing infinitesimally different paths, and supply no means to compare paths that cannot be bridged by a succession of infinitesimal variations.

The space of trajectories of identical particles, relevant to the question of quantum statistics, does fall into disconnected pieces. Suppose, for example, that we wish to construct the amplitude to have particles at positions $x_1, x_2, \ldots$ at time t_0 and again at time t_1. The total amplitude gets contributions not only from trajectories such that the particle originally at x_1 winds up at x_1, but also from trajectories where this particle winds up at some other x_k, and its place is taken up by a particle that started from some other position. All permutations of identity between the particles in the initial and final configurations, are possible. Clearly, trajectories that result in different permutations cannot be continuously deformed into one another. Thus we have the situation mentioned above, that the space of trajectories falls into disconnected pieces.

Although the classical limit cannot guide us in the choice of weights, there is an important consistency condition from quantum mechanics that severely limits the possibilities. We must respect the rule, that if we follow a trajectory α_{01} from t_0 to t_1 by a trajectory α_{12} from t_1 to t_2, then the amplitude assigned to the combined trajectory α_{02} should be the product of the amplitudes for α_{01} and α_{12}. This rule is closely tied up with the unitarity and linearity of quantum mechanics — i.e., with the probability interpretation and the principle of superposition — and it would certainly be very difficult to get along without it. The rule is automatically obeyed by the usual Feynman expression for the amplitude as the exponential of i times the classical action.

So let us determine the disconnected pieces, into which the space of identical particle trajectories falls. We need consider only closed trajectories, that is trajectories with identical initial and final configurations, since these are what appear in inner products. To begin with, let us focus on just two particles.

In *two* spatial dimensions, but not in any higher number, we can unambiguously define the angle through which one particle moves with respect to the other, as they go through the trajectory. It will be a multiple of π; an odd multiple if the particles are interchanged, an even multiple if they are not. Clearly the angle adds, if we follow one trajectory by another. Thus a weighting of the trajectories, consistent with the basic rule stated in the preceding paragraph, is

$$\rho(\alpha) = e^{i\theta\phi/\pi}, \tag{31}$$

where ϕ is the winding angle, and θ is a new parameter. As defined, θ is periodic modulo 2π. In three or more dimensions, the change in the angle ϕ cannot be defined unambiguously. In these higher dimensions it is only defined modulo 2π. In three or more dimensions, then, we must have $e^{i\theta\phi/\pi} = e^{i\theta\phi'/\pi}$ if ϕ and ϕ' differ by a multiple of 2π. So in three or more dimensions we are essentially reduced to the two cases $\theta = 0$ and $\theta = \pi$, which give a factor of unity or a minus sign respectively for trajectories with interchange. Thus in three dimensions we find only the familiar cases — bosons and fermions — of quantum statistics, and demonstrate that they exhaust the possibilities.

In two space dimensions, however, we see that there are additional possibilities for the weighting of identical particle paths. Particles carrying the new forms of quantum statistics, are called generically *anyons*.

Passing to N particles, we find that in three or more dimensions the disconnected pieces of trajectory space are still classified by permutations. With the obvious natural rule for composing paths (as used in our statement of the consistency requirement for quantum mechanics, above), we find that the disconnected pieces of trajectory space correspond to elements of the permutation group P_n. Thus the consistency rule, for three or more dimensions, requires that the weights assigned amplitudes from different disconnected classes must be selected from some representation of the group P_n.

In two dimensions there is a much richer classification, involving the so-called braid group B_n. The braid group is a very important mathematical object. The elements of the braid group are the disconnected pieces of trajectory space. The multiplication law, which makes it a group, is simply to follow one trajectory from the first piece, by another from the second piece — their composition lands in a uniquely determined piece of trajectory space, which defines the group product. The "braid" in braid group evidently refers to the interpretation of the disconnected pieces of trajectory space as topologically distinct methods of styling coils of hair.

It may be shown that the braid group for n particles is generated by $n-1$ generators σ_k satisfying the relations

$$\sigma_j \sigma_k = \sigma_k \sigma_j, \quad |j-k| \geq 2$$

$$\sigma_j \sigma_{j+1} \sigma_j = \sigma_{j+1} \sigma_j \sigma_{j+1}, \quad 1 \leq j \leq n-2. \tag{32}$$

The σs generate counterclockwise permutations of adjacent particles (with respect to some fixed ordering). Thus in formulating the quantum mechanics of identical particles, we are led to consider representations of P_n — or, in two spatial dimensions, B_n. The simplest representations are the one-dimensional ones. These are anyons with parameter θ, as previously defined. Higher-dimensional representations correspond to particles with some sort of internal degree of freedom, intimately associated with their quantum statistics.

This discussion of fractional statistics has been at the level of quantum particle kinematics. Their implementation in quantum field theory uses the so-called Chern–Simons construction. This was spelled out for the first time in the paper of Arovas, Schrieffer, Wilczek and Zee [17].

5. Fractional Quantum Numbers with Abstract Vortices

For reasons mentioned before, two-dimensional systems provide an especially fertile source of fractionalization phenomenon. In this section I'll discuss an idealized model that exhibits the salient phenomena in stripped-down form.

Consider a $U(1)$ gauge theory spontaneously broken to a discrete Z_n subgroup. In other words, we imagine that some charge ne field ϕ condenses, and that there are additional unit charge particles, produced by a field ψ, in the theory. The case $n=2$ is realized in ordinary BCS superconductors, where the doubly charged Cooper pair field condenses, and there are additional singly charged fields to describe the normal electron (pair-breaking) excitations.

Such a theory supports vortex solutions [18, 19], where the ϕ field behaves asymptotically as a function of the angle θ as

$$\phi(r,\theta) \to v e^{i\theta}, \; r \to \infty \tag{33}$$

where v is the value of ϕ in the homogeneous ground state. With this asymptotics for ϕ we must have for the gauge potential

$$A_\theta(r,\theta) \to \frac{1}{ne} \tag{34}$$

in order that the covariant derivative $D_\theta \phi = (\partial_\theta \phi - ineA_\theta)$, which appears (squared) in the energy density will vanish at infinity. Otherwise the energy would diverge.

In this set-up the magnetic field strength $B = \nabla \times A$ vanishes asymptotically. Indeed, since the fields transform as

$$\phi'(x) = \exp(iQ\Lambda(x))\phi(x) = \exp(ine\Lambda(x))\phi(x)$$
$$A'_\mu(x) = A_\mu(x) + \partial_\mu \Lambda(x) \tag{35}$$

under a gauge transformation we can, by choosing $\Lambda = -\theta/ne$, remove the space dependence of ϕ and make A_θ vanish altogether. We have, it appears, transformed back to the homogeneous ground state. However Λ is not quite a kosher gauge transformation, because the angle θ is not a legitimate, single-valued function.

The correct formulation is that the vortex asymptotics is trivial, and can be gauged away, locally but *not* globally. Since we can pick a well defined branch of θ in any patch that does not surround the origin, all local gauge invariant quantities must reduce to their ground state values (this explains why $D\phi$ and F vanish). But the line integral $\oint A \cdot dl$ of A around a closed loop surrounding the origin, which by Stokes' theorem measures the flux inside, cannot be changed by any legitimate gauge transformation. And it is definitely *not* zero for the vortex; indeed we have the enclosed flux $\Phi = \oint A \cdot dl = \frac{2\pi}{ne}$.

Another aspect of the global non-triviality of the vortex, is that our putative gauge transformation $\Lambda = -\theta/ne$ transforms a unit charge field ψ into something that is not single-valued. Since

$$\psi'(x) = \exp(i\Lambda(x))\psi(x) \tag{36}$$

we deduce

$$\psi'(\theta + 2\pi) = \exp\left(-\frac{2\pi i}{n}\right)\psi'(\theta). \tag{37}$$

Now let us discuss angular momentum. Superficially, vortex asymptotics of the scalar order parameter seems to trash rotational invariance. Indeed a scalar field like ϕ should be unchanged by a rotation, but $ve^{i\theta}$ acquires a phase. However we must remember that the phase of ϕ is gauge dependent, so we can't infer that any *physical* property of the vortex violates rotation symmetry. Indeed, it is easy to verify that if we supplement the naive rotation generator J_z with an appropriate gauge transformation

$$K_z = J_z - \frac{Q}{ne} \tag{38}$$

then K_z leaves both the action, and the asymptotic scalar field configuration of the vortex invariant.

Thus, assuming that the core is invariant, K_z generates a true rotation symmetry of the vortex. If the core is not invariant, the solution will have a finite moment of inertia, and upon proper quantization we will get a spectrum of rotational excitations of the vortex, similar to the band spectrum of an asymmetric molecule. That step, of course, does not introduce any fractional angular momentum.

For present purposes, the central point is that passing from J to K modifies the quantization condition for angular momentum of quanta orbiting the vortex. In general, their orbital angular momentum becomes fractional. The angular momentum of quanta with the fundamental charge e, for example, is quantized in units of $-\frac{1}{n}$ + integer.

In two space dimensions the object consisting of a vortex together with its orbiting electron will appear as a particle, since its energy-momentum distribution is well localized. But it carries a topological charge, of boundary type. That is the secret of its fractional angular momentum.

The general connection between spin and statistics suggests that objects with fractional angular momentum should likewise carry fractional statistics. Indeed there is a very general argument, the ribbon argument of Finkelstein and Rubenstein [20], which connects particle interchange and particle rotation. The space-time process of creating two particle-antiparticle pairs, interchanging the two particles, and finally annihilating the re-arranged pairs, can be continuously deformed into the process of creating a pair, rotating the particle by 2π, and finally annihilating. Therefore, in a path integral, these two processes must be accompanied by the same non-classical phase. This reasoning leads to the equation

$$P = e^{2\pi i S}, \tag{39}$$

where S is the spin and P is the phase accompanying (properly oriented) interchange. This gives the ordinary spin-statistics connection in $3+1$ space-time dimensions, in a form that generalizes to anyons. For our vortex-ψ composites, it is easy to see how the funny phase arises. It is a manifestation of the Aharonov–Bohm effect [21]. Transporting charge e around flux $1/ne$ — or, for interchange, half-transporting two such charges around one another's fluxes — accumulates non-classical phase $2\pi/n$.

6. Fractional Quantum Numbers in the Quantum Hall Effect

Microscopic understanding of the fractional quantum Hall effect has been built up from Laughlin's variational wave function, just as microscopic understanding of superconductivity was built up from the BCS variational wave function [22, 23]. To be concrete, let us consider the $\frac{1}{3}$ state. The ground state wave function for N electrons in a droplet takes the form

$$\psi(z_1, \ldots, z_N) = \prod_{i<j}(z_i - z_j)^3 \prod_i \exp(-|z_i|^2/l^2) \tag{40}$$

where $l^2 \equiv \frac{4}{eB}$ defines the magnetic length, and we work in symmetric gauge $A_x = -\frac{1}{2}By$, $A_y = +\frac{1}{2}Bx$.

The most characteristic feature of this wave function is its first factor, which encodes electron correlations. Through it, each electron repels other electrons, in a very specific (holomorphic) way that allows the wave function to stay entirely within the lowest Landau level. Specifically, if electron 1 is near the origin, so $z_1 = 0$, then the first factor contributes $\prod_{1<i} z_i^3$. This represents, for each electron, a boost of three units in its angular momentum around the origin. (Note that in the lowest Landau level the angular momentum around the origin is always positive.)

Such a universal kick in angular momentum has a simple physical interpretation, as follows. Consider a particle of charge q orbiting around a thin solenoid located along the $\hat{z}$ axis. Its angular momentum along the $\hat{z}$ axis evolves according to

$$\frac{dL}{dt} = qrE_\phi$$
$$= \frac{q}{2\pi}\frac{d\Phi}{dt} \qquad (41)$$

where E_ϕ is the value of the azimuthal electric field and Φ is the value of the flux through the solenoid; the second equation is simply Faraday's law. Integrating this simple equation we deduce the simple but profound conclusion that

$$\Delta L = \frac{1}{2\pi}\Delta(q\Phi). \qquad (42)$$

The change in angular momentum is equal to the change in the flux times charge. All details about how the flux was turned on cancel out.

From this point of view, we see that in the $\frac{1}{3}$ state each electron implements correlations as if it were a flux tube with flux $3\frac{2\pi}{q}$. This is three times the minimal flux. Now let us follow Schrieffer's idea, as previously discussed for polyacetylene, and remove the electron. Doing that produces a hole-like defect, but one that evidently, as in polyacetylene, begs to be broken into more elementary pieces. Either from the flux point of view, or directly from the wave function, it makes sense to break consider an elementary quasi-hole of the type

$$\psi(z_2,\ldots,z_N) = \prod z_i \prod_{i<j}(z_i - z_j)^3 \prod_i \exp(-|z_i|^2/l^2). \qquad (43)$$

(Note that electron 1 has been removed.) The first factor represents the defect. By adding three defects and an electron at the origin we get back to the ground state. Thus the elementary quasihole will carry charge $-e/3$.

Here again the Schrieffer counting argument is correct and utterly convincing, but a microscopic derivation adds additional insight. It is given in the paper by Arovas, Schrieffer, and Wilczek [24], through an orchestration of Berry's phase and the Cauchy integral theorem.

At another level of abstraction, we can use statistical transmutation (the Chern–Simons) construction to model electrons as vortices of a fictitious gauge field. This startling reinvention of the prototypical "elementary particle" leads to a profound insight into the nature of the quantum Hall effect. It ties together most of what we've discussed, and provides an appropriate climax.

A constant magnetic field frustrates condensation of electrically charged particles, because the gradient energy

$$\int |\partial \eta - iq_{\text{el.}} Ae\eta|^2 \sim (qe)^2 |\langle \eta \rangle|^2 \int A^2 \qquad (44)$$

grows faster than the volume, due to the growth of A, and therefore cannot be sustained. This is the deepest theoretical explanation of the expulsion of magnetic field, which is the soul of superconductivity (the Meissner effect). If each particle acts as a source of fictitious charge and flux, however, then the long-range part of the total potential $q_{\text{el.}}eA + q_{\text{fict.}}a$ will vanish, and the frustration will be removed, if

$$q_{\text{el.}}eB + q_{\text{fict.}} n_\eta \Phi_{\text{fict.}} = 0, \qquad (45)$$

where n_η is the number-density of η quanta and $\Phi_{\text{fict.}}$ is the fictitious flux each carries. Given $\frac{q_{\text{el.}}eB}{n}$ — that is to say, a definite filling fraction — a definite value of $q_{\text{fict.}}\Phi_{\text{fict.}}$ is implied. But it is just that parameter which specifies how the effective quantum statistics of the η quanta have been altered by their fictitious gauge charge and flux. Condensation will be possible if — and only if — the altered statistics is bosonic. Identifying the η quanta as electrons, i.e. fermions, we therefore require

$$q_{\text{fict.}}\Phi_{\text{fict.}} = (2m+1)\pi \qquad (46)$$

with m integral, to cancel the fermi statistics. We also have $\frac{q_{\text{el.}}eB}{n} = \frac{eB}{n} = \frac{\pi}{\nu}$, for filling fraction ν. Thus we derive

$$\frac{1}{\nu} = 2m+1, \qquad (47)$$

accounting for the primary Laughlin states.

These connections among superconductivity, statistical transmutation, and the quantum Hall effect can be extended to bring in anyon superconductivity [25, 26] and composite fermions [27, 28]; tightened into what I believe is a physically rigorous derivation of the quantum Hall complex, using adiabatic flux trading [29]; and generalized to multi-component systems (to describe multilayers, or states where both directions of spin play a role) [30], and more complicated orderings, with condensation of pairs [31, 32]. In this field, as in many others, the fertility of Bob Schrieffer's ideas has been invigorated, rather than exhausted, with the harvesting.

Acknowledgments

This work is supported in part by funds provided by the U.S. Department of Energy (D.O.E.) under cooperative research agreement No. DF-FC02-94ER40818. I would like to thank Reinhold Bertlmann and the University of Vienna, where this work was completed, for their hospitality.

References

1. R. Jackiw and C. Rebbi, *Phys. Rev.* **D13**, 3398 (1976).
2. W. P. Su, J. R. Schrieffer, and A. J. Heeger, *Phys. Rev. Lett.* **42**, 1698 (1979); *Phys. Rev.* **B22**, 2099 (1980).
3. W. P. Su and J. R. Schrieffer, *Phys. Rev. Lett.* **46**, 738 (1981).
4. R. Jackiw and J. R. Schrieffer, *Nucl. Phys.* **B190**, 253 (1981).
5. R. F. Peierls, *Quantum Theory of Solids* (Clarendon Press, Oxford, 1955).
6. S. Coleman and E. J. Weinberg, *Phys. Rev.* **D7**, 1888 (1973).
7. D. B. Kaplan, *Phys. Lett.* **B301**, 219 (1993).
8. T. L. Ho, J. R. Fulco, J. R. Schrieffer, and F. Wilczek, *Phys. Rev. Lett.* **52**, 1524 (1984).
9. J. Goldstone and F. Wilczek, *Phys. Rev. Lett.* **47**, 986 (1981).
10. H. Georgi and S. Glashow, *Phys. Rev. Lett.* **32**, 438 (1974); see also J. Pati and A. Salam, *Phys. Rev.* **D8**, 1240 (1973).
11. P. A. M. Dirac, *Proc. R. S. London* **A133**, 60 (1931).
12. G. 't Hooft, *Nucl. Phys.* **B79**, 276 (1974).
13. A. Polyakov, *JETP Lett.* **20**, 194 (1974).
14. T. H. R. Skyrme, *Proc. R. S. London* **260**, 127 (1961).
15. Reviewed in R. Rajaraman, *Solitons and Instantons* (North-Holland, 1982).
16. J. M. Leinaas and J. Myrheim, *Il Nuovo Cimento* **50**, 1 (1977); F. Wilczek, *Phys. Rev. Lett.* **48**, 1144 (1982); G. A. Goldin, R. Menikoff, and D. H. Sharp, *Phys. Rev. Lett.* **51**, 2246 (1983).
17. A. Arovas, J. R. Schrieffer, F. Wilczek, and A. Zee, *Nucl. Phys.* **B251**, 917 (1985).

18. A. A. Abrikosov, *Zh. Eksp. Teor. Fiz.* **32**, 1442 (1957).
19. H. B. Nielsen and P. Olesen, *Nucl. Phys.* **B61**, 45 (1973).
20. D. Finkelstein and J. Rubinstein, *J. Math. Phys.* **9**, 1762 (1968).
21. Y. Aharonov and D. Bohm, *Phys. Rev.* **115**, 485 (1959).
22. R. B. Laughlin, *Phys. Rev. Lett.* **50**, 1395 (1983).
23. Reviewed in R. Prange and S. Girvin, *The Quantum Hall Effect* (Springer Verlag, 1990).
24. D. Arovas, J. R. Schrieffer, and F. Wilczek, *Phys. Rev. Lett.* **53**, 2111 (1984).
25. R. B. Laughlin, *Science* **242**, 525 (1988); A. L. Fetter, C. B. Hanna, and R. B. Laughlin, *Phys. Rev.* **B39**, 9679 (1989).
26. Y.-H. Chen, F. Wilczek, E. Witten, and B. I. Halperin, *Int. J. Mod. Phys.* **B3**, 903 (1989).
27. J. K. Jain, *Phys. Rev. Lett.* **63**, 199 (1989).
28. B. I. Halperin, P. A. Lee, and N. Read, *Phys. Rev.* **B47**, 7312 (1993).
29. F. Wilczek, *Int. J. Mod. Phys.* **B5**, 1273 (1991); M. Greiter and F. Wilczek, *Mod. Phys. Lett.* **B4**, 1063 (1990).
30. X. G. Wen and A. Zee, *Phys. Rev.* **B46**, 2290 (1992).
31. G. Moore and N. Read, *Nucl. Phys.* **B360**, 362 (1991).
32. M. Greiter, X. G. Wen, and F. Wilczek, *Phys. Rev. Lett.* **66**, 3205 (1991).

Grand Occasions

Often awardees are asked to sing for their supper. It's a dangerous art form, poised to descend into ego-tripping and claim staking. But the more insidious danger is simply to bore with cliché and nostalgia. How to avoid these problems? Here's the advice I give myself: Look to the future as well as the past, submerge your ego into larger contexts, and keep it short.

"From Concept to Reality to Vision" was given in receipt of the high-energy physics prize of the European Physical Society. It contains brief narratives of the (unfinished) history of three lines of thought. The first, the fluidity of fundamental "constants," already featured in "Unification of Couplings." The second line of thought, narrowly viewed, concerns the deep structure of the proton: specifically, how the proton comes to appear, at the highest space-time resolutions, as a glob of colored glue. Of course other, superficially very different, pictures of what a proton is are used in chemistry, nuclear physics, and even hadron spectroscopy. That picture-album, I think, provides a most impressive example of the important philosophical insight that very different descriptions can apply to the same object, when different questions are asked of it. If I hadn't kept in mind the injunction to keep it short, I might have launched from there into a discussion of the possibility of religious versus scientific world-views; but I did, and so I didn't. The third line, technically, has to do with

QCD in extreme conditions. But here I'd like to liberate from technicalities a principle that I've found widely applicable and extremely useful. It is the Jesuit Credo. I learned from James Malley, S. J., who encountered it during his training as a seminarian. The Jesuit Credo is:

> It is more blessed to ask forgiveness than permission.

The "Nobel Biography" speaks for itself; it accompanies my acceptance lecture in the official book of the Nobel Foundation (*Les Prix Nobel 2004*). "Asymptotic Freedom: From Paradox to Paradigm" is the lecture. I had a very difficult time starting to prepare my Nobel lecture. It's an awesome assignment, and I kept changing my mind about how to approach it. As the delivery date approached I got polite but increasingly forceful requests for a title. Under that pressure I had to stop thinking Big Vague Thoughts and find a few specific words. That turned out to be very helpful, because the title imposed a natural, compelling structure on the material. In fact it became a joy to write. When Mark Twain wrote

> Science is a wonderful thing. You can squeeze so much conjecture from such a small deposit of fact.

he meant to be poking fun. But sometimes the conjectures pan out!

"Advice to Students" was given at the Academy of Achievement Summit in New York, where high-voltage luminaries were assembled to have a weeklong party and stir up a remarkable group of students (Rhodes and Marshall scholars, and the like). Yogi Berra, after being introduced by Roger Bannister (really!), gave the most profound advice, incorporating the many-worlds interpretation of quantum mechanics:

> When you come to a fork in the road, take it.

Many of the starriest luminaries were vague and windy, but as a lesser light I had the twin advantages of a very limited amount of time and no reputation to uphold, so I told a joke and gave some terse but genuinely useful marriage counseling.

From Concept to Reality to Vision[*]

I take a brief look at three frontiers of high-energy physics, illustrating how important parts of our current thinking evolved from earlier explorations at preceding frontiers.

Decoding the strong interaction was a vast enterprise to which many gifted scientists devoted their best efforts and made major contributions. While the subject is far from finished — the dramatic developments I'll be discussing this afternoon [1] bring that home! — I think it is clear that the foundations are secure. QCD, as the basic theory, is here to stay. It is a marvelous theory, which cleanly embodies mathematical ideas of great depth and beauty. Above all QCD demonstrates, in a unique way, the power of relativistic quantum field theory to produce an amazing wealth of phenomena (asymptotic freedom, jets, confinement, mass generation, resonance spectroscopy, chiral symmetry breaking, anomaly dynamics,…) in harmony with the observed facts of Nature.

David Gross has just described for you the whirlwind of events and discoveries that led us to propose this theory for the strong interaction, reinforced with concrete reasons to believe in it (and no other!), and packaged with proposals for critical, quantitative experimental tests. I don't want to repeat the details, but only want to endorse what David has already emphasized, that he and I were fortunate indeed to be in a position to leverage a vast accumulation of knowledge and technique built up

---

[*]Speech in acceptance of the European Physical Society prize for high-energy physics, Aachen, August 2003.

by a big international community of scientists over decades of dedicated work, much of it frustrating and not properly recognized. As members of this community we should all be proud of our joint achievement.

I'll freely admit that back in 1973 I didn't begin to anticipate the progress in experiment and theory that would bring our subject to the level where it is today. I had some hope that deep inelastic scattering experiments and perhaps measurements of electron-positron annihilation (the total cross section) would be made more precise, maybe precise enough that with careful analysis one would see hints of scaling deviations in the form we predicted, and thereby gradually build up a case for the correctness of QCD. Of course, reality has far outrun these expectations. One of the great joys of my life in physics has been to participate in the process — something like parenthood — whereby unshaped concepts mature in surprising ways into concrete realities, which then engender new visions. I'd like briefly to share with you three examples, in each case mixing a little nostalgia with pointers to the future.

1. From Running Coupling to Quantitative Unification to Supersymmetry

Running of gauge theory couplings, and in particular the peculiar antiscreening behavior we call asymptotic freedom, was first established by straight unguided calculation [2, 3]. It was first applied to renormalization group equations for deep Euclidean Green's functions and Wilson coefficients in operator product expansions, enabled through a rather cumbersome formalism to describe a very few physical processes [4, 5, 6]. Before long, antiscreening was understood in terms accessible to intuition [7]. And by now of course we've learned to exploit the concept much more boldly, and with great success. Quark and gluon degrees of freedom are identified directly in the energy-momentum flow of jets, and their basic couplings are made manifest in the iconic three-jet processes seen at LEP. The changing probability of such events, at different energies, exhibits the running in as clear and elementary a form as one could ask for.

The calculation of that running, of course, extends immediately to electroweak interactions. (Indeed, my original interest in it largely arose from that angle.) It was put to brilliant use in the famous work of Georgi, Quinn, and Weinberg [8], who indicated through its use how dreams about unification of interactions (Pati and Salam [9], Georgi and Glashow [10]) could be brought down to earth. One could check concretely whether the observed, unequal couplings might result from running a single coupling from ultra-short to accessible distances. A few years later Dimopoulos, Raby, and I realized [11] (to my great surprise, initially) that including the effects of low-energy supersymmetry, which is quite a drastic expansion of the physics, makes only comparatively small changes in the predictions that emerge from this sort of calculation. Precision experiments and improved calculations appear to endorse these dreams and ideas, in their supersymmetric version.

Unless this is a cruel tease on the part of Mother Nature, it means we can look forward to a lot of fun exploring supersymmetry, and maybe some aspects of unification, at the LHC. An especially poetic possibility is that other sorts of parameters, besides gauge couplings, derive by running from a unified value [12]. It is widely speculated that the masses of different sorts of gauginos, or of squarks and sleptons, might be related in this way.

2. From Dark Momentum to Gluonization to Higgs and Dark Matter

Feynman interpreted the famous SLAC experiments on deep inelastic scattering using an intuitive model of nucleons that postulated point-like particles (partons) as nucleon constituents and treated their dynamics in a crude impulse approximation, ignoring both interactions and quantum interference [13]. Identifying the partons as quarks, and building the weak and electromagnetic currents by minimal coupling to quarks, led to many successful predictions [14]. There was, however, one clear failure. The momentum carried by quarks inside a fast-moving proton does not add up to the total momentum of the proton, in fact it is less than half.

Today's "dark matter" problem in astronomy is reminiscent of this old "dark momentum" problem. In the mathematical treatment of deep inelastic scattering, the analogy becomes eerily precise. In that framework, a (failed) sum rule expresses the equality of the full energy-momentum tensor with the energy-momentum tensor constructed from quarks [5, 6]. Putting it differently where electroweak currents see just quarks, gravitons see more! We realized early on [5, 6] that including the color gluons of QCD, which are electroweak singlets but do carry energy-momentum, would enable us to keep the good predictions while losing the bad one. Evidently the gluons had to be major, though "dark" (or better: invisible), constituents of the proton.

Our analysis of deep inelastic scattering, which followed pioneering ideas of Wilson [15], and built on the insightful hard work of Christ, Hasslacher, and Mueller [16], went beyond the parton model in other ways too. Our equations predicted that an energetic quark appears, to probes at higher resolution (higher Q^2), to be composed of less energetic (smaller x) quarks, antiquarks and gluons, which in turn will resolve into more, still softer stuff. This changeable appearance seen experimentally as evolution of structure functions, is deeply characteristic of quantum field theory.

These evolution effects enhance the role of glue in the proton. Several of us worked out that there should be a dramatic pile-up of soft stuff, particularly soft glue, at small x [17]. Thus to a hard current (indirectly), or to a hard graviton (theoretically), the proton mostly looks like a blob of soft glue. Twenty years later, beautiful work at HERA confirmed these predictions in impressive detail [18].

Very soft or "wee" constituents of protons played a major role in Feynman's ideas about diffractive scattering [19]. His idea was that in diffractive scattering, by exchange of wee partons, the relative phases between different multiparton configurations in the proton wave function get disrupted, without much transfer of energy-momentum. These ideas are intuitively appealing, and have inspired some successful phenomenology, but as far as I know they haven't yet been firmly rooted in QCD.

Much better understood — I hope! — is the importance of gluonization for some frontier topics in high-energy physics, namely Higgs particle production and WIMP searches. (WIMP stands for dark matter made from Weak Interacting Massive Particle. The name's not my doing.) The primary, classical coupling of Higgs particles is to quarks, proportional to their mass. But because the u and d quarks we mainly find inside nucleons are so light, their direct coupling is heavily suppressed. Instead the most important coupling arises indirectly, as a quantum effect, through virtual top quark loops connecting to two gluons [20].

I was originally interested in this Higgs-gluon vertex for its potential to induce Higgs particle decays. Georgi, Glashow, Machacek, and Nanopoulos [21] quickly realized it could be exploited for production of Higgs particles, at hadron colliders, through "gluon fusion." That process, which of course relies completely on the glue content of protons, is expected to be the main production mechanism for Higgs particles at the LHC. It is important to calculate the production rate accurately, including good estimates of the gluon distribution functions, so that we will be able to interpret the observed production rate, and check whether the basic vertex is in fact what the standard model, in this intricate way, predicts.

The physical Higgs particle decays into very energetic gluons. In its decay we will see jets. Conversely, we can calculate its rate of production in proton-proton collisions, using gluon structure functions and perturbative QCD. When considering detection of the sorts of dark-matter candidates provided by models of low-energy supersymmetry (SUSY) we find ourselves a different domain. Since these WIMPs will be heavy and slowly moving (by particle physics standards), they will scatter at very small momentum transfer. The coupling of SUSY WIMPs to matter depends on poorly constrained details of the models, but in many realizations it is dominated by virtual Higgs exchange. Here the Higgs-gluon vertex comes in at essentially zero energy-momentum. Shifman, Vainshtein and Zakharov [22], in beautiful work, related the relevant gluon operator to the trace of the energy-momentum tensor, whose matrix elements are known. The very same operator appeared in the old dark momentum problem, bringing us full circle.

It is philosophically profound, and quite characteristic of modern physics, that even when viewing something so basic and tangible as a proton, what you see depends very much on how you choose to look. Low-energy electrons see point-like particles, the version described in old high-school textbooks; hard currents see an evolving pattern of quarks; gravitons see these plus lots of gluons as well; wee gluons see some complicated stuff we don't understand properly; real Higgs particles see gluons almost exclusively; and WIMPS, through exchange of virtual Higgs particles, see "The Origin of Mass." (The trace of the energy momentum tensor, to which they mainly couple, is on the one hand dominated by contributions from massless color gluons and nearly massless quarks, and on the other hand equal to the nucleon mass.) Each probe reveals different aspects of versatile reality.

3. From Asymptotic Simplicity to Quark-Gluon Plasma to Quark-Hadron Continuity

I mentioned earlier how we've learned to use the concept of asymptotic freedom more boldly over the years. To put it differently, we've learned fruitful ways to lower our standards. Instead of trying to prove directly from first principles that weak coupling applies, we usually content ourselves with consistency checks. That is, we tentatively assume that perturbative calculation of some quantity of interest starting with quark and gluon degrees of freedom will be adequate, and check whether the calculation makes sense [23]. This check is by no means trivial, since QCD is full of massless (color) charged particles that can lead to infinite answers (infrared divergences). So in cases where we find there are no infrared divergences we declare a well-earned victory, and anticipate that our calculations will approximate reality. This strategic compromise has supported a host of successful applications to describe jet processes, inclusive production, fragmentation, heavy quark physics, and more.

We aren't always forced to compromise. In some important applications, including low-energy spectroscopy, direct integration of the equations using the techniques of lattice gauge theory is practical. But as physicists hungry for answers, we properly regard strict mathematical rigor as a desirable luxury, not an indispensable necessity.

A particularly interesting and important application of the looser philosophy is to construct self-consistent descriptions of extreme states of matter, starting from quarks and gluons [24].

The high temperature, low baryon number regime is foundational for very early universe cosmology. It is also the object of an intense, international experimental program in relativistic heavy ion physics. The overarching theme is that a description of high-temperature matter, starting with free quarks and gluons, becomes increasingly accurate as the temperature increases. This can be seen, for the equation of state, from numerical simulation of the full theory [25]. After heroic calculations, which introduce several ingenious new techniques, controlled perturbative calculations (including terms up to sixth order in the coupling, and some infinite resummations) match the numerical work [26]. This is a milestone achievement in itself, and also promising for future developments, since the perturbative techniques are more flexible. They might be applied, for example, to calculate viscosity and energy loss, which experiments can probe. In this way, we can hope to do justice to the vision of quark-gluon plasma.

The regime of high baryon number density, and low temperature, is also intrinsically fascinating, and might be important for describing the inner dynamics of supernovae and the deep interior of neutron stars. The first fundamental result about QCD at high baryon number density is that many of its key properties, including for example the symmetry of the ground state and the energy and charge of the elementary excitations, can *not* be calculated to a good approximation starting from fermi balls of non-interacting quarks. The perturbation theory (for just about anything) contains infrared divergences [24].

Fortunately, the main source of these divergences is well understood. They signal instability toward the development of a condensate of quark pairs, similar to the Cooper pairs that occur in metallic superconductors. Whereas the phenomenon of superconductivity in metals is very delicate, because one must overcome the dominant Coulomb repulsion of like charges, color superconductivity is very robust, because there is a fundamentally attractive force between quarks (in the color and flavor antitriplet, spin singlet channel). We can construct an approximate ground state that accommodates the pairs, using the methods of BCS theory. Perturbation theory around this new ground state no longer has infrared divergences. Thus we find that strongly interacting matter at asymptotically high density can be calculated using weak coupling, but non-perturbative methods.

Color superconductivity has become an extremely active area of research over the past few years, and many surprises have emerged. Perhaps the most striking and beautiful result is the occurrence of color-flavor locking (CFL), a new form of symmetry breaking, in real-world (3 flavor) QCD at asymptotic densities [27]. The symmetry $SU(3)_C \times SU(3)_L \times SU(3)_R$ of local color times chiral flavor is broken down to the diagonal subgroup, a residual global $SU(3)$.

Color-flavor locking is a rigorous, calculable consequence of QCD at high density. It implies confinement and chiral symmetry breaking. The low-energy excitations are those created by the quark fields, those created by the gluon fields, and the collective modes associated with chiral symmetry breaking. Because CFL ordering mixes up color and flavor, the quarks form a spin-1/2 octet (plus a heavier singlet), the gluons form a vector octet, and the collective modes form a pseudoscalar octet under the residual $SU(3)$. Altogether there is an uncanny resemblance between the properties of dense hadronic matter one calculates for the CFL phase, and the properties one might anticipate for "nuclear matter" in a world with three massless quarks. A nice perspective on this arises if we consider coupling in the $U(1)$ of electromagnetism. Both the original color gauge symmetry and the original electromagnetic gauge symmetry are broken, but a combination survives. This is similar to what happens in the standard electroweak model, where both weak isospin and hypercharge are broken, but a certain combination survives (to provide electromagnetism). Just as in that case, also in CFL+QED the charge spectrum is modified. One finds that the quarks, gluons, and pseudoscalars acquire integral charges (in units of the electron charge). In fact, the charges precisely match those of the corresponding hadrons.

It is difficult to resist the conjecture that these two states are continuously related to one another, with no phase transition, as the density varies [28]. During this variation, degrees of freedom that are "obviously" three-quark baryons evolve continuously into degrees of freedom that are "obviously" single quarks! That nifty trick is possible because quark pairs can be borrowed from the condensate.

If the core of a neutron star is described by the CFL phase, which seems plausible, it will be a transparent insulator that partially reflects light — like a diamond! That particular consequence of the CFL phase is unlikely to be observed any time soon, but we are working toward defining indirect signatures in observable neutron star and supernova properties.

Unfortunately, existing numerical methods for calculating the behavior of QCD converge very slowly at high density and low temperature. They are totally impractical, even for the biggest and best modern computers. Developing usable algorithms for this kind of problem is a most important open challenge.

There are other stories linking the past with the future through asymptotic freedom and QCD, including a particularly interesting and potentially important one involving axions. But I'll stop here. Thanks again.

References

[1] R. Jaffe and F. Wilczek, hep-ph/0401187.
[2] D. Gross and F. Wilczek, *Phys. Rev. Lett.* **30** 1343–1346 (1973).
[3] H. Politzer, *Phys. Rev. Lett.* **30,** 1346–1349 (1973).
[4] D. Gross and F. Wilczek, *Phys. Rev.* **D8**, 3633–3652 (1973).
[5] D. Gross and F. Wilczek, *Phys. Rev.* **D9**, 980–993 (1974).
[6] H. Georgi and H. Politzer, *Phys. Rev.* **D9**, 416–420 (1974).
[7] N. Nielsen, *Am. J. Phys.* **49**, 1171 (1981); R. Hughes, *Nucl. Phys.* **B186**, 376 (1981).
[8] H. Georgi, H. Quinn, and S. Weinberg, *Phys. Rev. Lett.* **33**, 451 (1974).
[9] J. Pati and A. Salam, *Phys. Rev.* **D8**, 1240 (1973).
[10] H. Georgi and S. Glashow, *Phys. Rev. Lett.* **32**, 438 (1974).

[11] S. Dimopoulos, A. Raby, and F. Wilczek, *Phys. Rev.* **D24**, 1681 (1981).
[12] Reviewed in S. Martin hep-ph/9709356 (see especially his Ref. 75).
[13] R. Feynman, p. 773 in *The Past Decade in Particle Theory*, eds. E. Sudarshan and Y. Ne'eman (1970).
[14] J. Bjorken and E. Paschos, *Phys. Rev.* **185**, 1975 (1969).
[15] K. Wilson, *Phys. Rev.* **179**, 1499 (1969); *Phys. Rev.* **D3**, 1818 (1971).
[16] N. Christ, B. Hasslacher, and A. Mueller, *Phys. Rev.* **D6**, 3543 (1972).
[17] A. de Rujula, S. Glashow, H. Politzer, S. Treiman, F. Wilczek, and A. Zee, *Phys. Rev.* **D10**, 1649 (1974).
[18] Reviewed in G. Wolf, hep-ex/0105055.
[19] R. Feynman, *Phys. Rev. Lett.* **23**, 1415 (1969).
[20] F. Wilczek, *Phys. Rev. Lett.* **39**, 1304 (1977).
[21] H. Georgi, S. Glashow, M. Machacek, and D. Nanopoulos, *Phys. Rev. Lett.* **40**, 692 (1978).
[22] M. Shifman, A. Vainshtein, and V. Zakharov, *Phys. Lett.* **B78**, 443 (1978).
[23] G. Sterman and S. Weinberg, *Phys. Rev. Lett.* **39**, 1436 (1977).
[24] Reviewed by K. Rajagopal and F. Wilczek, p. 2061 in *Handbook of QCD*, ed. M. Shifman (2001).
[25] S. Katz, hep-lat/0310051, and references therein.
[26] Y. Schroder and A. Vuorinen, hep-ph/0311323, and references therein.
[27] M. Alford, K. Rajagopal, and F. Wilczek, *Nucl. Phys.* **B537**, 443 (1999).
[28] T. Schaefer and F. Wilczek, *Phys. Rev. Lett.* **82**, 3956 (1999).

Nobel Biography

The most deeply formative events of my scientific career long preceded my first contact with the research community; indeed, some of them preceded my birth.

My grandparents emigrated from Europe in the aftermath of World War I, as young teenagers; on my father's side they came from Poland and on my mother's side from Italy, near Naples. My grandparents arrived with nothing, and no knowledge of English. My grandfathers were a blacksmith and a mason, respectively. Both my parents were born on Long Island, in 1926, and they have lived there ever since. I was born in 1951, and grew up in a place called Glen Oaks, which is in the northeast corner of Queens, barely within the city limits of New York City.

I've always loved all kinds of puzzles, games, and mysteries. Some of my earliest memories are about the questions I "worked on" even before I went to school. When I was learning about money, I spent a lot of time trying out various schemes of exchanging different kinds of money (e.g., pennies, nickels, and dimes) in complicated ways back and forth, hoping to discover a way to come out ahead. Another project was to find ways of getting very big numbers in a few steps. I discovered simple forms of repeated exponentiation and recursion for myself. Generating large numbers made me feel powerful.

With these inclinations, I suspect I was destined for some kind of intellectual work. A few special circumstances led me to science, and eventually to theoretical physics.

My parents were children during the time of the Great Depression, and their families struggled to get by. This experience shaped many of their attitudes, and especially their aspirations for me. They put very great stock in education, and in the security that technical skill could bring. When I did well in school they were very pleased, and I was encouraged to think about becoming a doctor or an engineer. As I was growing up my father, who worked in electronics, was taking night classes. Our little apartment was full of old radios, early-model televisions, and the many books he was studying. It was the time of the Cold War. Space exploration was a new and exciting prospect, nuclear war a frightening one; both were ever-present in newspapers, TV, and movies. At school, we had regular air raid drills. All this made a big impression on me. I got the idea that there was secret knowledge that, when mastered, would allow Mind to control Matter in seemingly magical ways.

Another thing that shaped my thinking was religious training. I was brought up a Roman Catholic. I loved the idea that there was a great drama and a grand plan behind existence. Later, under the influence of Bertrand Russell's writings and my increasing awareness of scientific knowledge, I lost faith in conventional religion. A big part of my later quest has been trying to regain some of the sense of purpose and meaning that was lost. I'm still trying.

I went to public schools in Queens, and was fortunate to have excellent teachers. Because the schools were big, they could support specialized and advanced classes. At Martin van Buren High School there was a group of thirty or so of us who went to many such classes together, and both supported and competed with one another. More than half of us went on to successful scientific or medical careers.

I arrived at the University of Chicago with large but amorphous ambitions. I flirted with brain science, but soon decided that the central questions were not ready for mathematical treatment, and that I lacked the patience for laboratory work. I read voraciously in many subjects, but I wound up majoring in mathematics, largely because doing that gave me the most freedom. During my last term at Chicago, I took a course about the use of symmetry and group theory in physics from Peter Freund. He was an extremely enthusiastic and inspiring teacher, and I felt an instinctive resonance with the material. I went to Princeton University as a graduate student in the math department, but kept a close eye on what was going on in physics. I became aware that deep ideas involving mathematical symmetry were turning up at the frontiers of physics; specifically, the gauge theory of electroweak interactions, and the scaling symmetry in Wilson's theory of phase transitions. I started to talk with a young professor named David Gross, and my proper career as a physicist began.

The great event of my early career was to help discover the basic theory of the strong force, QCD. That is the subject of the following lecture. The equations of QCD are based on gauge symmetry principles, and we make progress with them using (approximate) scaling symmetry. It was very gratifying to find that the ideas I admired as a student could be used to get a powerful and accurate theory for an important part of fundamental physics. I continue to apply these ideas in new ways, and I am certain that they have a great future.

An aspect of my later work that is not much reflected in the lecture, has been to use insights and methods from "fundamental" physics to address "applied" questions, and vice versa. I'm not sure that fractional quantum numbers, transmuted quantum statistics, exotic superfluidities, or the gauge theory of swimming at low Reynolds number have really arrived as applied physics (yet?), but I've derived a lot of joy from my discoveries in these areas.

To me, the unity of knowledge is a living ideal and goal. I continue, as in my student days, to read voraciously in many subjects, and to think about them. I hope to further expand the horizons of my writing and work in the future.

I've been blessed with a wife, Betsy Devine, and two daughters, Amity and Mira, who've been an inexhaustible source of joy and entertainment.

Here are a few photographs.

Betsy from around the time I met her.

Me speaking at Fermilab in 1977.

My hero Feynman giving me grief at Murph Goldberger's birthday party in 1983; the devil horns are from Sam Treiman.

Finally, here is a family picture from 2001.

NOBEL BIOGRAPHY 399

Asymptotic Freedom: From Paradox to Paradigm*

1. A Pair of Paradoxes

In theoretical physics, paradoxes are good. That's paradoxical, since a paradox appears to be a contradiction, and contradictions imply serious error. But Nature cannot realize contradictions. When our physical theories lead to paradox we must find a way out. Paradoxes focus our attention, and we think harder.

When David Gross and I began the work that led to this Nobel Prize [1, 2, 3, 4], in 1972, we were driven by paradoxes. In resolving the paradoxes we were led to discover a new dynamical principle, asymptotic freedom. This principle in turn has led to an expanded conception of fundamental particles, a new understanding of how matter gets its mass, a new and much clearer picture of the early universe, and new ideas about the unity of Nature's forces. Today I'd like to share with you the story of these ideas.

1.1. *Paradox 1: Quarks Are Born Free, but Everywhere They Are in Chains*

The first paradox was phenomenological.

Near the beginning of the twentieth century, after pioneering experiments by Rutherford, Geiger and Marsden, physicists discovered that most of the mass and all of the positive charge inside an atom is

---

*Nobel Lecture, December 8, 2004.

concentrated in a tiny central nucleus. In 1932, Chadwick discovered neutrons, which together with protons could be considered as the ingredients out of which atomic nuclei could be constructed. But the known forces, gravity and electromagnetism, were insufficient to bind protons and neutrons tightly together into objects as small as the observed nuclei. Physicists were confronted with a new force, the most powerful in Nature. It became a major challenge in fundamental physics, to understand this new force.

For many years physicists gathered data to address that challenge, basically by bashing protons and neutrons together and studying what came out. The results that emerged from these studies, however, were complicated and hard to interpret.

What you would expect, if the particles were really fundamental (indestructible), would be the same particles you started with, coming out with just their trajectories changed. Instead, the outcome of the collisions was often many particles. The final state might contain several copies of the originals, or different particles altogether. A plethora of new particles was discovered in this way. Although these particles, generically called hadrons, are unstable, they otherwise behave in ways that broadly resemble the way protons and neutrons behave. So the character of the subject changed. It was no longer natural to think of it as simply as the study of a new force that binds protons and neutrons into atomic nuclei. Rather, a new world of phenomena had come into view. This world contained many unexpected new particles, that could transform into one another in a bewildering variety of ways. Reflecting this change in perspective, there was a change in terminology. Instead of the nuclear force, physicists came to speak of the strong interaction.

In the early 1960s, Murray Gell-Mann and George Zweig made a great advance in the theory of the strong interaction, by proposing the concept of quarks. If you imagined that hadrons were not fundamental particles, but rather that they were assembled from a few more basic types, the quarks, patterns clicked into place. The dozens of observed hadrons could be understood, at least roughly, as different ways of putting together just three kinds ("flavors") of quarks. You can have a given set of quarks in different spatial orbits, or with their spins aligned in different ways. The energy of the configuration will depend on these things, and so there will be a number of states with different energies, giving rise to particles with different masses, according to $m = E/c^2$. It is analogous to the way we understand the spectrum of excited states of an atom, as arising from different orbits and spin alignments of electrons. (For electrons in atoms the interaction energies are relatively small, however, and the effect of these energies on the overall *mass* of the atoms is insignificant.)

The rules for using quarks to model reality seemed quite weird, however.

Supposedly, quarks hardly noticed one another when they were close together, but if you tried to isolate one, you found that you could not. People looked very hard for individual quarks, but without success. Only bound states of a quark and an antiquark — mesons — or bound states of three quarks — baryons — are observed. This experimental regularity was elevated into The Principle of Confinement. But giving it a dignified name didn't make it less weird.

There were other peculiar things about quarks. They were supposed to have electric charges whose magnitudes are fractions (2/3 or 1/3) of what appears to be the basic unit, namely the magnitude of charge carried by an electron or proton. All other observed electric charges are known, with great accuracy, to be whole-number multiples of this unit. Also, identical quarks did not appear to obey the normal rules of quantum statistics. These rules would require that, as spin 1/2 particles, quarks should be fermions, with antisymmetric wave functions. The pattern of observed baryons cannot be understood using antisymmetric wave functions: it requires symmetric wave functions.

The atmosphere of weirdness and peculiarity surrounding quarks thickened into paradox when J. Friedman, H. Kendall, R. Taylor and their collaborators at the Stanford Linear Accelerator (SLAC) used energetic photons to poke into the inside of protons [5]. They discovered that there are indeed entities that look like quarks inside protons. Surprisingly, though, they found that when quarks are hit hard they seem to move (more accurately: to transport energy and momentum) as if they were free particles. Before the experiment, most physicists had expected that whatever caused the strong interaction of quarks would also cause quarks to radiate energy abundantly, and thus to dissipate their motion rapidly, when they got violently accelerated.

At a certain level of sophistication, that association of radiation with forces appears inevitable, and profound. Indeed, the connection between forces and radiation is associated with some of the most glorious episodes in the history of physics. In 1864, Maxwell predicted the existence of electromagnetic radiation — including, but not limited to, ordinary light — as a consequence of his consistent and comprehensive formulation of electric and magnetic forces. Maxwell's new radiation was subsequently generated and detected by Hertz, in 1883 (and over the twentieth century its development has revolutionized the way we manipulate matter and communicate with one another). Much later, in 1935, Yukawa predicted the existence of pions based on his analysis of nuclear forces, and they were subsequently discovered in the late 1940s. The existences of many other hadrons were predicted successfully using a generalization of these ideas. (For experts: I have in mind the many resonances that were first seen in partial wave analyses, and then later in production.) More recently the existence of W and Z bosons, and of color gluons, and their properties, was inferred before their experimental discovery. Those discoveries were, in 1972, still ahead of us, but they serve to confirm, retroactively, that our concerns were worthy ones. Powerful interactions ought to be associated with powerful radiation. When the most powerful interaction in Nature, the strong interaction, did not obey this rule, it posed a sharp paradox.

1.2. Paradox 2: Special Relativity and Quantum Mechanics Both Work

The second paradox is more conceptual. Quantum mechanics and special relativity are two great theories of twentieth-century physics. Both are very successful. But these two theories are based on entirely different ideas, which are not easy to reconcile. In particular, special relativity puts space and time on the same footing, but quantum mechanics treats them very differently. This leads to a creative tension, whose resolution has led to three previous Nobel Prizes (and ours is another).

The first of these prizes went to P. A. M. Dirac (1933). Imagine a particle moving on average at very nearly the speed of light, but with an uncertainty in position, as required by quantum theory. Evidently there will be some probability for observing this particle to move a little faster than average, and therefore faster than light, which special relativity won't permit. The only known way to resolve this tension involves introducing the idea of antiparticles. Very roughly speaking, the required uncertainty in position is accommodated by allowing for the possibility that the act of measurement can involve the creation of several particles, each indistinguishable from the original, with different positions. To maintain the balance of conserved quantum numbers, the extra particles must be accompanied by an equal number of antiparticles. (Dirac was led to predict the existence of antiparticles through a sequence of ingenious interpretations and re-interpretations of the elegant relativistic wave equation he invented, rather than by heuristic reasoning of the sort I've presented. The inevitability and generality of his conclusions, and their direct relationship to basic principles of quantum mechanics and special relativity, are clear only in retrospect.)

The second and third of these prizes were to R. Feynman, J. Schwinger, and S.-I. Tomonaga (1965) and to G. 't Hooft and M. Veltman (1999) respectively. The main problem that all these authors in one way or another addressed is the problem of ultraviolet divergences.

When special relativity is taken into account, quantum theory must allow for fluctuations in energy over brief intervals of time. This is a generalization of the complementarity between momentum and position that is fundamental for ordinary, non-relativistic quantum mechanics. Loosely speaking, energy can be borrowed to make evanescent virtual particles, including particle-antiparticle pairs. Each pair passes away soon after it comes into being, but new pairs are constantly boiling up, to establish an equilibrium distribution. In this way the wave function of (superficially) empty space becomes densely populated with virtual particles, and empty space comes to behave as a dynamical medium.

The virtual particles with very high energy create special problems. If you calculate how much the properties of real particles and their interactions are changed by their interaction with virtual particles, you tend to get divergent answers, due to the contributions from virtual particles of very high energy.

This problem is a direct descendant of the problem that triggered the introduction of quantum theory in the first place, i.e. the "ultraviolet catastrophe" of blackbody radiation theory, addressed by Planck. There the problem was that high-energy modes of the electromagnetic field are predicted, classically, to occur as thermal fluctuations, to such an extent that equilibrium at any finite temperature requires that there is an infinite amount of energy in these modes. The difficulty came from the possibility of small-amplitude fluctuations with rapid variations in space and time. The element of discreteness introduced by quantum theory eliminates the possibility of very small-amplitude fluctuations, because it imposes a lower bound on their size. The (relatively) large-amplitude fluctuations that remain will occur very rarely in thermal equilibrium, and cause no problem. But quantum fluctuations are much more efficient than are thermal fluctuations at exciting the high-energy modes, in the form of virtual particles, and so those modes come back to haunt us. For example, they give a divergent contribution to the energy of empty space, the so-called zero-point energy.

Renormalization theory was developed to deal with this sort of difficulty. The central observation that is exploited in renormalization theory is that although interactions with high-energy virtual particles appear to produce divergent corrections, they do so in a very structured way. That is, the same corrections appear over and over again in the calculations of many different physical processes. For example in quantum electrodynamics (QED) exactly two independent divergent expressions appear, one of which occurs when we calculate the correction to the mass of the electron, the other of which occurs when we calculate the correction to its charge. To make the calculation mathematically well-defined, we must artificially exclude the highest energy modes, or dampen their interactions, a procedure called applying a cut-off, or regularization. In the end we want to remove the cut-off, but at intermediate stages we need to leave it in, so as to have well-defined (finite) mathematical expressions. If we are willing to take the mass and charge of the electron from experiment, we can identify the formal expressions for these quantities, including the potentially divergent corrections, with their measured values. Having made this identification, we can remove the cutoff. We thereby obtain well-defined answers, in terms of the measured mass and charge, for everything else of interest in QED.

Feynman, Schwinger, and Tomonaga developed the technique for writing down the corrections due to interactions with any finite number of virtual particles in QED, and showed that renormalization theory worked in the simplest cases. (I'm being a little sloppy in my terminology; instead of saying the number of virtual particles, it would be more proper to speak of the number of internal loops in a Feynman graph.) Freeman Dyson supplied a general proof. This was intricate work, that required new mathematical techniques. 't Hooft and Veltman showed that renormalization theory applied to a much wider class of theories, including the sort of spontaneously broken gauge theories that had been used by Glashow, Salam, and Weinberg to construct the (now) standard model of electroweak interactions. Again, this was intricate and highly innovative work.

This brilliant work, however, still did not eliminate all the difficulties. A very profound problem was identified by Landau [6]. Landau argued that virtual particles would tend to accumulate around a real particle as long as there was any uncancelled influence. This is called screening. The only way for this screening process to terminate is for the source plus its cloud of virtual particles to cease to be of interest to additional virtual particles. But then, in the end, no uncancelled influence would remain — and no interaction!

Thus all the brilliant work in QED and more general field theories represented, according to Landau, no more than a temporary fix. You could get finite results for the effect of any particular number of virtual particles, but when you tried to sum the whole thing up, to allow for the possibility of an arbitrary number of virtual particles, you would get nonsense — either infinite answers, or no interaction at all.

Landau and his school backed up this intuition with calculations in many different quantum field theories. They showed, in all the cases they calculated, that screening in fact occurred, and that it doomed any straightforward attempt to perform a complete, consistent calculation by adding up the contributions of more and more virtual particles. We can sweep this problem under the rug in QED or in electroweak theory, because the answers including only a small finite number of virtual particles provide an excellent fit to experiment, and we make a virtue of necessity by stopping there. But for the strong interaction that pragmatic approach seemed highly questionable, because there is no reason to expect that lots of virtual particles won't come into play, when they interact strongly.

Landau thought that he had destroyed quantum field theory as a way of reconciling quantum mechanics and special relativity. Something would have to give. Either quantum mechanics or special relativity might ultimately fail, or else essentially new methods would have to be invented, beyond quantum field theory, to reconcile them. Landau was not displeased with this conclusion, because in practice quantum field theory had not been very helpful in understanding the strong interaction, even though a lot of effort had been put into it. But neither he, nor anyone else, proposed a useful alternative.

So we had the paradox, that combining quantum mechanics and special relativity seemed to lead inevitably to quantum field theory; but quantum field theory, despite substantial pragmatic success, self-destructed logically due to catastrophic screening.

2. Paradox Lost: Antiscreening, or Asymptotic Freedom

These paradoxes were resolved by our discovery of asymptotic freedom.

We found that some very special quantum field theories actually have anti-screening. We called this property asymptotic freedom, for reasons that will soon be clear. Before describing the specifics of the theories, I'd like to indicate in a rough, general way how the phenomenon of anti-screening allows us to resolve our paradoxes.

Antiscreening turns Landau's problem on its head. In the case of screening, a source of influence — let us call it charge, understanding that it can represent something quite different from electric charge — induces a canceling cloud of virtual particles. From a large charge, at the center, you get a small observable influence far away. Antiscreening, or asymptotic freedom, implies instead that a charge of intrinsically small magnitude catalyzes a cloud of virtual particles that enhances its power. I like to think of it as a thundercloud that grows thicker and thicker as you move away from the source.

Since the virtual particles themselves carry charge, this growth is a self-reinforcing, runaway process. The situation appears to be out of control. In particular, energy is required to build up the thundercloud, and the required energy threatens to diverge to infinity. If that is the case, then the source could never be produced in the first place. We've discovered a way to avoid Landau's disease — by banishing the patients!

At this point our first paradox, the confinement of quarks, makes a virtue of theoretical necessity. For it suggests that there *are* in fact sources — specifically, quarks — that cannot exist on their own. Nevertheless, Nature teaches us, these confined particles can play a role as building blocks. If we have, nearby to a source particle, its antiparticle (for example, quark and antiquark), then the catastrophic growth of the antiscreening thundercloud is no longer inevitable. For where they overlap, the cloud of the source can be canceled by the anticloud of the antisource. Quarks and antiquarks, bound together, can be accommodated with finite energy, though either in isolation would cause an infinite disturbance.

Because it was closely tied to detailed, quantitative experiments, the sharpest problem we needed to address was the paradoxical failure of quarks to radiate when Friedman, Kendall, and Taylor subjected them to violent acceleration. This too can be understood from the physics of antiscreening. According to this mechanism, the color charge of a quark, viewed up close, is small. It builds up its power to drive the strong interaction by accumulating a growing cloud at larger distances. Since the power of its intrinsic color charge is small, the quark is actually only loosely attached to its cloud. We can jerk it away from its cloud, and it will

— for a short while — behave almost as if it had no color charge, and no strong interaction. As the virtual particles in space respond to the altered situation they rebuild a new cloud, moving along with the quark, but this process does not involve significant radiation of energy and momentum. That, according to us, was why you could analyze the most salient aspects of the SLAC experiments — the inclusive cross-sections, which only keep track of overall energy-momentum flow — as if the quarks were free particles, though in fact they are strongly interacting and ultimately confined.

Thus both our paradoxes, nicely dovetailed, get resolved together through antiscreening.

The theories that we found to display asymptotic freedom are called nonabelian gauge theories, or Yang-Mills theories [7]. They form a vast generalization of electrodynamics. They postulate the existence of several different kinds of charge, with complete symmetry among them. So instead of one entity, "charge," we have several "colors." Also, instead of one photon, we have a family of color gluons.

The color gluons themselves carry color charges. In this respect the nonabelian theories differ from electrodynamics, where the photon is electrically neutral. Thus gluons in nonabelian theories play a much more active role in the dynamics of these theories than do photons in electrodynamics. Indeed, it is the effect of virtual gluons that is responsible for antiscreening, which does not occur in QED.

It became evident to us very early on that one particular asymptotically free theory was uniquely suited as a candidate to provide the theory of the strong interaction. On phenomenological grounds, we wanted to have the possibility to accommodate baryons, based on three quarks, as well as mesons, based on quark and antiquark. In light of the preceding discussion, this requires that the color charges of three different quarks can cancel, when you add them up. That can occur if the three colors exhaust all possibilities; so we arrived at the gauge group $SU(3)$, with three colors, and eight gluons. To be fair, several physicists had, with various motivations, suggested the existence of a three-valued internal color label for quarks years before [8]. It did not require a great leap of imagination to see how we could adapt those ideas to our tight requirements.

By using elaborate technical machinery of quantum field theory (including the renormalization group, operator product expansions, and appropriate dispersion relations) we were able to be much more specific and quantitative about the implications of our theory than my loose pictorial language suggests. In particular, the strong interaction does not simply turn off abruptly, and there is a non-zero probability that quarks will radiate when poked. It is only asymptotically, as energies involved go to infinity, that the probability for radiation vanishes. We could calculate in great detail the observable effects of the radiation at finite energy, and make experimental predictions based on these calculations. At the time, and for several years later, the data was not accurate enough to test these particular predictions, but by the end of the 1970s they began to look good, and by now they're beautiful.

Our discovery of asymptotic freedom, and its essentially unique realization in quantum field theory, led us to a new attitude towards the problem of the strong interaction. In place of the broad research programs and fragmentary insights that had characterized earlier work, we now had a single, specific candidate theory — a theory that could be tested, and perhaps falsified, but which could not be fudged. Even now, when I re-read our declaration [3]

> Finally let us recall that the proposed theories appear to be uniquely singled out by Nature, if one takes both the SLAC results and the renormalization-group approach to quantum field theory at face value.

I re-live the mixture of exhilaration and anxiety that I felt at the time.

3. A Foursome of Paradigms

Our resolution of the paradoxes that drove us had ramifications in unanticipated directions, and extending far beyond their initial scope.

3.1. *Paradigm 1: The Hard Reality of Quarks and Gluons*

Because, in order to fit the facts, you had to ascribe several bizarre properties to quarks — paradoxical dynamics, peculiar charge, and anomalous statistics — their "reality" was, in 1972, still very much in question. This despite the fact that they were helpful in organizing the hadrons, and even though Friedman, Kendall, and Taylor had "observed" them! The experimental facts wouldn't go away, of course, but their ultimate significance remained doubtful. Were quarks basic particles, with simple properties, that could be used to in formulating a profound theory — or just a curious intermediate device, that would need to be replaced by deeper conceptions?

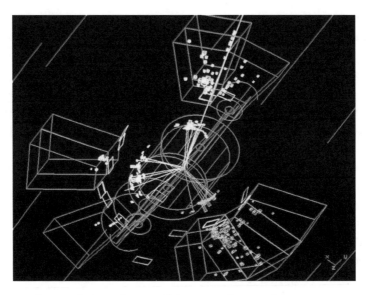

Figure 1. A photograph from the L3 collaboration, showing three jets emerging from electron-positron annihilation at high energy [9]. These jets are the materialization of a quark, antiquark, and gluon.

Now we know how the story played out, and it requires an act of imagination to conceive how it might have been different. But Nature is imaginative, as are theoretical physicists, and so it's not impossible to fantasize alternative histories. For example, the quasiparticles of the fractional quantum Hall effect, which are not basic but rather emerge as collective excitations involving ordinary electrons, also cannot exist in isolation, and they have fractional charge and anomalous statistics! Related things happen in the Skyrme model, where nucleons emerge as collective excitations of pions. One might have fantasized that quarks would follow a similar script, emerging somehow as collective excitations of hadrons, or of more fundamental preons, or of strings.

Together with the new attitude toward the strong interaction problem, that I just mentioned, came a new attitude toward quarks and gluons. These words were no longer just names attached to empirical patterns, or to notional building blocks within rough phenomenological models. Quarks and (especially) gluons had become ideally simple entities, whose properties are fully defined by mathematically precise algorithms.

You can even see them! Here's a picture, which I'll now explain.

Asymptotic freedom is a great boon for experimental physics, because it leads to the beautiful phenomenon of jets. As I remarked before, an important part of the atmosphere of mystery surrounding quarks arose from the fact that they could not be isolated. But if we change our focus, to follow flows of energy and momentum rather than individual hadrons, then quarks and gluons come into view, as I'll now explain.

There is a contrast between two different kinds of radiation, which expresses the essence of asymptotic freedom. Hard radiation, capable of significantly re-directing the flow of energy and momentum, is rare. But soft radiation, that produces additional particles moving in the same direction, without deflecting the overall flow, is common. Indeed, soft radiation is associated with the build-up of the clouds I discussed before, as it occurs in time. Let's consider what it means for experiments, say to be concrete the sort of experiment done at the Large Electron Positron collider (LEP) at CERN during the 1990s, and contemplated for the International Linear Collider (ILC) in the future. At these facilities, one studies what emerges from the annihilation of electrons and positrons that collide at high energies. By well-understood processes that belong to QED

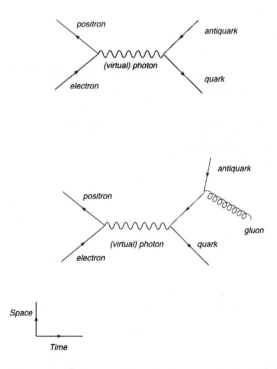

Figure 2. These Feynman graphs are schematic representations of the fundamental processes in electron-positron annihilation, as they take place in space and time. They show the origin of two-jet and three-jet events.

or electroweak theory, the annihilation proceeds through a virtual photon or Z boson into a quark and an antiquark. Conservation and energy and momentum dictate that the quark and antiquark will be moving at high speed in opposite directions. If there is no hard radiation, then the effect of soft radiation will be to convert the quark into a spray of hadrons moving in a common direction: a jet. Similarly, the antiquark becomes a jet moving in the opposite direction. The observed result is then a 2-jet event. Occasionally (about 10% of the time, at LEP) there will be hard radiation, with the quark (or antiquark) emitting a gluon in a significantly new direction. From that point on the same logic applies, and we have a 3-jet event, like the one shown in Figure 1. The theory of the underlying space-time process is depicted in Figure 2. And roughly 1% of the time 4 jets will occur, and so forth. The relative probability of different numbers of jets, how it varies with the overall energy, the relative frequency of different angles at which the jets emerge and the total energy in each — all these detailed aspects of the "antenna pattern" can be predicted quantitatively. These predictions reflect the basic couplings among quarks and gluons, which define QCD, quite directly.

The predictions agree well with very comprehensive experimental measurements. So we can conclude with confidence that QCD is right, and that what you are seeing, in Figure 1, is a quark, an antiquark, and a gluon — although, since the predictions are statistical, we can't say for sure which is which!

By exploiting the idea that hard radiation processes, reflecting fundamental quark and gluon interactions, control the overall flow of energy and momentum in high-energy processes, one can analyze and predict the behavior of many different kinds of experiments. In most of these applications, including the original one to deep inelastic scattering, the analysis necessary to separate out hard and soft radiation is much more involved and harder to visualize than in the case of electron-positron annihilation. A lot of ingenuity has gone, and continues to go, into this subject, known as perturbative QCD. The results have been quite successful and gratifying. Figure 3 shows one aspect of the success. Many different kinds of experiments, performed at many different energies, have been successfully described by QCD predictions, each in terms of the one relevant parameter of the theory, the overall coupling strength. Not only must each experiment, which may involve hundreds of independent measurements, be fit consistently, but one can then check whether the values of the coupling change with the energy scale in the way we predicted. As

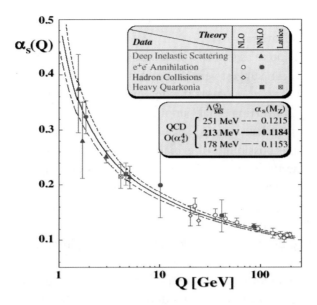

Figure 3. Many quite different experiments, performed at different energies, have been successfully analyzed using QCD. Each fits a large quantity of data to a single parameter, the strong coupling α_s. By comparing the values they report, we obtain direct confirmation that the coupling evolves as predicted [10].

you can see, it does. A remarkable tribute to the success of the theory, which I've been amused to watch evolve, is that a lot of the same activity that used to be called *testing* QCD is now called *calculating backgrounds*.

As a result of all this success, a new paradigm has emerged for the operational meaning of the concept of a fundamental particle. Physicists designing and interpreting high-energy experiments now routinely describe their results in terms of producing and detecting quarks and gluons: what they mean, of course, is the corresponding jets.

3.2. Paradigm 2: Mass Comes from Energy

My friend and mentor Sam Treiman liked to relate his experience of how, during World War II, the U.S. Army responded to the challenge of training a large number of radio engineers starting with very different levels of preparation, ranging down to near zero. They designed a crash course for it, which Sam took. In the training manual, the first chapter was devoted to Ohm's three laws. Ohm's first law is $V = IR$. Ohm's second law is $I = V / R$. I'll leave it to you to reconstruct Ohm's third law.

Similarly, as a companion to Einstein's famous equation $E = mc^2$ we have his second law, $m = E/c^2$.

All this isn't quite as silly as it may seem, because different forms of the same equation can suggest very different things. The usual way of writing the equation, $E = mc^2$, suggests the possibility of obtaining large amounts of energy by converting small amounts of mass. It brings to mind the possibilities of nuclear reactors, or bombs. Stated as $m = E/c^2$, Einstein's law suggests the possibility of explaining mass in terms of energy. That is a good thing to do, because in modern physics energy is a more basic concept than mass. Actually, Einstein's original paper does not contain the equation $E = mc^2$, but rather $m = E/c^2$. In fact, the title is a question: "Does the Inertia of a Body Depend Upon its Energy Content?" From the beginning, Einstein was thinking about the origin of mass, not about making bombs.

Modern QCD answers Einstein's question with a resounding "Yes!" Indeed, the mass of ordinary matter derives almost entirely from energy — the energy of massless gluons and nearly massless quarks, which are the ingredients from which protons, neutrons, and atomic nuclei are made.

The runaway build-up of antiscreening clouds, which I described before, cannot continue indefinitely. The resulting color fields would carry infinite energy, which is not available. The color charge that threatens to induce this runaway must be cancelled. The color charge of a quark can be cancelled either with an antiquark of the opposite color (making a meson), or with two quarks of the complementary colors (making a baryon). In either case, perfect cancellation would occur only if the particles doing the canceling were located right on top of the original quark — then there would be no uncanceled source of color charge anywhere in space, and hence no color field. Quantum mechanics does not permit such perfect cancellation, however. The quarks and antiquarks are described by wave functions, and spatial gradients in these wave function cost energy, and so there is a high price to pay for localizing the wave function within a small region of space. Thus, in seeking to minimize the energy, there are two conflicting considerations: to minimize the field energy, you want to cancel the sources accurately; but to minimize the wave function localization energy, you want to keep the sources fuzzy. The stable configurations will be based on different ways of compromising between those two considerations. In each such configuration, there will be both field energy and localization energy. This gives rise to mass, according to $m = E/c^2$, even if the gluons and quarks started out without any non-zero mass of their own. So the different stable compromises will be associated with particles that we can observe, with different masses; and metastable compromises will be associated with observable particles that have finite lifetimes.

To determine the stable compromises concretely, and so to predict the masses of mesons and baryons, is hard work. It requires difficult calculations that continue to push the frontiers of massively parallel processing. I find it quite ironical that if we want to compute the mass of a proton, we need to deploy something like 10^{30} protons and neutrons, doing trillions of multiplications per second, working for months, to do what one proton does in 10^{-24} seconds, namely figure out its mass. Maybe it qualifies as a paradox. At the least, it suggests that there may be much more efficient ways to calculate than the ones we're using.

In any case, the results that emerge from these calculations are very gratifying. They are displayed in Figure 4. The observed masses of prominent mesons and baryons are reproduced quite well, stating from an extremely tight and rigid theory. Now is the time to notice also that one of the data points in Figure 3, the one labeled "Lattice," is of a quite different character from the others. It is based not on the perturbative physics of hard radiation, but rather on the comparison of a direct integration of the full equations of QCD with experiment, using the techniques of lattice gauge theory.

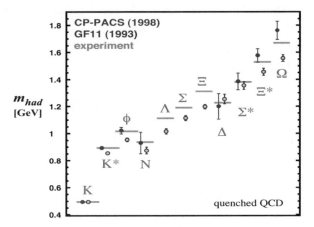

Figure 4. Comparison of observed hadron masses to the energy spectrum predicted by QCD, upon direct numerical integration of the equations, exploiting immense computer power [11]. The small remaining discrepancies are consistent with what is expected given the approximations that were necessary to make the calculation practical.

The success of these calculations represents the ultimate triumph over our two paradoxes:

- The calculated spectrum does not contain anything with the charges or other quantum numbers of quarks; nor of course does it contain massless gluons. The observed particles do not map in a straightforward way to the primary fields from which they ultimately arise.
- Lattice discretization of the quantum field theory provides a cutoff procedure that is independent of any expansion in the number of virtual particle loops. The renormalization procedure must be, and is, carried out without reference to perturbation theory, as one takes the lattice spacing to zero. Asymptotic freedom is crucial for this, as I discussed — it saves us from Landau's catastrophe.

By fitting some fine details of the pattern of masses, one can get an estimate of what the quark masses are, and how much their masses are contributing to the mass of the proton and neutron. It turns out that what I call QCD Lite — the version in which you put the u and d quark masses to zero, and ignore the other quarks entirely — provides a remarkably good approximation to reality. Since QCD Lite is a theory whose basic building blocks have zero mass, this result quantifies and makes precise the idea that most of the mass of ordinary matter — 90% or more — arises from pure energy, via $m = E/c^2$.

The calculations make beautiful images, if we work to put them in eye-friendly form. Derek Leinweber has made some striking animations of QCD fields as they fluctuate in empty space. Figure 5 is a snapshot from one of his animations. Figure 6 from Greg Kilcup, displays the (average) color fields, over and above the fluctuations, that are associated with a very simple hadron, the pion, moving through space-time. Insertion of a quark-antiquark pair, which we subsequently remove, produces this disturbance in the fields.

These pictures make it clear and tangible that the quantum vacuum is a dynamic medium, whose properties and responses largely determine the behavior of matter. In quantum mechanics, energies are associated with frequencies, according to the Planck relation $E = h\nu$. The masses of hadrons, then, are uniquely associated to tones emitted by the dynamic medium of space when it is disturbed in various ways, according to

$$\nu = mc^2/h \tag{1}$$

Figure 5. A snapshot of spontaneous quantum fluctuations in the gluon fields [12]. For experts: what is shown is the topological charge density in a typical contribution to the functional integral, with high frequency modes filtered out.

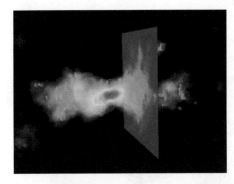

Figure 6. The calculated net distribution of field energy caused by injecting and removing a quark-antiquark pair [13]. By calculating the energy in these fields, and the energy in analogous fields produced by other disturbances, we predict the masses of hadrons. In a profound sense, these fields are the hadrons.

We thereby discover, in the reality of masses, an algorithmic, precise Music of the Void. It is a modern embodiment of the ancients' elusive, mystical "Music of the Spheres."

3.3. *Paradigm 3: The Early Universe was Simple*

In 1972 the early universe seemed hopelessly opaque. In conditions of ultrahigh temperatures, as occurred close to the Big Bang singularity, one would have lots of hadrons and antihadrons, each one an extended entity that interacts strongly and in complicated ways with its neighbors. They'd start to overlap with one another, and thereby produce a theoretically intractable mess.

But asymptotic freedom renders ultra-high temperatures friendly to theorists. It says that if we switch from a description based on hadrons to a description based on quark and gluon variables, and focus on quantities like total energy, that are not sensitive to soft radiation, then the treatment of the strong interaction, which was the great difficulty, becomes simple. We can calculate to a first approximate by pretending that the quarks, antiquarks and gluons behave as free particles, then add in the effects of rare hard interactions. This makes it quite practical to formulate a precise description of the properties of ultrahigh temperature matter that are relevant to cosmology.

We can even, over an extremely limited volume of space and time, reproduce Big Bang conditions in terrestrial laboratories. When heavy ions are caused to collide at high energy, they produce a fireball that briefly attains temperatures as high as 200 MeV. "Simple" may not be the word that occurs to you in describing the explosive outcome of this event, as displayed in Figure 7, but in fact detailed study does permit us to reconstruct aspects of the initial fireball, and to check that it was a plasma of quarks and gluons.

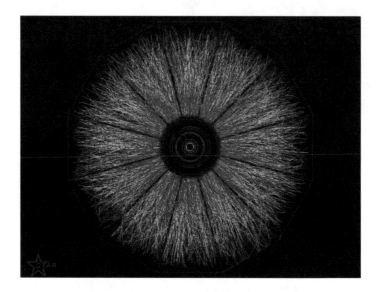

Figure 7. A picture of particle tracks emerging from the collision of two gold ions at high energy. The resulting fireball and its subsequent expansion recreate, on a small scale and briefly, physical conditions that last occurred during the Big Bang [14].

3.4 Paradigm 4: Symmetry Rules

Over the course of the twentieth century, symmetry has been immensely fruitful as a source of insight into Nature's basic operating principles. QCD, in particular, is constructed as the unique embodiment of a huge symmetry group, local $SU(3)$ color gauge symmetry (working together with special relativity, in the context of quantum field theory). As we try to discover new laws, that improve on what we know, it seems good strategy to continue to use symmetry as our guide. This strategy has led physicists to several compelling suggestions, which I'm sure you'll be hearing more about in future years! QCD plays an important role in all of them — either directly, as their inspiration, or as an essential tool in devising strategies for experimental exploration.

I will discuss one of these suggestions schematically, and mention three others telegraphically.

3.4.1. Unified Field Theories

Both QCD and the standard electroweak standard model are founded on gauge symmetries. This combination of theories gives a wonderfully economical and powerful account of an astonishing range of phenomena. Just because it is so concrete and so successful, this rendering of Nature can and should be closely scrutinized for its aesthetic flaws and possibilities. Indeed, the structure of the gauge system gives powerful suggestions for its further fruitful development. Its product structure $SU(3) \times SU(2) \times U(1)$, the reducibility of the fermion representation (that is, the fact that the symmetry does not make connections linking all the fermions), and the peculiar values of the quantum number hypercharge assigned to the known particles all suggest the desirability of a larger symmetry.

The devil is in the details, and it is not at all automatic that the superficially complex and messy observed pattern of matter will fit neatly into a simple mathematical structure. But, to a remarkable extent, it does.

Most of what we know about the strong, electromagnetic, and weak interactions is summarized (rather schematically!) in Figure 8. QCD connects particles horizontally in groups of 3 ($SU(3)$), the weak interaction connects particles vertically in groups of 2 ($SU(2)$) in the horizontal direction and hypercharge ($U(1)$) senses the little subscript numbers. Neither the different interactions, nor the different particles, are unified. There are three different interaction symmetries, and five disconnected sets of particles (actually fifteen sets, taking into account the threefold repetition of families).

We can do much better by having more symmetry, implemented by additional gluons that also change strong into weak colors. Then everything clicks into place quite beautifully, as displayed in Figure 9.

$$\begin{pmatrix} u & u & u \\ d & d & d \end{pmatrix}^L_{1/6}$$

$$\begin{pmatrix} \nu \\ e \end{pmatrix}^L_{-1/2}$$

$$(u \quad u \quad u)^R_{2/3}$$

$$SU(3) \times SU(2) \times U(1)$$
mixed, not unified

$$(d \quad d \quad d)^R_{-1/3}$$

$$(e)^R_{-1}$$

No ν^R

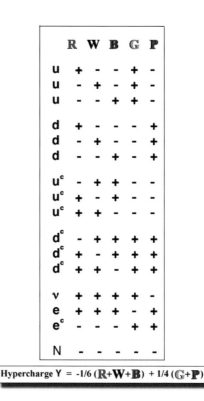

Hypercharge Y = -1/6 (**R**+**W**+**B**) + 1/4 (**G**+**P**)

Figure 8. A schematic representation of the symmetry structure of the Standard Model. There are three independent symmetry transformations, under which the known fermions fall into five independent units (or fifteen, after threefold family repetition). The color gauge group $SU(3)$ of QCD acts horizontally, the weak interaction gauge group $SU(2)$ acts vertically, and the hypercharge $U(1)$ acts with the relative strengths indicated by the subscripts. Right-handed neutrinos do not participate in any of these symmetries.

Figure 9. The hypothetical enlarged symmetry $SO(10)$ [15] accommodates all the symmetries of the Standard Model, and more, into a unified mathematical structure. The fermions, including a right-handed neutrino that plays an important role in understanding observed neutrino phenomena, now form an irreducible unit (neglecting family repetition). The allowed color charges, both strong and weak, form a perfect match to what is observed. The phenomenologically required hypercharges, which appear so peculiar in the Standard Model, are now theoretically determined by the color and weak charges, according to the formula displayed.

There seems to be a problem, however. The different interactions, as observed, do not have the same overall strength, as would be required by the extended symmetry. Fortunately, asymptotic freedom informs us that the observed interaction strengths at a large distance can be different from the basic strengths of the seed couplings viewed at short distance. To see if the basic theory might have the full symmetry, we have to look inside the clouds of virtual particles, and to track the evolution of the couplings. We can do this, using the same sort of calculations that underlie Figure 3, extended to include the electroweak interactions, and extrapolated to much shorter distances (or equivalently, larger energy scales). It is convenient to display inverse couplings and work on a logarithmic scale, for then the evolution is (approximately) linear. When we do the calculation using only the virtual particles for which we have convincing evidence, we find that the couplings do approach each other in a promising way, though ultimately they don't quite meet. This is shown in the top panel of Figure 10.

Interpreting things optimistically, we might surmise from this near-success that the general idea of unification is on the right track, as is

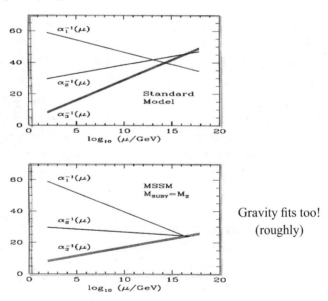

Figure 10. We can test the hypothesis that the disparate coupling strengths of the different gauge interactions derive a common value at short distances, by doing calculations to take into account the effect of virtual particle clouds [16]. These are the same sort of calculations that go into Figure 3, but extrapolated to much higher energies, or equivalently shorter distances. Top panel: using known virtual particles. Bottom panel: including also the virtual particles required by low-energy supersymmetry [17].

our continued reliance on quantum field theory to calculate the evolution of couplings. After all, it is hardly shocking that extrapolation of the equations for evolution of the couplings beyond their observational foundation by many orders of magnitude is missing some quantitatively significant ingredient. In a moment I'll mention an attractive hypothesis for what's missing.

A very general consequence of this line of thought is that an enormously large energy scale, of order 10^{15} GeV or more, emerges naturally as the scale of unification. This is a profound and welcome result. It is profound, because the large energy scale — which is far beyond any energy we can access directly — emerges from careful consideration of experimental realities at energies more than ten orders of magnitude smaller! The underlying logic that gives us such leverage is a synergy of unification and asymptotic freedom. If evolution of couplings is to be responsible for their observed gross inequality then, since this evolution is only logarithmic in energy, it must act over a very wide range.

The emergence of a large mass scale for unification is welcome, first, because many effects we might expect to be associated with unification are observed to be highly suppressed. Symmetries that unify $SU(3) \times SU(2) \times U(1)$ will almost inevitably involve wide possibilities for transformation among quarks, leptons, and their antiparticles. These extended possibilities of transformation, mediated by the corresponding gauge bosons, undermine conservation laws including lepton and baryon number conservation. Violation of lepton number is closely associated with neutrino oscillations. Violation of baryon number is closely associated with proton instability. In recent years neutrino oscillations have been observed; they correspond to miniscule neutrino masses, indicating a very feeble violation of lepton number. Proton instability has not yet been observed, despite heroic efforts to do so. In order to keep these processes sufficiently small, so as to be consistent with observation, a high scale for unification, which suppresses the occurrence of the transformative gauge bosons as virtual particles, is most welcome. In fact, the unification scale we infer from the evolution of couplings is broadly consistent with the observed value of neutrino masses, and that encourages further vigorous pursuit of the quest to observe proton decay.

The emergence of a large mass scale for unification is welcome, secondly, because it opens up possibilities for making quantitative connections to the remaining fundamental interaction in Nature: gravity. It is notorious that gravity is absurdly feebler than the other interactions, when they are compared acting between fundamental particles at accessible energies. The gravitational force between proton and electron, at any macroscopic distance, is about $Gm_e m_p/\alpha \sim 10^{-40}$ of the electric force. On the face of it, this fact poses a severe challenge to the idea that these forces are different manifestations of a common source — and an even more severe challenge to the idea that gravity, because of its deep connection to space-time dynamics, is the primary force.

By extending our consideration of the evolution of couplings to include gravity, we can begin to meet these challenges.

- Whereas the evolution of gauge theory couplings with energy is a subtle quantum mechanical effect, the gravitational coupling evolves even classically, and much more rapidly. For gravity responds directly to energy-momentum, and so it appears stronger when viewed with high-energy probes. In moving from the small energies where we ordinarily measure to unification energy scales, the ratio GE^2/α ascends to values that are no longer absurdly small.
- If gravity is the primary force, and special relativity and quantum mechanics frame the discussion, then Planck's system of physical units, based on Newton's constant G, the speed of light c, and Planck's quantum of action h, is privileged. Dimensional analysis then suggests that the value of naturally defined quantities, measured in these units, should be of order unity. But when we measure the proton mass in Planck units, we discover

$$m_p \sim 10^{-18} \sqrt{\frac{hc}{G}} \qquad (2)$$

On this hypothesis, it makes no sense to ask "Why is gravity so feeble?". Gravity, as the primary force, simply is what it is. The right question is the one we confront here: "Why is the proton so light?" Given our new, profound understanding of the origin of the proton's mass, which I've sketched for you today, we can formulate a tentative answer. The proton's mass is set by the scale at which the strong coupling, evolved down from its primary value at the Planck energy, comes to be of order unity. It is then that it becomes worthwhile to cancel off the growing color fields of quarks, absorbing the cost of quantum localization energy. In this way, we find, quantitatively, that the tiny value of the proton mass in Planck units arises from the fact that the basic unit of color coupling strength, g, is of order 1/2 at the Planck scale! Thus dimensional reasoning is no longer mocked. The apparent feebleness of gravity results from our partiality toward the perspective supplied by matter made from protons and neutrons.

3.4.2. *Supersymmetry*

As I mentioned a moment ago, the approach of couplings to a unified value is suggested, but not accurately realized, if we infer their evolution by including the effect of known virtual particles. There is one particular proposal to expand the world of virtual particles, which is well motivated on several independent grounds. It is known as low-energy supersymmetry [18].

As the name suggests, supersymmetry involves expanding the symmetry of the basic equations of physics. This proposed expansion of symmetry goes in a different direction from the enlargement of gauge symmetry. Supersymmetry makes transformations between particles having the same color charges and different spins, whereas expanded gauge symmetry changes the color charges while leaving spin untouched. Supersymmetry expands the space-time symmetry of special relativity.

In order to implement low-energy supersymmetry, we must postulate the existence of a whole new world of heavy particles, none of which has yet been observed directly. There is, however, a most intriguing indirect hint that this idea may be on the right track: If we include the particles needed for low-energy supersymmetry, in their virtual form, in the calculation of how couplings evolve with energy, then accurate unification is achieved! This is shown in the bottom panel of Figure 10.

By ascending a tower of speculation, involving now both extended gauge symmetry and extended space-time symmetry, we seem to break though the clouds, into clarity and breathtaking vision. Is it an illusion, or reality? This question creates a most exciting situation for the Large Hadron Collider (LHC), due to begin operating at CERN in 2007, for this great accelerator will achieve the energies necessary to access the new world of of heavy particles, if it exists. How the story will play out, only time will tell. But in any case I think it is fair to say that the pursuit of unified field theories, which in past (and many present) incarnations has been vague and not fruitful of testable consequences, has in the circle of ideas I've been describing here attained entirely new levels of concreteness and fruitfulness.

3.4.3. Axions [19]

As I have emphasized repeatedly, QCD is in a profound and literal sense constructed as the embodiment of symmetry. There is an almost perfect match between the observed properties of quarks and gluons and the most general properties allowed by color gauge symmetry, in the framework of special relativity and quantum mechanics. The exception is that the established symmetries of QCD fail to forbid one sort of behavior that is not observed to occur. The established symmetries permit a sort of interaction among gluons — the so-called θ term — that violates the invariance of the equations of QCD under a change in the direction of time. Experiments provide extremely severe limits on the strength of this interaction, much more severe than might be expected to arise accidentally.

By postulating a new symmetry, we can explain the absence of the undesired interaction. The required symmetry is called Peccei-Quinn symmetry after the physicists who first proposed it. If it is present, this symmetry has remarkable consequences. It leads us to predict the existence of new very light, very weakly interacting particles, *axions*. (I named them after a laundry detergent, since they clean up a problem with an axial current.) In principle axions might be observed in a variety of ways, though none is easy. They have interesting implications for cosmology, and they are a leading candidate to provide cosmological dark matter.

3.4.4. *In Search of Symmetry Lost* [20]

It has been almost four decades since our current, wonderfully successful theory of the electroweak interaction was formulated. Central to that theory is the concept of spontaneously broken gauge symmetry. According to this concept, the fundamental equations of physics have more symmetry than the actual physical world does. Although its specific use in electroweak theory involves exotic hypothetical substances and some sophisticated mathematics, the underlying theme of broken symmetry is quite old. It goes back at least to the dawn of modern physics, when Newton postulated that the basic laws of mechanics exhibit full symmetry in three dimensions of space despite the fact that everyday experience clearly distinguishes 'up and down' from 'sideways' directions in our local environment. Newton, of course, traced that asymmetry to the influence of Earth's gravity. In the framework of electroweak theory, modern physicists similarly postulate that the physical world is described by a solution wherein all space, throughout the currently observed Universe, is permeated by one or more (quantum) fields that spoil the full symmetry of the primary equations.

Fortunately this hypothesis, which might at first hearing sound quite extravagant, has testable implications. The symmetry-breaking fields, when suitably excited, must bring forth characteristic particles: their quanta. Using the most economical implementation of the required symmetry breaking, one predicts the existence of a remarkable new particle, the so-called Higgs particle. More ambitious speculations suggest that there should be not just a single Higgs particle, but rather a complex of related particles. Low-energy supersymmetry, for example, requires at least five "Higgs particles."

Elucidation of the Higgs complex will be another major task for the LHC. In planning this endeavor, QCD and asymptotic freedom play a vital supporting role. The strong interaction will be responsible for most of what occurs in collisions at the LHC. To discern the new effects, which will be manifest only in a small proportion of the events, we must understand the dominant backgrounds very well. Also, the production and decay of the Higgs particles themselves usually involves quarks and gluons. To anticipate their signatures, and eventually to interpret the observations, we must use our understanding of how protons — the projectiles at LHC — are assembled from quarks and gluons, and how quarks and gluons show themselves as jets.

4. The Greatest Lesson

Evidently asymptotic freedom, besides resolving the paradoxes that originally concerned us, provides a conceptual foundation for several major insights into Nature's fundamental workings, and a versatile instrument for further investigation.

The greatest lesson, however, is a moral and philosophical one. It is truly awesome to discover, by example, that we humans can come to comprehend Nature's deepest principles, even when they are hidden in remote and alien realms. Our minds were not created for this task, nor were appropriate tools ready at hand. Understanding was achieved through a vast international effort involving thousands of people working hard for decades, competing in the small but cooperating in the large, abiding by rules of openness and honesty. Using these methods — which do not come to us effortlessly, but require nurture and vigilance — we can accomplish wonders.

5. Postscript: Reflections

That was the conclusion of the lecture as I gave it. I'd like to add, in this written version, a few personal reflections.

5.1. *Thanks*

Before concluding I'd like to distribute thanks.

First I'd like to thank my parents, who cared for my human needs and encouraged my curiosity from the beginning. They were children of immigrants from Poland and Italy, and grew up in difficult circumstances during the Great Depression, but managed to emerge as generous souls with an inspiring admiration for science and learning. I'd like to thank the people of New York, for supporting a public school system that served me extremely well. I also got a superb undergraduate education, at the University of Chicago. In this connection I'd especially like to mention the inspiring influence of Peter Freund, whose tremendous enthusiasm and clarity in teaching a course on group theory in physics was a major influence in nudging me from pure mathematics toward physics.

Next I'd like to thank the people around Princeton who contributed in crucial ways to the circumstances that made my development and major work in the 1970s possible. On the personal side, this includes especially my wife Betsy Devine. I don't think it's any coincidence that the beginning of my scientific maturity, and a special surge of energy, happened at the same time as I was falling in love with her. Also Robert Shrock and Bill Caswell, my fellow graduate students, from whom I learned a lot, and who made our extremely intense life-style seem natural and even fun. On the scientific side, I must of course thank David Gross above all. He swept me up in his drive to know and to calculate, and through both his generous guidance and his personal example started and inspired my whole career in physics. The environment for theoretical physics in Princeton in the 1970s was superb. There was an atmosphere of passion for understanding, intellectual toughness, and inner confidence whose creation was a great achievement. Murph Goldberger, Sam Treiman, and Curt Callan especially deserve enormous credit for this. Also Sidney Coleman, who was visiting Princeton at the time, was very actively interested in our work. Such interest from a physicist I regarded as uniquely brilliant was inspiring in itself; Sidney also asked many challenging specific questions that helped us come to grips with our results as they developed. Ken Wilson had visited and lectured a little earlier, and his renormalization group ideas were reverberating in our heads.

Fundamental understanding of the strong interaction was the outcome of decades of research involving thousands of talented people. I'd like to thank my fellow physicists more generally. My theoretical efforts have been inspired by, and of course informed by, the ingenious persistence of my experimental colleagues. Thanks, and congratulations, to all. Beyond that generic thanks I'd like to mention specifically a trio of physicists whose work was particularly important in leading to ours, and who have not (yet?) received a Nobel Prize for it. These are Yoichiro Nambu, Stephen Adler, and James Bjorken. Those heroes advanced the cause of trying to understand hadronic physics by taking the concepts of quantum field theory seriously, and embodying them in specific mechanistic models, when doing so was difficult and unfashionable. I'd like to thank Murray Gell-Mann and Gerard 't Hooft for not quite inventing everything, and so leaving us something to do. And finally I'd like to thank Mother Nature for her extraordinarily good taste, which gave us such a beautiful and powerful theory to discover.

This work is supported in part by funds provided by the U.S. Department of Energy (D.O.E.) under cooperative research agreement DE-FC02-94ER40818.

5.2. *A Note to Historians*

I have not, here, given an extensive account of my personal experiences in discovery. In general, I don't believe that such accounts, composed well after the fact, are reliable as history. I urge historians of science instead to focus on the contemporary documents; and especially the original papers, which by definition accurately reflect the understanding that the authors had at the time, as they could best articulate it. From this literature, it is I think not difficult to identify where the watershed changes in attitude I mentioned earlier occurred, and where the outstanding paradoxes of strong interaction physics and quantum field theory were resolved into modern paradigms for our understanding of Nature.

References

[1] In view of the nature and scope of this write-up, its footnoting will be light. Our major original papers [2, 3, 4] are carefully referenced.//
[2] D. Gross and F. Wilczek, *Phys. Rev. Lett.* **30**, 1343 (1973).//
[3] D. Gross and F. Wilczek, *Phys. Rev.* **D8**, 3633 (1973).//
[4] D. Gross and F. Wilczek, *Phys. Rev.* **D9**, 980 (1974).//
[5] J. Friedman, H. Kendall, and R. Taylor received the Nobel Prize for this work in 1990.//
[6] L. Landau, in *Niels Bohr and the Development of Physics*, ed. W. Pauli (McGraw-Hill, New York, 1955).//
[7] C.-N. Yang and R. Mills, *Phys. Rev.* **96**, 191 (1954).//
[8] An especially clear and insightful early paper, in which a dynamical role for color was proposed, is Y. Nambu, in *Preludes in Theoretical Physics*, ed. A. De-Shalit, H. Feshbach and L. van Hove (North-Holland, Amsterdam, 1966).//
[9] Figure courtesy L3 collaboration, CERN.//
[10] Figure courtesy S. Bethke, hep-ex/0211012.//
[11] Figure courtesy Center for Computational Physics, University of Tsukuba.//
[12] Figure courtesy D. Leinweber, http://www.physics.adelaide.edu.au/theory/staff/leinweber/VisualQCD/Nobel/.//
[13] Figure courtesy G. Kilcup, http://www.physics.ohio-state.edu/~kilcup.//
[14] Figure courtesy STAR collaboration, Brookhaven National Laboratory.//
[15] Unification based on $SO(10)$ symmetry was first outlined in H. Georgi, in *Particles and Fields — 1974*, ed. C. Carlson (AIP, New York, 1975).//
[16] H. Georgi, H. Quinn, and S. Weinberg, *Phys. Rev. Lett.* **33**, 451 (1974).//
[17] S. Dimopoulos, S. Raby, and F. Wilczek, *Phys. Rev.* **D24**, 1681 (1981).//
[18] A standard review is H. P. Nilles, *Phys. Rep.* **110**, 1 (1984).//
[19] A standard review is J. Kim, *Phys. Rep.* **150**, 1 (1987). I also recommend F. Wilczek, hep-ph/0408167.//
[20] I treat this topic more amply in F. Wilczek, *Nature* **433**, 239 (2005).

Advice to Students

Dear students,

In preparing my advice for you I asked myself "What would Einstein say?" And it occurred to me that Einstein, being an intelligent fellow, would probably start with a joke. Fortunately, I happen to know Einstein's favorite joke. It turns out to be quite relevant. Here goes.

> A man is having trouble with his car; it frequently stalls. So he goes to a garage, and asks them to fix it. They replace the transmission and put in new spark plugs. But his car still doesn't run right, so he takes it to another garage. At this second garage, the mechanic pokes around for ten minutes, then pulls a screwdriver out of his belt and tightens a screw. And now the car runs perfectly.
>
> But the man is irate when he gets a bill in the mail for $200. He storms back to the mechanic, and says, "This is outrageous! All you did was tighten a screw, and you ask for $200! I want an itemized bill!" So the mechanic takes out a pad and pencil, and writes down an itemized bill, as follows:
>
> Labor: turning screw $5
> Knowing which screw to turn: $195

My first piece of advice is to consider very carefully the possibilities for what you can do, before choosing. This principle works on several levels. You should consider many different possibilities for what general sort of work you want to do, before settling into one. And when you have finished one project, you should think about many different possibilities for what to do next. And when you encounter a problem, you should consider various possible approaches, before investing heavily in any one.

It's easy to give vague advice, but I will break new ground, and give you an algorithm. Many of you are probably thinking about getting married, and naturally you would like to maximize your chance of finding the best possible mate. I'll give you an algorithm for that.

You have to estimate the number N of suitors that you can expect to deal with over your career in courtship. We'll assume that you evaluate them one at a time, and that once you've broken up with one, then that one is gone forever. Then what you should do is this. Evaluate, but do not accept, each of the first N/e suitors. Here e is a number, the base of natural logarithms, approximately 2.7. Then accept the first subsequent suitor who is better than all the earlier ones. That is how to maximize your chance of getting the best possible mate.

For example, if N is 10, then you should evaluate but reject each of the first 4 suitors, and accept the first one after that who is better than them. In my own case, I estimated $N = 3$. I dutifully broke up with my first serious girlfriend, but the second was better, and I married her. It worked out fine.

Of course the precise assumptions that underlie this particular algorithm might not always be appropriate, but the underlying lesson is much more general. You should put considerable effort into gathering information before choosing what to invest in. The great mathematician Henri Poincaré, when asked how he came up with such good creative ideas, responded, "I generate a lot of ideas, and discard most of them." This is also Nature's trick, in natural selection.

My second piece of advice is to learn about the history of your endeavor. This has many advantages. By reading masterworks you come in contact with great minds, and get to feel how they operate. Often the original works are well expressed, and you can learn valuable lessons about how to express yourself. Most important, you can begin to see yourself and your work as part of a continuing narrative, that started before you entered, and that will continue after you leave. That is a beautiful thing to realize.

Breaking into Verse

Dirac had a low opinion of poetry. He said

> When I write, I try to express difficult ideas in a simple way. In poetry, it is just the opposite.

That seems a little harsh. In any case, there's some poetry I've enjoyed a lot. Mostly it's very old-fashioned, like Homer, Lucretius, Chaucer, Donne; but I also got a great kick out of Vikram Seth's *Golden Gate*, not to mention the rhyming duels in *Curse of Monkey Island*.

When I was in elementary school I was a very enthusiastic versifier, but in the fourth grade I had a crushing experience. We were asked to write poems, and I wrote a very long one. I admit, in retrospect, that some of the meter may have been dubious, and a few of the rhymes were slightly forced. Still, I was unprepared for the teacher's remarks when she returned our papers:

> Frank is extremely enthusiastic about writing poetry. Janice is really good at it.

Though my confidence never entirely recovered, I've occasionally found reasons to break into verse. Herewith, a sampling.

"Virtual Particles" and "Gluon Rap" were written for *Longing for the Harmonies*. "Virtual Particles" also appeared in the *Norton Anthology of Light Verse*, edited by Russell Baker. I'm pretty sure it's the only poem in the anthology that has anything to do with science. I'm very pleased to be an anthologized poet, and I encourage other guardians of the canon to consider following Baker's inspired lead.

I'm quite proud of "Reply in Sonnet Form." Response to high-minded criticism can be a dreary business both for the responder and for his readers (if any). I managed to make it fun for myself, at least. I also suspect that I broke new poetic ground in rhyming "hysterical" with "numerical," "aviation" with "simulation," and "unsound" with "background." (Speaking of background, the letter I was responding to complained that numerical results, which I touted, could never lead to truly reliable conclusions.)

"From Beneath An e-Avalanche" was composed under the pressure of a deluge of congratulatory e-mail following announcement of my Nobel Prize. I could not hope to respond to each message individually in any reasonable time frame, yet I wanted to avoid a canned, cliché response.

"Frog Sonnet" was a drunken improvization. Halfway into the ceremony of the Order of the Ever Smiling and Jumping Green Frog, I was informed I'd have to give a closing speech. This was a somewhat terrifying prospect, given my internal state and the high literary standard set by the drinking songs and liturgical elements of the ceremony. With napkin and pen I jotted down seven rhyming couplets, and came forth with what's here.

"Archaeopteryx" appears here for the first time. It was inspired by a phrase in Olaf Stapledon's *Odd John*, one of my favorite books. John, the superchild, tells the conventional human narrator

> You are the archaeopteryx of the spirit.

As a teenager I identified with Odd John, but life has taught me, as it would have taught Odd John, to identify with archaeopteryx.

Virtual Particles

Beware of thinking nothing's there —
Remove all you can, despite your care
Behind remains a restless seething
Of mindless clones beyond conceiving.

They come in a wink, they dance about,
Whatever they touch is seized by doubt:
What am I doing here? What should I weigh?
Such thoughts often lead to rapid decay.

Fear not! The terminology's misleading;
Decay is virtual particle breeding
Their ferment, though mindless, does serve noble ends:
Those clones, when exchanged, make a bond between friends.

To be or not? The choice seems clear enough,
But Hamlet vacillated. So does this stuff.

Gluon Rap

O! O! O! You eight colorful guys!
You won't let quarks materialize.
You're tricky, but now we realize
You hold together our nucleis.

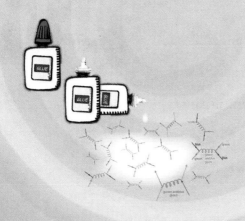

Reply In Sonnet Form

Won't you admit it's a trifle hysterical
To disbelieve *every* result that's numerical?

How, then, could you use modern aviation?
For the planes are designed by simulation.
And are experiments at accelerators all unsound,
Because they *simulate* the QCD background?

O why do you recoil in terror
From calculations that control their error?
Give it up! The symmetry's surely broken,
The order parameter (its token)
Refuses, by 20 σ, to go away.
What's that, *a coincidence*? No way!

No offense, but it's silly to avert your eyes
After 10^{18} floating point multiplies.

From Beneath An e-Avalanche

I don't suppose that colored quarks and glue
Think overmuch about what they're up to;
They just do whatever comes naturally
And leave the worrying to you and me.

Free spirits! They seemed blithely unconcerned
With sacred lessons we'd with effort learned.
But by invoking then heretical
Wild hypotheses theoretical
I found their workings could be understood:
So the world makes sense, as it damn well should.

The Prize recalls those days of search and find,
Warm notes from friends bring human joy to mind;
My heart is full, as is my thanks to you
My In box also, I'm afraid — adieu!

Frog Sonnet

("Thank you" speech at the ceremony of the Ever Jumping and Smiling Green Frog, 13 Dec. 2004.)

I'm well aware that the primary virtue of a speech like this is brevity. I couldn't come up with a worthy haiku, so you'll have to settle for a sonnet. The sonnet is a strict form, that requires fourteen lines. Here goes.

When studying physics, who would have thunk,
It'd lead to public sonnetizing, quasi-drunk
After days of partying and eating like a hog
Climaxed by jumping around like a frog.

Through it all I've found that I'm taking a liking
For living life large, like a Viking.
So, when I return to my peaceful local village
I'll be tempted to indulge in some looting and pillage.

But you've taught me how Swedes got past doing wrongs:
By investing time, instead, in learning drinking songs.
On the surface everyone seems calm and nice
But beneath, in reserve, there is plenty of spice.

For this lesson, frog, and fun: Thanks a lot!
That's fourteen lines, it's all I've got.

Archaeopteryx

A heron stirs with easy majesty
Spreads strong outreaching wings and grasps the air
As if it were a solid thing, to push
His massy body, from water derived
Upward, where vision grows far and clear.

A hummingbird flails with frenzied fury
But she is calm, serene, and fixed in space
As she accepts her bribe from flowery depths:
Our Sun's fierce energy, made tame and sweet
To power the pollen-bearer's delight.

Archaeopteryx is not a virtuoso of the air.
He is a feathered reptile, too heavy for his strength;
He moves with vigor, but his motion is crude;
His stability is precarious;
He suspends his descent, and only that.

The proto-bird does not know what he is:
A trial, an experiment, a bridge.
I know my spiritual irony:
Archaeopteryx I aspire to be.

Another Dimension

Betsy Devine, as you may already have guessed, is my wife. I'm a lucky guy.

One aspect of my good fortune is that Betsy captures the memory of so much of our life together. For many years she did this through photographs and journals. In retrospect, it's clear to me that Betsy was a blog waiting to happen. Sure enough, for the past few years Betsy has kept an online journal/diary — in short, a blog. The popularity of *Betsy Devine: Funny Ha Ha or Funny Peculiar?* is such that if you Google simply "Betsy" it usually appears as the third or fourth item — after Betsy Ross, who she probably won't catch, and two others she's got a shot at.

Betsy began her Nobel adventure a minute or so before I did, since she picked up the phone. (As she doesn't fail to inform you, I was in the shower.) Throughout our many voyages and exploits, she took time to make a blog entry almost every day, and often more than one. Her take on things was invariably interesting, always amusing, and mostly accurate. Betsy made history by becoming the first, and so far the only, blog to be linked at nobelprize.org.

I don't have any secrets from Betsy (well, not very many). Fortunately, she has discretion and a sweet nature. So go ahead and read.

The big news: October, 2004

Wow and super-wow...

So this morning the phone woke me up at 5:30 and it was a lady with a beautiful Swedish accent. Frank was already in the shower, but he got out and dripped all over the floor while she informed him that he and his thesis advisor David Gross, and a third physicist named David Politzer, just won this year's Nobel Prize in Physics!

Awesome! I'm so pleased! Frank Wilczek rulez!

Posted 10/5/04; Department 'Nobel'

What it's like...

Heh, not like what you might think.

Exhibit A: After Frank gave me a thumbs-up that this morning's 5:30 phone call was what we both hoped, I ran out to the kitchen to listen on an extension as a series of very distinguished and impressive Nobel Committee members congratulated him.

It wasn't until the second or third Swedish voice (a Lars or a Niels or a Sven, I was losing track) that I realized — Frank was still totally dripping water from the shower he climbed out of. So I ran back and grabbed a big bathrobe to drape around him. Not that he probably noticed, but I felt good about it.

Exhibit B: I'm not used to having people ring the doorbell at 6 a.m. who turn out to be photographers from Reuters. But Brian was very nice, and took some good pictures.

Exhibit C: Lame remarks that are fun in retrospect: "Could you let me park in this parking lot because my husband just won the Nobel Prize and I'm late for his press conference?"

That's how I wound up with a huge cardboard special permit with a handwritten endorsement. No, it didn't say what you might think,[*] it said, "Vendor."

[*]What you might think it said: "Crazy."

Posted 10/5/04; Department 'Nobel'

The Nobel Prize from our dog's point of view

Photo courtesy of Reuters/Brian Snyder.

Here's my favorite photo from yesterday, taken by Brian Snyder of Reuters as Frank walked out the door for his usual 3-mile walk to MIT.

Can you see the West Highland terrier, down by my knee?

Marianne, age 16, or 102 in dog years, had the best time of anyone yesterday morning — her breakfast and morning walk arrived hours early! Getting to sniff Brian Snyder's photo equipment was an extra treat.

She was by no means finished sniffing it when he left.

Posted 10/6/04; Department 'Nobel'

Thanks for the funny e-card, this means you!

It's been great getting email and phone calls from old friends. Frank went out to dinner tonight with some physicists — when he got home he had 110 new messages. (Mostly not Nigerian sex-aid offers, because MIT has a pretty good spam-filter system.)

We've heard from former professors, long-ago babysitters, sixth-grade school friends, plus several members of the "Princeton Eulers," a softball team I organized at the Institute for Advanced Study. And my blog friends have been generous with their link-love!* Our kids, parents, siblings, etc. are getting similar deluges of congrats.

If I'd known 15 minutes of fame would be this much fun, I would have wished harder for it to arrive. OTOH, if I'd known 15 minutes of fame would be this much work, I might have decided to hide under the bed.

*Thanks, Adam, JR, Julie, Enoch, Frank, Yvonne, Dervala, Lisa, Judith, Scott, Susan, Peter, Paul, Zoe... .

Posted 10/6/04; Department 'Nobel'

Big Nobel downside: One thousand thank-you letters!
Even my optimistic husband Frank was taken aback by the job of replying to all the old friends and others who congratulated him on his Nobel Prize.

Frank — who doesn't like to do a not-very-good job on anything — solved the problem of replying-quickly versus replying-personally in a characteristically outside-the-box-Frank way.

He wrote a sonnet he could send everyone fast. [*See page 437.*]

Then he organized his in-box so he could write real replies one day at a time — first to all the As — then to all the Bs — then, on Thursday night, his hard disk crashed. It's off at Disk Doctors being (hopefully) salvaged.

But if you wrote to Frank and your first name begins with C, you're top on his list when he gets his computer back.

Posted 10/23/04; Department 'Nobel'

Getting ready

Web research on other people's prizes

Nobel's formal events require white-tie-and-tails or (there is an exception!) "national costume." Unfortunately, the national costume of physics is anything-goes-topped-with-a-funny-logo-on-your-Tshirt.

I've been doing online "research" on what people wear to all the various parties in Sweden — an economist's family blog was a great resource. The Engles went to three balls in one week…wow!

As a direct result of this research, I got Frank to come shopping with me and buy some non-sneakers.

Posted 10/15/04; Department 'Nobel'

Nobel Prize and math

There is no Nobel Prize for mathematics[*] — but there's lots of math involved in Nobel Prizes. Word problems… .

1. If 7 of us fly to Stockholm on the redeye, plot our best distribution onto airplane seats, bearing in mind that Amity's husband Colin has very long legs and neither of Frank's parents should have to sit immobilized for too long.
2. Which will be harder and take more time: to find the required white-tie-and-tails Nobel outfit in Boston and lug it to Stockholm, or to figure our how to take 8 different measurements of my husband and then convert them all into metric so that someone in Stockholm can rent the outfit for him?
3. Rank these four events in order of probabability: Lightning will strike Mel Gibson, Lightning will strike Mel Brooks, Betsy Devine will have triplets nine months from now, Frank Wilczek will need to wear white-tie-and-tails to some event unrelated to Nobel Prizes.

Show all calculations, and remember, neatness counts.

[*]According to urban legend, Alfred Nobel cold-shouldered math because his wife was sleeping with a mathematician. Nobel was a bachelor, so it's hard to guess how this rumor got started — though I have no doubt it was by a mathematician.

Posted 10/8/04; Department 'Nobel'

"Manchester family proud... ."

My old hometown paper just ran a piece on Frank's Nobel Prize. The headline gives you the flavor of the whole thing: "Manchester Family Proud of Newest Nobel Prize Winner." Manchester NH has 100,000 people but every one of them knows it as a small town.

In the 1950s, the *Manchester Union Leader* was run by a right-wing nut[*] named William Loeb who gained some nationwide notoriety during the 1972 primaries.

My mother — a Rockefeller-type Republican from Massachusetts — refused to allow "that filthy rag" in our house, which meant we had to look in a neighbor's copy to find out which movies were playing. She warned us never to talk to strangers because they might be *Union Leader* reporters trying to dig up dirt about my dad. (My dad held various unpaid offices in the NH Democratic Party and attracted some nasty attacks from Mr Loeb.)

In those days, I imagined reporters hung out at playgrounds all over the city, so they could ask unsuspecting children about their fathers.

Times change. Mr Loeb left the planet long ago, and the *Union Leader*, though still right-wing, shows some admirable attempts at balance. Its editorials strongly favor George Bush, but the paper's star political columnist John DiStaso played a major role in exposing Republican wrongdoing in the NH phone jamming scandal.

And now the paper has kind words to say about some Devines. I just wish my mother were here — she'd be smiling too.

[*] A front-page editorial titled "Kissinger the K*ke" comes to mind.

Posted 10/9/04; Department 'Nobel'

The Supreme Order of the Ever Jumping and Smiling Green Frog

Stockholm University students have their own Nobel tradition — new laureates are invited to a December 13 ball, then inducted into "The Supreme Order of the Ever Jumping and Smiling Green Frog." Their actual paper snailmail invitation contained no email address or URL. For information about this unique event, it had to be Google once again to the rescue!

The guest of one 1987 Nobel-winner describes a banquet with drinking-songs followed by "hi-jinks."

From a detailed diary of 1996 Nobel-guest adventures kept by David Mermin: "between midnight and 1 the culminating event of the week takes place… details of the lunatic ceremony are lost in the fog, but somehow it manages to culminate in all six of the 1996 Nobel Laureates in Physics and Chemistry lined up together and uttering cries of "Rivet, rivet" while squatting on their haunches and yumping up and down. A fitting end to a can-you-top-this week."

Stockholm University science students explain the frog insignia: "The Frog is made out of an alloy of lead with bismuth, frogish green and tied in a very special way. The Insignia is supposed to be weared with tail-coat for the gentlemen and evening-dress for the ladies. When a Frog-member dies, the relatives are asked to send the Insignia back to the Order or to destroy it."

Frog-members also get to drink pea soup with brandy in March — the group's only other official event. Thank you, Google — jumping and smiling will be on Frank's Nobel week schedule!

Posted 11/10/04; Department 'Nobel'

Swedish TV crew…

Three very nice young TV-crew-men are downstairs getting ready to interview Frank, so of course I decided it would be fun to live-blog this novel event.

Roland, the interviewer from Sweden, is young and TV-good-looking — he has already interviewed two US Nobel laureates before he came here — visiting Minneapolis and Santa Barbara.

Guy, the cameraman/lighting guru, lives in Wellesley. Rob, who does sound, is also a local freelancer.

They're setting up lights and cameras and microphones all over our house's familiar spaces, while Roland and Frank chit-chat about physics and travel.

Rob dangles a huge microphone in a thick fur sweater (a "Rykote windscreen") over one of our dining room chairs.

Guy pulls down the shades in our dining room so that he can take full control of the ambient light.

More soon… .

Posted 11/12/04; Department 'Nobel'

"More schmutz"

Frank is sitting in the dining room chair, getting lit for the camera.

"More schmutz," says Guy. Schmutz turns out to be two pieces of waxy-looking paper he hangs in front of a light that is shining on Frank.

"Why is that schmutz?" I ask.

"That's just what we call it," says Guy. "The real name, the brand name for it, is actually Opal."

Something with a pretty name like Opal gets nicknamed "schmutz"? It must be a guy thing. Definitely, a guy thing.

Posted 11/12/04; Department 'Nobel'

Hail and farewell….

…or at least, snow and farewell. It's pouring down snow, and roofs are getting white.

Frank and the camera crew left for MIT after shooting footage of just about everything you can imagine.

I am wondering what Swedish tv audiences will think of all the events this crew took pictures of:

- ❖ Snow falling on our garden, as seen from our bedroom window.
- ❖ Frank reading in his favorite chair.
- ❖ Frank playing the piano.
- ❖ Frank and Betsy sitting in two chairs talking. Have you ever tried to talk with your spouse non-stop for 10 minutes, pretending that just the two of you are there?
- ❖ Marianne the dog, watching Frank and Betsy as if we were a Wimbledon tennis match.

- ❖ Frank and Betsy being interviewed by Roland about *Longing for the Harmonies*, which W. W. Norton now plans to re-publish.
- ❖ Betsy blogging, with Frank looking on, bemused.

The crew then, intrepidly, shot some more footage outdoors. Frank and I were instructed to shut the front door, count to ten, then open it and stroll out, going for a walk. We are not allowed to look at Guy or Rob or Roland or the camera. So, we do some walking following these instructions. Then we are asked to do some more walking, starting from a point outside the house and going together around the corner.

I said to Guy, "Maybe Roland would like to walk with us?" Guy said, not unkindly, "You haven't done a lot of this, have you?"

So we walked to the corner and past it and kept on going. Snow fell on us. Frank was wearing his brand-new black shoes, bought for Sweden, and flakes of snow were melting all over them. I was having a great time, maybe because, as Guy noted, I haven't done a lot of this.

Now our house is quiet. Marianne, in her 102 dog years, has rarely seen or smelled so much excitement. She's conked out in her fuzzy dogbed now, so fast asleep that she's not even snoring.

Marianne has got the right idea.

Posted 11/12/04; Department 'Nobel'

Harry-met-Sally boots

I once spent a week on jury duty with John Nash. I'm glad I was smoking back then[*], because it meant I got to hang out with him next to the gritty Trenton courthouse. (No lawyer wanted either of us for a jury.)

This was after he won the Nobel Prize in Economics but before *A Beautiful Mind* was written (let alone filmed), so he was a celebrity to nobody there but me, which suited us both just fine.

I wish I remembered more of the things he told me about his Nobel adventures. I asked him if he had bought something great with the money — he said the best thing was a cordless phone! He loved being able to have the phone with him when he was taking a bath. (This was before anyone had mobile phones.) On the way home that night, I bought Frank a cordless phone.

My favorite luxury since Frank won the Nobel Prize has been a pair of Merrell Yeti high boots. And two more pairs, one for each of our daughters.

On the weekend my wonderful sister Marie came to help us all buy ball gowns, the Tannery shoe store was having a two-for-one sale. This required a quick detour from ball gowns...

Amity, trying on Merrell Yeti boots, said, "Oh, these feel so good. I love walking in them. They feel so warm and so soft and so comfortable..."

Marie looked at me and said, "I'll have what she's having."

Caveat — if you get these beautiful boots, wear socks inside them. The black fleece lining leaves your ankles and feet a dusty gray. Not the best look to go with party shoes!

*And I'm even gladder that I quit later that year.

Posted 11/20/04; Department 'Nobel'

How many suitcases for the Ice Hotel?

Packing is always a challenge, but for Nobel Week? Its many events line up under three different dress codes:

- ❖ "casual";
- ❖ business suit/cocktail dress; and
- ❖ white tie/ballgown (three different events require these).

Frank can get away with a couple of suits and two pairs of black shoes — it's easy to rent the white-tie-and-tails in Stockholm. I'm bringing a few more options for all this dress-up. We'll each need a grown-up black coat, plus mittens, boots, mufflers, longjohns, etc. for sub-zero Sweden.

But then, in the early morning of December 14 (after the late-late-night party with Jumping Frogs), we've been invited to visit the Arctic Circle for a week that requires layers and layers more clothing! Forget the wimpy lightweight Capilene longjohns, so usefully invisible under a suit. Bring on some expedition-weight fleece, and never mind the bulges! The dress code is stuff for hiking or riding in dogsleds — for hanging out in the midnight anti-Sun and chasing auroras!

Now this is starting to add up to way too much luggage...

And then, we schlep it all to the Ice Hotel? Guests sleep in a room made of ice, on top of a bed made of ice (sleeping bags and padding in between you and the ice), surrounded by sculptures made of ice... . The door and the toilet are not made of ice, because the room has neither. It also has no space for luggage, which gets stored all night in a locker somewhere. That's too bad, because I might want some ballgowns round about 2 a.m., to pile up on top of my sleeping bag to keep warm...

Posted 11/21/04; Department 'Nobel'

Pre-Nobel Thanksgiving

The first Thanksgiving I cooked for was the hardest. Amity was a tiny baby — Frank and I both had flu. I managed to stagger into the kitchen and heat up a can of Campbell's chicken soup with rice for us to celebrate with. We were both thankful we could keep the soup down that day, a sign that we were finally getting better.

I remember the Thanksgiving when I was 10, when my Aunt Mary let me help make the giblet gravy. It was delicious. She and I kept tasting it in the kitchen, and when it was time to serve it we* had none left.

I remember the many holiday meals we shared with Frank's grandparents. Grandma Wilczek would cook an authentic Polish feast with lots of kielbasa. Then we would all drive over to Grandma Cona's for an Italian super-spectacular — turkey plus pans of lasagne, meatballs, and sausage. It's a miracle we have any arteries left.

I remember when I realized, 10 years ago, that my computer could help me stage-manage Thanksgiving. I created timetables, lists of dishes and recipes. I don't know how people did this before they had printers.

I remember last year, when I blogged the universal Thanksgiving prayer ("Oh lord, you know I don't know how to cook this ugly bird...")

This year, we'll be 12 around the table (pardon my elbow!) No canned soup, but plenty of veggie pot pie and killer brownies along with the turkey and gravy for carnivores like me. Just about all of us are headed for Sweden to cheer for Frank — less than two weeks away. Eek!

Now, I better get cooking!

*Heh. Almost 50 years later, I just realized — Aunt Mary was quick to claim half the blame for the missing gravy, but she probably didn't drink more than a tablespoon of it. No wonder everybody loved Aunt Mary!

Posted 11/26/04; Department 'Life, the Universe, and Everything'

Prelude to the Arctic

Once Nobel Week is over, Frank and I will head north for a week of auroras and anti-midnight non-sun. Swedish physicist Sverker Fredriksson has been sending us fascinating email about what to expect — today I got permission to post some. So here he is, my first-ever guest-blogger.

Hi Betsy and Frank,

I am mailing you from the Gothenburg airport while waiting for the bus to Boras. I missed the previous one because we arrived one hour late due partly to the snow yesterday, partly to the ice today. There is only one shuttling aircraft between Luleå and Gothenburg, and it collects delays during the day. And this time even overnight. The last one yesterday was so late that they had to let the crew sleep this morning, and took off one hour late. Then they had to de-ice the wings, which normally takes just a few minutes. However, 7 am is rush-hour from Luleå, with four airlines, so we had to stand in line for another 30 minutes, waiting to be sprayed.

The good thing is that this brought me one hour at the airport internet café. The bad thing is that this gives me one hour less with my wife.

Anyway, we "dog people" seem to have the same view of our darlings. My wife always jokes that Qrispin is reading the newspaper's gossip column when he checks out which dog has done what since he read it last time. Qrispin is not the least interested in a shoot-out on TV, or someone making a noise in our apartment house. But he can stand still for minutes to sort out all details around one spot on the lawn. And that is even if he can clearly see the other dog 50 meters ahead.

As you know dogs have some 250,000 times better "noses" than we have. It is even claimed that they can detect one single molecule. I try to train Qrispin to tell one single free quark, but he probably knows that they are only asymptotically free, and not worth the effort.

I am amazed that your dog is 16 years old, and still so cute! My wife and I are a little depressed over Qrispin's 12.5 years, because in the books most races are predicted to live only until the age of 14. We hope that these predictions are as wrong as the one we heard when we bought a little rabbit 20 years ago (Stumper). We counted on "5–7 years" but he became 12 until he died of high age. We think this had to do with his habit of eating bacon chips, After Eight mints, art books and electric cables.

I have seen balloon launches* only on TV, but sometimes we see them in the northern sky from Luleå, when they have reached their operating altitudes of some 35 km. There they look like very bright stars. Every time they are sent up, people phone the authorities and report UFOs. I do not know yet who owns the balloon of mid-December, but there is a fair chance that it is NASA. They have decided to use Kiruna for scientific balloons that will fly to Alaska.

Normally the wind here blows to the east, so that balloons are taken down just before the Ural mountains. I guess that NASA balloons have to wait for specific weather types.

About weather: We had a one-week period of unusually cold weather in both Luleå and Kiruna — down to −22°C where I live, and even −27°C in Kiruna. Now it is up to a more normal — 3 to −5. Here on the west coast it is +8 just now, which is comparable to a cold summer night in northern Sweden.

The year my wife and I moved to Luleå, August had a mean temperature of +8°C, which was the coldest summer for decades. On the other hand, two years ago we had a week of more than +30°C, with a record +35°C one day. Remember that this is as far north as northern Alaska or northern Greenland. An advantage is that the air here is always dry, which is a comfort both with cold and warm weather. In other words we do not have New York weather.

That's all for this internet café!

Best regards,
Sverker

*We are hoping to see a stratospheric balloon launched from Esrange during our visit.

Posted 11/26/04; Department 'Nobel'

My Swedish language emergency kit — and Frank's

My Swedish* "language emergency kit" covers essentials like finding bathrooms or an Internet cafe. Frank will be facing a much, much bigger challenge.

I just found out this morning he's agreed to give a "little talk" to some schoolchildren, ages 7 and up. I'm sure he'll come up with something interesting to say — in English, that is. The challenge will then be to extract a few key words or a sentence that he can also say in Swedish.

Translating "gluon" or "asymptotic freedom" wouldn't make the words any less Greek to a bunch of young Swedes.

At the opposite extreme, Frank could memorize a Swedish version of Churchill's "Never give up speech."†

One can never go wrong with "thanks very much," but it isn't very surprising.

I look forward to seeing what Frank comes up with!

*We spent time in Uppsala last summer, thank heaven — so I already got past my original Muppet-Swedish-chef-Swedish.
†According to legend, Churchill's speech to the students at Harrow School ran as follows: "Never, ever, ever, ever, ever, ever, ever, give up. Never give up. Never give up. Never give up." Then he sat down again.

Posted 11/27/04; Department 'Nobel'

We are in a metastable state

Physicists know about lots of phase transitions — "an abrupt sudden change of one or more physical properties," says Wikipedia — for example, ice melts into water — water boils up into steam — some supercooled substances turn into superconductors... .

I'm getting ready for the phase transition from pre-traveling-Betsy to Betsy-in-transit.

I just finished putting our garden to bed for the winter — I know I won't have another free minute before December 18. Pre-traveling-Betsy is wearing blue jeans, with one sneaker dripping water all over the floor. (As I coiled up a garden hose, it soaked my hosiery.)

In-transit-Betsy, I hope, will look cool and serene. She'll be wearing a suit in shades of black, silver, and gray, plus her favorite red walking shoes. She'll definitely have dry socks on both feet.

Now, to get past the intermediate state, which seems to be blogging-Betsy... .

Posted 11/30/04; Department 'Nobel'

Hello, Mr. President

One-party Washington, DC

Elephant neckties but no donkey neckties are for sale in the Wyndham Hotel gift shop. Ninety-nine kinds of George Bush buttons or cute little heart-shaped buttons with Laura Bush on them.

Well, I guess Democrats don't have much reason to come here unless, for example, the Swedish Ambassador asks them to drop in for dinner.

I realized something on the tiny plane that bounced us from Boston into Ronald Reagan Airport. The reason Republicans were "conservative" for years was that they were out of power. Democrats were running things and spending government money on programs Republicans didn't want.

Once Republicans started running the country, after 2000 but with power-ups in 2002 and 2004, they used their new power to make some big changes of their own.

Now it's Democrats wanting things back the way they were.

Posted 11/30/04; Department 'Life, the Universe, and Everything'

Most of my letter to Mia

Hi Mia!

It's a tradition that the Swedish ambassador to DC invites Nobel laureates to dinner and arranges a White House reception for them beforehand. The date is set by the White House and you can see (it was Dec. 1) they put it off until the last possible moment. Six laureates and spouses in turn shook hands with Bush — we had all* made up our minds to be polite, because being otherwise wouldn't help anybody but could embarrass the Swedes, who have been so nice.

Ambassador Jan Eliasson is very impressive, with a background in diplomacy and mediation: http://www.swedenabroad.se/pages/general__7038.asp

The announcement that he'd be President of the UN General Assembly is very recent: http://www.swedenabroad.se/pages/news__29882.asp&root=6989

In person he is charming, a good speaker, thoughtful, etc. His wife Kerstin is Deputy Minister for Culture — she comes over from Stockholm to be part of the DC event each year — he joked that this was one of its many benefits.

He is also so modest that we didn't hear about his new UN position until Justice Sandra Day O'Connor congratulated him in an after-dinner speech. I also met the more-palatably-politicked Justice Ruth Bader Ginsburg — for some reason she was wearing tiny black crocheted gloves. Let's see, other DC gossip — Condi Rice should buy bigger or better shoes; she was very trimly dressed but with little white "ouch" pads peeking over her high heeled shoes where they rubbed the back of her ankles.

I wish the US had the good luck to be led by somebody like Jan Eliasson.

Love and xxx
Betsy

*I think at least one laureate supported Bush. Other points of view expressed by "Nobel Laureates Other Than Frank Wilczek" (NLOTFW): "It's the office, not the man. It's the office, not the man." "I'll be polite but I'm not making any small talk." As it turned out, small talk was not required. See my next post for more…

Posted 12/3/04; Department 'Nobel'

Meanwhile, back at the Oval Office

The White House account of the Oval Office visit of six Nobel laureates puts "Frank Wilczek" in the center of the photo — which sounds better than it is, because the person they think is Frank Wilczek is economics laureate Finn Kydland.

Filling in blanks, from my December notebook:

At two o'clock, all in suits and most in coats, we piled into a limousine-minibus hybrid to go to the White House. One NLOTFW* had forgotten his photo ID, which delayed our departure. From Wyndham Hotel to White House, about 20 minutes.

For security reasons, the bus stopped outside WH gates. We walked past huge stages with canvas-wrapped lights and podiums where news anchors stand when doing exterior shots. We all got white plastic tags that said 'A' (appointment).

As we entered the West Wing, I heard a helicopter's whep-whep-whep getting closer. The President, arriving from Canada, said a young blonde aide. Then we went into the Teddy Roosevelt Room, where a young blond aide, this one a Navy officer with lots of medals, explained the drill: He had a list, and laureates should line up in that order. Family members, stick close to your own laureates. When the door to the Oval Office opened, the President would be ready to begin. Don't go in until the Navy guy announces the laureate's name. Laureate goes in, is greeted by Bush, then laureate introduces family. Then photo of Bush with laureate, then photo of Bush with laureate and spouse. Then laureate and spouse move away and keep going counterclockwise to stand by the fireplace as Bush greets the next laureate. Then a group photo of Bush with all laureates by White House photographers. Then press photographers come in to take that exact same photo. Then it's time to leave for the reception.

It was a very formal choreography, given its meaning by cameras recording it all.

Navy Guy: Frank Wilczek, Nobel Prize in physics. (Frank walks in, trailed by Betsy. President Bush is standing just inside the door. Bush shakes Frank's hand.)

Bush: "Congratulations. We're very proud of you. [Frank probably said something intelligent and polite here, but I didn't catch it.] Who's that with you?

Frank: This is my wife, Betsy Devine.

Bush: (shaking my hand and smiling) Congratulations.

Betsy: That was very nice of you to jump right out of your helicopter and come see us.

Bush: (looking pleased, thinking what to reply) It's windy out there.

Betsy: Good thing we're in here then.

Bush (grabbing Frank) Now, if you'll excuse us, we'll take a picture with just the two of us. (They pose, **F∗l∗a∗s∗h**) Now a picture with all three of us. (**F∗l∗a∗s∗h**) Thank you so much.

Frank and I moved off toward the fireplace and Bush really hustles toward the door to greet his next guest (and after the next guest, the next one after that) with word-for-word exactly the same script. "Congratulations … We're very proud … Who's that with you?" etc.

I imagine anybody who "meets" people in batches has to have some routines worked out in advance … though campaign donors probably get "Thank you, appreciate your support" instead of "Congratulations, we're very proud." "Who's that with you?" has a friendly, informal sound — and works equally well for one wife or fifteen grandchildren.

But I digress — then the press photographers arrived, in two groups, each one racing and leaping to set up in a good position. (**F∗l∗a∗s∗h**) (∗!∗F∗l∗a∗s∗h∗!∗) (!∗!∗F∗L∗A∗S∗H∗!∗!)

Then, as we had been told, it was time to leave for the reception.

BTW, I really did think it showed admirable self-discipline for Bush to hustle into our little ceremony right after his trip — especially considering that he probably didn't want to meet us at all. But I was surprised to hear myself saying it. It's funny the things that tumble out of our mouths at those moments we don't know what to say.

*Nobel Laureate Other Than Frank Wilczek

Posted 1/7/05; Department 'Nobel'

Nobel adventures in Stockholm

Lingonberry juice

Stockholm's beautiful Grand Hotel is full up with Nobel laureates and their gene pools. David Gross and his kids arrived with baseball caps that say "Gross: The Strong Force." Edward Prescott's cute grandchildren were in the lobby when we arrived. I haven't seen Kydlands or Politzers or Axels yet, but it's just a matter of time...er, scratch that, yes I have.

And we're all drinking metaphorical lingonberry juice. All day long.

Swedish hospitality involves lots of lingonberry juice. It's a kind of a national "welcome to Sweden" beverage. The Nobel officials who meet jet-lagged travelers at the SAS gateway whisk you away to drink lingonberry juice. The Ice Hotel brings hot lingonberry juice to guests waking up on a fur-covered ice bed. And so on — IKEA dispenses it by the imperial gallon.

Sweden is so lovely and everyone is so kind that my tongue will be metaphorically blue for a while. Now I have to get off this computer so that Finn Kydland can use it.

Posted 12/5/04; Department 'Nobel'

Nobel Museum and the tippe top

Spin a "tippe top" on its hemisperical base — it wobbles until it flips over to spin on its stem.

Physicist Anders Bárány (Nobel Museum) is one of the world's top experts...on tippe tops! Frank and I first met him in 2003, and within minutes Anders and Frank were on the floor taking a bunch of tippe tops out for a spin. It was with great pleasure we saw him again today at a Nobel Museum reception for the newest prize-winners.

The reception was part of a long, topsy-turvy day that began at 8 a.m. with the arrival of Frank's aunt and uncle. Their hotel room wasn't ready, but the Grand Hotel kindly furnished them with a "resting room" until it was. We strolled off into some glorious Nordic sunshine and found a small ice-skating rink already open for business. Sanity and jet lag soon led us home again to the hotel.

A few more fragments:
- ❖ Frank looks handsome in a tailcoat, and the people who rent them can tell just by looking at you what coat will fit you.
- ❖ Stockholm looks lovely from the top of a television tower, but cell phone reception is surprisingly bad there.

❖ The Nobel Museum store sells beautifully detailed "Nobel medals" made of gold foil with chocolate inside.
❖ Lots of people here ask Nobel laureates for autographs.

Our day ended with great pleasure and some more sanity at the hotel's Franska Matsalen, where we managed to have a wonderful family dinner that included Frank's "Nobel attendant" Cecilia Ekholm (in her real life, a foreign service officer) and Roland Zuiderveld of SVT 2. (Remember Roland, the Swedish-cultural-television interviewer whose arrival at our house I live-blogged?)

Frank needs the computer and I need some sleep. (I'm luckier than Frank here.)

And tomorrow's schedule is even more of a whirl.

Posted 12/6/04; Department 'Nobel'

My husband the kvarkmastare

Reading the Swedish newspapers online this morning — I like the picture of Frank in *Dagens Nyheter*.

I can't tell you, however, whether Frank or the reporter made this statement: "Teorin var den sista pusselbiten som behövdes för att förklara hur naturens krafter verkar."

Posted 12/7/04; Department 'Nobel'

Lunch at Ghost Castle

In Stockholm, I seem to have stepped through the looking glass into a sensible world where journalists have better things to write about than "Is blogging journalism?" For example, a bunch of new Nobel laureates with good things to say about more interesting questions.

I'll say this for real journalists — how many bloggers would jump out of bed early enough to interview my husband at 7:15 a.m.? Well, some nice young man from Swedish radio did.

Later, the Royal Swedish Academy of Sciences fed the science laureates breakfast in their tower, interrupted by a rare astronomical event — an occultation of Jupiter. (Alas, the sky was too bright for us to see it, though we could all see the slender crescent moon.)

Another press conference and a rehearsal later, the President of Stockholm University treated us to a delicious lunch in "Spökslottet", a 17th century mansion said to be haunted — hence its name "Ghost castle."

Lars Bergstrom, the Stockholm University physicist who gave such a good explanation of the strong interaction, told me that Ingmar Bergmann got the chess-playing-Death motif in *Seventh Seal* from a mural in the church in his small hometown. Even scarier, he told me that the final decision about physics Nobel Prizes, can be reached as late as the morning of the announcement!

David Gross, on my other side, gave me some Nobel-worthy wisdom about the festivities. (I had asked him if one of the laureates ought to propose a thank-you toast to our host.) David said, "This whole event is so well-planned that if we were supposed to do it, someone would have told us. And if nobody told us to do it, we probably shouldn't."

Frank and David are now on their third or maybe fourth interview. But I think bloggers blogging about journalists is even more boring than journalists journalizing about bloggers, don't you?

Posted 12/7/04; Department 'Nobel'

Music hath charms…

Tomorrow (December 8) come the Nobel lectures. To keep laureates from worrying too much, the Nobel Committee creates a giant smorgasbord of activities all day long December 7.

Then there was the moment when Frank thought the lecture notes inside his computer had vanished — but they hadn't. Whew.

Tonight, the Swedish Royal Academy of Science treated us all to dinner in yet another lovely light-filled 18th-century space. After dinner, some young Stockholm singers (Riltons Vänner (Friends of Rilton)) charmed us with *a cappella* singing, sometimes funny and other times movingly beautiful.

We were so spellbound we all forgot to worry about anything, even Nobel Prize lectures to be given tomorrow.

Posted 12/7/04; Department 'Nobel'

Sled dogs, Swedish royalty, and more from Sverker

How brilliant is this? Nobel Prize winners get to invite 16 of their dearest family, friends, and colleagues to Stockholm for the party — then find themselves with such a heavy schedule that you see all these people mostly at huge receptions or during random encounters in hotel lobbies.

Laureate: Uncle Walter! Aunt Billie! Hi! How are you? Er, sorry, the car is waiting...

That's one more reason I'm grateful to Nobel-guest-blogger Sverker Fredriksson for some more e-mail I know readers will enjoy.

Late Thursday night:

Hi Betsy and Frank,

I guess you are asleep now before the first serious duty in Sweden, to deliver the Nobel talk tomorrow. I saw you two on TV tonight, and observed that your appearance on TV fits very well the picture I already have, that you enjoy life and are easy-going and charming.

This means that Frank has a good chance to break an old tradition of somewhat boring Nobel talks. Most speakers seem to forget the audience of mainly non-experts, eager to get some knowledge about the Nobel Laureate and the discovery, and instead try to deliver their talks straight into the History Book. Just like with Academy Award Winners, the main content of the talks are long acknowledgments.

I guess Thursday is for rehearsing, before "Good Friday".

Tonight I heard David Gross tell that he looks forward to meet our King. The King is a very nice chap, who could joke and laugh when out of serious duty, but also be formal and a bit shy while doing traditional things, like inaugurating something.

The King is tremendously popular in Sweden. Being a "Republican" here does not give any votes in an election (unlike in your country — unfortunately).

Our ruling party has some footnote somewhere that Sweden should become a republic, but the government has avoided the issue successfully for about a century.

Anyway, when you get your five minutes with the Royal family on Friday night, don't forget to tell that you are going to the Arctic and will stay overnight in the Ice Hotel and visit space organizations and the mine. I guarantee that the King's attention will be on top.

You, Frank, will also probably be called to deliver a two-minute talk at the banquet. These are normally light-hearted and well fitted to an audience in a good mood, including us in front of TV. Last year's literature winner devoted his talk to "all mothers", without whose support no one would ever qualify for a Nobel prize.

This morning our main newspaper told that most American winners are concerned about those strange "tails" that you have to carry on Friday. In Sweden they symbolize the uniform of the academic world, and demonstrate that in science one is judged from ideas and arguments, and not from how fancy one looks. In folklore the entrees into the various ceremonies are called penguin parades.

Finally some news: Ingrid Sandahl [physicist and expert on auroras] is a great person, who knows almost everyone in Kiruna. She has arranged a dog sled for you on Thursday evening, after our visit to the Space High-School. She has a friend, Lena, who owns so many dogs that she plans to start up commercial tours for tourists. It would be a great thing for her upcoming business to mention you as her first "passengers". Of course, the tour is free of charge. It will not be as long as the existing commercial ones, because they can take up to a few hours, and stop for coffee somewhere (and cost a fortune).

Since you are dog-lovers, you will be able to chat with the dogs and tell them about your own beloved one at home. If you have seen sled-tours on TV, you know that the problem is not to make the dogs run — it is to make them stop. They love to run, and one normally has to throw out an anchor around a tree to get them to halt. Therefore, I cannot promise that you will get a chance to pilot them yourselves — it is too tricky. Let's hope for both Northern Lights and some moon next Thursday!

Even later Thursday night, Sverker responded to my response (telling him I've been seated next to the Swedish Prince at the Nobel banquet):

Congratulations Betsy to a nice and handsome partner at the banquet. Traditionally, the oldest physics winner's wife will sit next to the King. That was a bit of a problem when Salam got the prize, because he brought both wives. The oldest lady got the honour while the younger one had to sit with the students. This situation was not quite foreseen by Mr. Nobel, whose will is still used for how to place people at the tables.

All three Royal kids are nice and cute. And above all, our Royal family is 100% scandal-free, unlike the British one, which is 100% scandal-burdened.

Now I will go to bed too. I have spent the night with preparing a lecture later today — about alien abduction and earth radiation, as case studies within a new course in "good and bad science", where I provide examples from physics and space.

I will probably oversleep the lecture now, and all work will be wasted. Qrispin fell asleep many hours ago.

Good luck for the rest of the week, or break a leg, as the saying goes here among superstitious colleagues.

Best wishes from a half-sleeping Sverker

P.S. Also the King and the Prince are dog-lovers. The King once had a labrador named Ali. But the Swedish Muslim community objected, so he renamed it Charlie. The Prince recently walked his dog unleashed in central Stockholm, and the tabloid press immediately pointed out that this is illegal. That's the nearest a royal scandal we have been since the early 1950s! Maybe some of them have also walked against red light once or twice?

The Nobel lectures should soon be online at http://nobelprize.org. I hope the online version includes the great story Frank told (heard from Sam Treiman) about "Ohm's Three Laws."

Posted 12/8/04; Department 'Nobel'

Separating two quarks with his bare hands...

Not really, but giving a very dynamic talk about what it means that quarks won't let you do that.

Picture of Frank Wilczek giving his Nobel Lecture in Physics, Dec. 8, 2004, taken by Jonas Förare, talented photographer and biologist, now science editor and press officer of the The Royal Swedish Academy of Sciences. Thanks, Jonas!

Posted 12/9/04; Department 'Nobel'

Lots more people named Wilczek than you would imagine

How many families end up with a name that means "wolfcub"? More than you'd think, and we've had e-mail from many!

Frank's Wilczek roots are Polish — his father's dad and mom came from Warsaw and Galicia respectively, arriving sometime between the two World Wars. (Frank's mom is Italian, but that's another story.)

This morning, 5 copies of *Przeglad* Weekly Magazine were delivered to our hotel from Warsaw, Poland — you can see a shorter version of Frank's interview at their website — they called it "Lowca kwarkow," which means "Quarks hunter," I'm told.

Here, at the request of their editor, is a photo of Frank, in a rare moment of Nobel downtime, reading *Przeglad*.

Thanks to Waldemar Piasecki for making it happen!

Posted 12/8/04; Department 'Nobel'

Santa Lucia and the laureate

On December 13, Swedish families celebrate a festival of light, where young boys and girls in white robes "surprise" adult sleepers with songs and saffron cookies. The traditional headdress for girls was a crown with lit candles, but that can create surprises that nobody wants.

Needless to say, the Grand Hotel also offers Lucia wake-ups on December 13, although I'm told their Lucias are beautiful Swedish twenty-somethings.

I am eternally grateful to one of my dinner partners last night for this piece of useful information: that traditional costume for adults being surprised is "your most dignified pyjamas."

Our most dignified pyjamas. Hmmmm. Unfortunately I don't seem to have packed these. But I'm glad to have 4 days notice that they'll be needed.

Posted 12/8/04; Department 'Nobel'

Before and after Nobel Week

The American Ambassador to Sweden hosted a lunch for laureates today, including a magical chocolate dessert with a lighted "sparkler" for Frank. (Many thanks to Ambassador and Mrs. Bivins!)

Pre-Nobel-Prize Frank Wilczek with Betsy Devine. A few more banquets and the Wilczeks will be much too much "broadened" by our experience.

Posted 12/9/04; Department 'Nobel'

Wow, a Nobel honor for blogging!

This is a first — the official Nobel Prize website now links to this blog from its page on "Other Resources" for Frank Wilczek!

Now (Friday morning, December 10) we have to go practice marching to our places in the Stockholm Concert Hall. No fancy clothes needed for this exercise, although white tie and tails for all the men were delivered to the Grand Hotel last night.

Posted 12/10/04; Department 'Nobel'

Up to and including the Nobel Prize ceremony

Someone described Nobel Week as a series of "can you top this?" experiences. December 10 was a full day of such events.

The first big event was rehearsing the Nobel Prize ceremony. That is, Frank and the other laureates rehearsed marching and getting medals from the King while wives and reporters rehearsed sitting in the audience.

The part of the King was played by a very distinguished Nobel Board member — Michael Sohlman, I think — and I took several photos of the pre-event for this blog but was asked not to post them in fairness to newspapers, who had been told the rehearsal was off-limits to photos.

Then Harald drove us back to the Grand Hotel, whose fitness center serves a great salad lunch. Frank did some exercising, took a sauna, and walked to the Nobel Museum to buy more toys while the two girls and I had our hair done at the hotel…so much fun, many thanks to Morgan Johansson and his team.

Ten big black limos lined up at the Grand Hotel to ferry laureates to Stockholm Concert Hall. We rode through the twilight streets and into a courtyard illuminated by lighted torches that some pre-teen scouts were waving to welcome us.

Then, after clambering around various backstage corridors, Frank went to wait with the other laureates while I went to sit in the audience with other family. (Non-family guests, friends, and colleagues sat somewhere else.)

Frank's Uncle Walter and Aunt Billie were already sitting down when I arrived. We had great seats, in the second row from the front. I'd already showed Frank my assigned seat that morning — #84, right in the middle. For the actual ceremony, I was sitting between our daughter Amity and the Austrian ambassador. (He was there on behalf of literature laureate Elfrida Jelinek.)

I won't describe the ceremony — it's 3 a.m., and newspaper accounts (or online videos) will do better. All the laureates shook hands with the king properly, and nobody forgot to do all the right bows. The music was lovely, especially a Rossini aria sung by the young Swedish soprano Susanna Andersson. (Both the Prince and the King, when I met them later on, mentioned how very much they'd enjoyed her singing.)

After the final music and formal exits, family members rush up on stage to reclaim their own laureates. I said to Frank, "Show us your medal," and he said, "Here's the box, you open it and show us."

Wow.

Posted 12/10/04; Department 'Nobel'

Marble stair anxiety

The Nobel banquet's menu and music are secrets, right up to the day itself. But the seating arrangement — and the processional order for going in — are featured in Swedish newspapers days beforehand.

I was a bit nervous to hear that I would be walking down three flights of marble stairs on the arm of His Royal Highness Prince Carl Philip.

Long-time readers of this blog no doubt expect here some jokes about the handsome (and charming) princes in fairy tales. On the other hand, I can imagine the kind of teasing a twenty-something young man would get from friends if he were the topic of such a blogpost, no matter how admiringly I would mean it.

I had a wonderful time at dinner, between the charming and computer-graphics-savvy Prince on my left and my fellow opera-fanatic Richard Axel on my right — for even more Nobel gossip, see *Svenska Dagblad* and *Dagens Nyheter*.

Posted 12/11/04; Department 'Nobel'

Food "presentation" that will be hard to live up to

Why settle for garnishing food with sprigs of spices or with paper-thin slices of dark, dark chocolate?

The Nobel banquet staff dessert "presentation" requires a Mozart mini-opera prelude, plus hundreds of waitstaff marching in unison, holding trays high with one hand and ringing small bells with the other.

Those three flights of marble stairs add dramatic tension, and I'm told waiters aren't allowed to help serve dessert until they are veterans of previous Nobel banquets.

Posted 12/11/04; Department 'Nobel'

A happy Nobel-ified birthday December 11

It's traditional for the King and Queen to invite new Nobel laureates to dine at the palace on the next day after prize awards — so today's birthday dinner will be hard to live up to.

I might add that in addition to a charming prince who remembered today was my birthday, Sweden also has a very charming prime minister. Thanks to my table companions Göran Persson and Finn Kydland for sharing my enthusiasm for speculating about economic issues, a much safer pastime than speculating in currency.

The Grand Hotel, which does nothing in a small way, sent a huge vase full of flowers with birthday wishes. (This advantage of having hotels check your passport had never occurred to me before.)

Frank gave me a beautiful necklace of amber beads — Amity gave me a cellphone holder with a huge script 'B' picked out in cute diamond-oids — Mira gave me a huge chocolate cigar full of marzipan — mmm — and Joi Ito's birthday best wishes linked to my blog.

This was such a good birthday, I think I'll just have to stop getting older. No more birthdays for me!

Posted 12/11/04; Department 'Nobel'

Nobel ballgown advice

My childhood idea of a "ballgown" was based on Disney's cartoon Cinderella — a gleaming display of fine shoulders but no naughty bits. My Google researches before buying gowns of my own showed that Sweden's beautiful Queen and the Royal Aunt wear long flowing dresses with cap sleeves or even with long sleeves.

For the traditional royal dinner at the palace, I wore my own long-sleeved long velvet dress so that my new birthday necklace would show. I was glad of the sleeves, because Stockholm's Royal Palace turns out to be much colder than City Hall!

Posted 12/14/04; Department 'Nobel'

Other people's pictures: Swedish TV

The Swedish television Nobel page links to lots of fun pop-up pages that don't have independent URLs of their own. For the benefit of my non-Swedish readers, I've mis-translated a few of the links you might click.

To see: Still shots of Nobel ceremony and banquet.

   Click *Nobelpriser och festmingel* (Nobel ceremony and party-mingle)

To see: Two photos of Günter Grass jitterbugging, past Nobel ice cream, and more.

   Click *Snillen och glitter på tidigare feste* (Remembrance of past glitter/literati)

To see: Royalty and others walk downstairs to banquet hall.

Click *Vem bar snyggaste klänningen?* (Who is wearing the most gorgeous clothing?)

Posted 12/11/04; Department 'Feedster'

Keystone Cops version of Nobel banquet video

The Nobel Foundation has also posted some videos of the award ceremony and dinner, including a two-minute time-lapse film of the 2002 banquet, with Keystone Cops music, which is either hilariously funny or else a deeply tragic modern equivalent of Percy Bysshe Shelley's *Ozymandias*.

Posted 1/11/05; Department 'Nobel'

Prelude to Lucia Day

Last night before dinner, Laetitia and Alice showed us their Lucia costumes.

My understanding of the Lucia celebration has progressed far beyond the need for "dignified pajamas." It's now 6:15 a.m. Frank and I are both up getting toothbrushed, combed, bathed, etc. so that when Lucia arrives we can climb back in bed and look, of course, very suitably surprised.

Posted 12/12/04; Department 'Nobel'

We are no longer Lucia Day novices

At 6:50 a.m., the sparkling-clean Wilczeks jumped back into bed and turned off the light. We were, of course, wearing "dignified pajamas." (Since I'd had no time to buy new pajamas in Stockholm, we had put on our heaviest and still-unwrinkled long underwear.) Soon, we heard the Lucia knock at our door... .

The Lucia singers were escorted by two ladies from the Grand Hotel's "guest services," so we didn't have to get up to unlock the door. Soft singing and candlelight slowly progressed into our darkened bedroom. The Lucia girl had a headdress with real lighted candles — her attendants all carried candles in their hands. (I later found out these were students from a local music college — their voices were lovely!)

The Lucia attendants wore tall pointy hats (the "star boys") or green wreaths with flowers ("maids of honor"). They sang "Santa Lucia" (in Swedish, that's pronounced "Loo-see-ah" rather than "Loo-chee-ah"), a bit more Swedish Christmas music, then slowly filed out singing "Santa Lucia" again.

David Gross told me later that one year a laureate was really surprised by this Swedish custom. The sleepy laureate woke up to melodious singing by handsome young blondes in long flowing robes and jumped to the conclusion he'd died and gone to heaven.

The Grand Hotel Lucia singers also brought us coffee, saffron buns, and a wrapped Lucia gift that turned out to be a ceramic Lucia. At the time, the coffee was much less exciting than their singing. Which, for coffee-addicts like Frank and me, suggests some kind of miracle.

Later, I saw the same singers down in the lobby and snapped a few photos. The singers are serious but the people they sing to are smiling from ear to ear — I know we were.

Posted 12/13/04; Department 'Nobel'

Jumping back in time, ribbet, ribbet, to Dec. 13

I want to jump back to Stockholm University's ceremony of the Ever Jumping and Smiling Green Frog on December 13 — one of the most fun parties we've ever enjoyed.

As the youngest laureate, Frank discovered (after the first few drinking songs and toasts with aquavit) that he would be making a midnight thank-you speech. He scribbled 7 pairs of rhyming words on a paper napkin and delivered the following:

I'm well aware that the primary virtue of a speech like this is brevity. I couldn't come up with a worthy haiku, so you'll have to settle for a sonnet. The sonnet is a strict form, that requires fourteen lines. Here goes.

> *When studying physics, who would have thunk,*
> *It'd lead to public sonnetizing, quasi-drunk*
> *After days of partying and eating like a hog*

Climaxed by jumping around like a frog.
Through it all I've found that I'm taking a liking
For living life large, like a Viking.
So, when I return to my peaceful local village
I'll be tempted to indulge in some looting and pillage.
But you've taught me how Swedes got past doing wrongs:
By investing time, instead, in learning drinking songs.
On the surface everyone seems calm and nice
But beneath, in reserve, there is plenty of spice.
For this lesson, frog, and fun: Thanks a lot!
That's fourteen lines, it's all I've got.

Footnote: Remembering this night from home, in our peaceful village, I now want sleep a lot more than looting and pillage...

Posted 12/18/04; Department 'Nobel'

Ohhhh, 5 a.m.!

Hmmmm... the football-helmet-hairstyle of seventies TV-icon Patty Duke? Or the drat-those-orcs tangled coiffure of her son Sean Astin, playing Sam Gamgee in Lord of the Rings?

Last night was great fun and Frank Wilczek is now a green frog. We went to bed happily smiling at 1 a.m. — now it's 5 and we are taking last showers before packing up for the north.

I have to decide what to do with my Patty-Duke helmet-hair. Don't blame my elegant young hairdresser Morgan Johansson for creating this time-honored style twice during our stay — I talked him into it, both times. Without lots of smoothing and fluffing and serious hairspray, my hair likes to curl itself up into tiny dreadlocks.

So tempted as I am to wash it this morning, I'd much rather look like Sean Astin's mom than like third-reel Frodo or Sam. And I'll regret my helmet hair if I fall off that dogsled!

Posted 12/13/04; Department 'Nobel'

Nobelbesok above the Arctic Circle

Goodbye to Nobel glitter

Today about 1 p.m. we saw our last daylight — at least until we fly south again on December 18. We're pausing in Luleå for one night — their church was built in 1492 — tomorrow we head north again, past the Arctic Circle.

We've said fond good-byes now to Stockholm and to Nobel Week. We're saying hello to life without Cecilia, who solved all our problems with unmatchable efficiency and good humor. We're wondering how we'll survive without Harald Gustavsson, who drove us everywhere in his stretch Volvo limousine.

The Arctic Hotel of Luleå has comfortable large rooms and even free wireless Internet. But I think Frank and I will spend some time pining for Stockholm's incredible Grand Hotel…

- Golden elevators lined with sparkling mirrors.
- Fresh flowers and even chocolates for laureates.
- White-clad maidens who deliver breakfasts or pick up your laundry as if they were taking a pleasant break from performing complex brain surgery elsewhere.

Still, the future lies ahead, including the Ice Hotel tomorrow!

Posted 12/14/04; Department 'Nobel'

Ice Hotel, surprisingly, not too overcrowded with British tourists…

Hans from Stockholm sent this email to you through Betsy Devine: Funny Ha-Ha or Funny Peculiar?

Hi Betsy, what a fantastic blog you write. I really enjoy reading about your Nobel adventures. But it seems like you have restricted yourself a bit since the link came up on nobelprize.org. Anyway I saw this little article and thought you might be interested in where the other hotel guests will come from.

http://www.reuters.co.uk/newsArticle.jhtml?type=oddlyEnoughNews&storyID=7067483*

Keep up the good work, no answer expected...

Thanks, Hans!

The article:

British dream of Yuletide sunshine

LONDON (Reuters) — Britons favour turkey with all the trimmings at Christmas over finer foods like lobster or caviar but half of the nation would swap a traditional snowy setting for sun, a seasonal survey has shown...

But people were split over the perfect destination for their holiday alternative — 50 percent chose the Caribbean island of Mustique, while just more than a quarter favoured Sweden's Ice Hotel.

Posted 12/16/04; Department 'Nobel'

Email I hope to respond to, very soon

Hi mom —

How's the frozen north? The frog ball? I'm dying to hear more about the secret rituals and bismuth amphibians.

love
mickey

I'm blogging from the conference center at Esrange, just about to go watch a stratospheric balloon launch. And I will not neglect the frog ball but first a quick core-dump of this morning's notes, as we drove from the Ice Hotel (loved it!) to the space station:

Kiruna is home to about 500 people who make a living by herding reindeer and an equal number of space scientists...

Some of whom are waiting for me now — more later!

Posted 12/16/04; Department 'Nobel'

I love you, Esrange!

What a glorious place — more later, as we have to dash off to visit the Space High School…

Posted 12/16/04; Department 'Nobel'

More about Sweden's unique "space high school"

I mentioned our plans to ride in a Kiruna dogsled, but didn't say enough about Sweden's space high school. Sverker Fredriksson told us:

It has national recruitment, and VERY good pupils. Many of them come from southern Sweden, 2000 km away. It takes real enthusiam to move that far from their parents at the age of 16, in order to study space for three years.

I then got some interesting blog-email with more information and links to photos:

Odd Minde [1] sent this email to you through Betsy Devine: Funny Ha-Ha or Funny Peculiar? [2] regarding this page [3].

Hello! I found on Web in your Nobel-blogging: "…Finally some news: Ingrid Sandahl [physicist and expert on auroras] is a great person, who knows almost everyone in Kiruna. She has arranged a dog sled for you on Thursday evening, after our visit to the Space High-School…"

Here you have a picture of that Space High School. The name of the school: Rymdgymnasiet in Kiruna. Our astronomical observatory is named Bengt Hultquist observatory. Picture you find here:

http://www.malmgruppen.com/t1pubhttpdocs/temp/282422793974968_BHO_mini.jpg

Odd Minde

Project Manager
Teacher

[1] http://w1.171.telia.com/~u17106184/skolan.JPG
[2] http://BetsyDevine.weblogger.com/
[3] http://betsydevine.weblogger.com/

Thanks to Odd Minde for letting me share this email!

Posted 12/11/04; Department 'Nobel'

Nobel trip redux: Alien encounter (Dec. 16)

En route to the space high school, Sverker told us that the format of this visit would be ... unusual. Instead of a lecture, there would be a panel discussion. And the others on the panel would be...aliens!

These aliens, Sverker told, had planned to destroy all life on the planet Earth. Fortunately, Odd Minde had persuaded them to meet first with the visiting Nobel Prizewinner. If our answers to their questions satisfied them, the entire planet would be saved!

Very polite young students escorted us into the conference room — where, it turned out, the plan was also to question the Nobel spouse-cum-blogger. They'd even set up our table with two water glasses. After we were seated, "aliens" filed in, wearing long robes and striking green or blue makeup. The green leader carried a skull (which Frank later signed for him.) Never have I seen a more enraptured audience!

Their questions were a good mix of profound and jokey. And, fortunately, they liked the answers we gave (one of which was "42"). Instead of destroying the planet, they showered us with booty, including a puzzle showing the Northern Lights and a pin of a snowgoose ("kiruna") that's now on my jacket.

A good time was had by all — what a wonderful format! We came away very, very impressed with the Rymdgymnasiet students and their teachers.

Posted 1/30/05; Department 'Nobel'

Nobel trip redux: Digging LKAB in Kiruna (Dec. 17)

I've been getting e-mail from readers frustrated by the many Nobel stories I started to tell but never found time to finish. One person called it "blogus interruptus."

For example, I started several times to blog about Kiruna but never got much past telling you that it has as many space scientists as reindeer herders (about 500 of each).

On December 17, Frank and I spent the morning with Sverker Fredriksson and Ingrid Sandahl visiting the huge LKAB iron mine, which employs another 2,000 of Kiruna's 24,000 non-space-scientists and non-reindeer-herders.

Here are a few LKAB facts I jotted down:

- ❖ You could build 12 new Eiffel towers every day with the iron ore dug up at LKAB.
- ❖ Every night around 2 a.m., the mine does its blasting — which makes the town of Kiruna vibrate slightly.
- ❖ Iron ore is easier to ship and grade if you process it into uniform pellets.
- ❖ LKAB backs up its goals (quality and safety) by paying company-wide bonuses based on how well the goals are met.
- ❖ People who work in LKAB mine agree that theirs are the best jobs in the best company in the best town in the world.

That last item, however, seemed pretty widespread among people in Kiruna. I just hope space scientists and reindeer herders don't ever get into fights with the people of LKAB about who does, in fact, have the best job ever.

Posted 1/5/05; Department 'Nobel'

The adventure continues

Picturing Nobel festivities…

A delicate snow is falling on Cambridge right now — I love Halley's flour-sifter image. Good-bye, stick-figure skeletons of summer mint plants! I don't want my memories to have similar fates…

The Swedish television tape of Frank with the Space High School students who dressed up as aliens will probably vanish by Thursday, Dec. 30 (but thanks to Odd Minde for sending the link to me).

Of course, I can still ask Frank to sing Abba tunes in real life…

Anyway, I've started posting my own Nobel photo albums.

And thanks to DigitalaBonder — someone I don't know — who Flickr-ed a photo of me with Prince Carl Philip.

Posted 12/20/04; Department 'Nobel'

Unpacking the frog

One good thing about losing a suitcase in transit — it brings your unpacking into sharper focus.

Just a few things I'm so glad we didn't lose — a blue tippe top from the Nobel Museum, Ella Carlsson's thesis on Martian water, a postcard of the Kiruna Lappish church, information about space exploration at Esrange and iron mining at LKAB…and of course Frank's green bismuth jumping frog.

I'm told our lost suitcase isn't really lost, by the way. It just missed a connection in Stockholm and will arrive via Reykjavik later today — no doubt with stories of its own to tell.

Posted 12/19/04; Department 'Nobel'

Författare Betsy Devine and other Nobel Prize-winning goofs

One of my original goals in Nobel-blogging was to help future laureates dodge those "if only I had known" moments. So, here in one handy location, are just a few:

1. When you fill out a form with everyone's name, profession, and title — "title" doesn't mean "job title" as in the US, it means titles like Professor, Dr., Mrs., maybe HRH or even Duke of Earl. Oops! I was surprised but not unhappy to end up with "Författare" (Swedish for "writer") on my Nobel banquet place setting — but I wonder what happened to friend Naomi, whose dual career I'd summed up as "Film director and hotel owner"…
2. Laureates and spouses don't have to figure out how to get to the Grand Hotel from Arlanda Airport — or how to get anywhere else, for that matter — because the Nobel Committee delivers a huge Volvo limousine with a great driver (thank you, Harald!) almost to the door of the airplane.

3. When you arrive at the Royal Palace for dinner on December 11, don't draw the conclusion from previous banquets that you should start shaking hands along the huge reception line to your left. Your job is to stand in a reception line on the right.
4. Don't worry about making my mistakes, or new ones of your own, because your "Nobel attendant" (thank you, Cecilia!) will rescue you with good-humored clever kindness.

Posted 12/21/04; Department 'Nobel'

Joy to the airport

I spent seven hours of Christmas-Eve-Eve-day (today) trying to find our lost suitcase in Boston's Logan Airport.

I hadn't been waiting long when three men appeared — not Magi on camels, but young men in silly red hats who carried a trumpet, a trombone, and a big golden French horn. They sat down on folding chairs, rustled sheet music, and started to play, rather softly, "Joy to the World." They gave it an oompah bass line that was somehow funny and touching at the same time.

The music transformed the way I experienced all the people around me — and the airport was packed with holiday travelers. I saw them as people who wanted to be with their families. I remembered my brother Mark's great delight in driving children around to see Christmas lights. I remembered my mother's pride in her Yorkshire pudding.

Looking for my suitcase today meant I had to bother a whole bunch of busy, tired people. Every one of them treated me with kindness and concern.

Some of the time one of us said "Merry Christmas" to the other and some of the time one we didn't, and some people say "Happy holidays" instead, and it amazes me that people who think of themselves as Christian can get angry about "Happy holidays" or because you can't see baby Jesus in the White House crèche.

Anyway, I didn't find my suitcase (drat!) but I did find the spirit of Christmas in the kind and caring way people were treating each other, all over that crowded airport. And also, of course, in the oompah backbeat to "Silent Night."

Posted 12/23/04; Department 'Pilgrimages'

Non-Christmas non-dinner and other Dec. 25 gossip

Our non-Christmas non-dinner today was open-faced California grilled veggie sandwiches. Mmmmm, delicious.

We're saving our family Christmas hullabaloo for December 30, when both daughters will be around. Stockings will hang by the chimney with care on the night of December 29, while New-Year's-Eve-Eve will feature more gifts and big festive meal.

There's something comforting about shifting "Christmas" (the name we give our family's own pagan festival of gifting and getting together) away from December 25, the official birthday of the Christ child since 336 A.D. It somehow gives us a little more leverage against the thousand and differing voices trying each year to tell other people how Christmas "should" be celebrated.

The commercial push to buy and spend, for example, doesn't push so hard on people who don't share the same "last-minute shopping days" with everyone else.

Our first early Christmas was in 1999 or so. Because we were scheduled to spend Dec. 25 on an airplane bound for Chile, we moved Christmas to December 21. Amazingly, the next four days felt calmer for us than for anyone else we knew — getting ready for 3 weeks on the other side of the Equator and the International Date Line was much less stressful than gearing up for Christmas.

Of course, our family might just cheat a tiny bit. As I type this, Frank is thumping out "Joy to the World" on the piano. And we had some Christmas stollen for lunch dessert.

Anyway, I wish all my readers a good celebration of any kind you fancy, and a happy 2005 to follow!

Posted 12/25/04; Department 'Life, the Universe, and Everything'

Tidings of joy and a squishy dodecahedron

The five platonic solids, left to right: octahedron, icosahedron, cube, tetrahedron, and the squishy one is a dodecahedron.

In other words, our lost suitcase was found — almost exactly a week after disappearing. (Thank you, Scandinavian Air, for finding it, and thanks to Continental for delivering it to our house at 1:30 a.m.)

Now, I was extremely pleased — though the next morning was soon enough to check on it — that my winter hat, two pairs of shoes, et cetera, were once again part of my life. Frank, however, was much more excited about his magnetic "stix and balls"* from the Nobel Museum, which he has now turned into five platonic solids.

"The Platonic solids were known to the ancient Greeks," according to Steve Wolfram. Furthermore, "Plato equated the tetrahedron with the "element" fire, the cube with earth, the icosahedron with water, the octahedron with air, and the dodecahedron with the stuff of which the constellations and heavens were made (Cromwell 1997)."

Of course, if the heavens were made of magnets and balls, they'd also squish out of shape from their own weight. But I know Frank has some tricky un-squishing planned for the immediate future…

*Magnetic stix and balls are made in the UK, and various people sell them online, if you want some. Frank has 3 kits — they cost about $10 each.

Posted 1/2/05; Department 'Nobel'

What came out of the schmutz on November 11

I just found out that the TV camera crew I live-blogged on November 11 produced quite a good short movie with that footage.

It includes a lot of the stuff I wrote about here (our piano really needs a tuning!) as well as some intriguing physics clips:

- ❖ BANG! A man shot from a cannon flies into a net (at least 20 times).
- ❖ Crash! An old truck gets bashed into rubble (only once), and
- ❖ Hmmmm…The night sky opens up to reveal…David Gross!

Roland Zuiderveld does a good job on the science of the strong force, quarks, and gluons — all in 15 minutes and in Swedish, with time for Frank on the accordion and the two of us slowly walking near our house.

Our dog Marianne doesn't get any screen time — neither does kitchen-blogging with my computer on top of the stove — but despite these small flaws, I really enjoyed the movie — so thank you, Roland! And thanks to Dennis, mysterious friend of GD, for emailing me the link.

Posted 1/8/05; Department 'Nobel'

Hard to believe, Nobel parties were a month ago...

I'm not complaining — I'm enjoying my real life one heck of a lot. And W.W. Norton just re-published the paperback version of *Longing for the Harmonies*, a book Frank and I wrote together that a whole generation of kids (maybe you?) used to learn about fun non-math physics.

Norton mailed us some paperbacks — the cover is new, but the text is an old friend. I love re-seeing the graphics I did with MacPaint and my dot matrix printer. I'll post some another day.

Meantime, I even got an Amazon Associate link to help me track its statistics. If you buy a book by clicking on a link in my blog, I might ultimately get 50 cents.

Messing with html and hoping to get 50 cents does somehow feel more like the real me than dressing up in an evening gown every night — but the real me enjoyed that too, a month ago!

Posted 1/14/05; Department 'Nobel'

Postmature ignitulation (Loud sirens, but no kaboom)

Our old furnace choked up today while I was shopping. I got home to a house full of dark smells and weird heat — and the basement, when I checked it, was billowing smoke!

Three loudly-sirenaceous-red-firetrucks full of kind-and-cheerful-firemen later, the firebox of my furnace was still smoldering but at least not blowing out smoke all over the house.

"You need to get the furnace cleaned every year," said the fireman. Well, I usually do, but, dang, 2004 was busy with both daughters graduating, both daughters moving (twice), one daughter getting married, and just as things are starting to settle down Frank gets that Nobel phone call, and one of the things that slipped my mind was the furnace.

Not that I'm complaining, especially since our house did not blow up.

At least not yet.

Posted 1/20/05; Department 'Life, the Universe, and Everything'

First blogger with DNA for sale on eBay...

One small step for blogging, one giant surprise for the blogger: the blue and green aliens at the Swedish Space High School have teamed up with eBay to offer "Unique Nobel Prize Winner 2004 Memorabilia": two drinking glasses left unwashed since we used them.

If you happen to feel like bidding on what would indeed make a truly unique gift for that special someone…

***

Bidding on Wilczek and Devine DNA continues, thanks to link-love not only from me but also from Greenpass Zoe ("People will do anything!") and a lovely post by Julie titled "Clone Betsy Devine".

Frank, who is as we know a genius, points out that bidder "betsythedevine" is probably not too anonymous. Said bidder's excuse is that he or she would enjoy a souvenir of an unforgettable encounter with blue and green painted aliens at the Rymdgymnasiet.

Opposing bidder "oceanman" no doubt represents a secret evil clone-lab somewhere in the mid-Pacific.

Posted 1/25/05; Department 'Nobel'

"275 kronor for genius DNA"

The eBay auction is over! I outbid "oceanman" to win a unique souvenir of our alien encounter in Kiruna.

One surprise side benefit has been oodles of emails from our friends in Sweden!

> *Dear Betsy and Frank, Today you were mentioned in the newscast on national Swedish radio and there was also a story in the newspaper. You probably guess why. The story about the DNA bidding…*

Still making waves…I saw a short article in today's paper about the glass that the students tried to sell at an auction on the internet — and that you bought it! It was great fun to see your names in the paper again.

Hello Betsy! I followed the bidding at eBay with "Frank's DNA". Swedish biggest newspapers, local papers and media write about it today, i.e. http://www.aftonbladet.se/vss/telegram/0,1082,64554414_852_,00.html…

Heh! Here's my rough mini-translation: "One drinking glass as used by a Nobel Prizewinner in physics was offered on eBay by pupils of Space High School in Kiruna. Nobel Prizewinner Frank Wilczek drank from glass while visiting the school, and thereby deposited some genius DNA. Said DNA will never be used for a future clone, however, as it was his wife who bought it for 275 kronor."

Even when you add an extra $25 for the novel experience of having said DNA delivered by Airborne Express — that's a genuine bargain on my favorite DNA!

One thing I'm pretty sure of — this will never replace the good-old-fashioned method of DNA transfer.

Posted 2/4/05; Department 'Nobel'

Unique Nobel DNA souvenir in my email and elsewhere…

I just got email from Elizabeth Thomson at the MIT News Office…it seems she just got email from Odd Minde of Kiruna, Sweden….

From: Elizabeth Thomson

Date: February 25, 2005 9:29:27 AM EST

Subject: Fwd: DNA from Wilczek

Betsy and Frank!

Received the following note, and EVERYBODY over here has thoroughly enjoyed it. Ultra-bizarre! Just noticed that bidding has now stopped, and the winner is…betsythedevine. Betsy: did you indeed recover the glasses?

Do tell!

Elizabeth

>*Hello!*

>*In connection with your article:*

>*MIT's Wilczek wins 2004 Nobel Prize in physics*

>*Elizabeth A. Thomson, News Office*

>*October 5, 2004; updated October 6, 2004*

>*I suggest you to look at:*

>*http://cgi.ebay.com/ws/eBayISAPI.dll?ViewItem&item=6149217230*

>*Best wishes*

>*Odd Minde*

>*Rymdgymnasiet*

>*Kiruna*

>*Sweden*

My reply:

Hi Elizabeth — yes, the package from Sweden arrived yesterday!

The Rymdgymnasiet students had packaged each glass in layers of bubble-wrap with labels "Betsy Devine" and "Frank Wilczek." I hesitate to un-bubble-wrap them now, and maybe the bubble wrap is part of the story.

As you can see, I put the bubble-wrapped glasses over our fireplace, where some of Frank's robots and Lego constructions seem to be happy to see them.

All best,

Betsy

P.S. Do you mind if I include your nice email when blogging this?

Posted 2/25/05; Department 'Nobel'

Nobel prize for ghostbusting?

I'm racing around getting ready to drive to MIT where I hope to live-blog Frank's latest adventure — he's appearing on Penn and Teller as one of their experts to talk about ghosts!

Penn and Teller started out as magicians with hugely successful and very funny stage show. They now have a TV show on Showtime called, er, "*Tishllub*." (I'm assuming the censors of Google can't spell that backward, but I'm sure you can.) Each episode pokes fun at some popular fakery — like UFO abductions or talking to the dead — showing the audience how its "results" can be fudged.

It's a pretty amusing show, spiked with pretty surprising language. As Teller (I think it was Teller) explained, "We could get sued if we call somebody a liar, but our lawyers say calling them &#@@!! or **&^%$??!! is fine." (Do their lawyers know about Michael Powell?)

Anyway, Frank won't have to swear — at least, none of the experts on other shows did — and he has some pretty cool physics demos set up. This particular show will be about ghosts, a topic on which I do have more to say, but in another blogpost.

Posted 3/8/05; Department 'Nobel'

My personal run-in with Albert Einstein's ghost

Years ago, when our now-grown-up daughters were smaller, Frank and I moved into Einstein's house, where we would live for the next 8 or so years, surrounded by Einstein's furniture.

Of course, we two planned to sleep in Einstein's bedroom, a tiny room that was dwarfed by Einstein's study right next to it. Einstein's big Biedermeier bed was not really big enough to hold two people, we later decided, but on our first night there we didn't know that yet.

Now, I don't believe in ghosts, not even in Einstein's. But it was very strange, in the middle of that first night, to wake up and hear the sound of slow, heavy breathing that was not Frank's breathing or my breathing or the breathing of one of our daughters.

Whish–pause–whoosh. Whish–pause–whoosh.

It's one thing not to believe in ghosts, and it's another thing not to be spooked by strange midnight noises!

One of the good (and bad) things about a small bed is that if you are awake in the middle of the night, you don't have to do something active to wake up your partner. Your partner will automatically wake up anyway. Here's how it played out, at least in my recollections:

Mysterious noise: Whish–pause–whoosh. Whish–pause–whoosh.

Frank: (Sleepily) Betsy, is something wrong?

Mysterious noise: Whish–pause–whoosh. Whish–pause–whoosh.

Betsy: (very tiny voice) What is that noise?

Mysterious noise: Whish–pause–whoosh. Whish–pause–whoosh.

Frank: It's the steam radiator.

Radiator noise: Whish–pause–whoosh. Whish–pause–whoosh.

Betsy: Oh.

So Penn and Teller may not realize it, but when they got Frank, they got a real ghostbuster!

Posted 3/8/05; Department 'Nobel'

Tishllub TV: *Virtual Penn and an ethereal Teller*

Inside Frank's office, the interview filming has started.

I'm not sure if I'm happy or sad that the giant, big-gestured Penn and the fast-fingered Teller aren't here in person to ask Frank their emailed questions. (They're in Las Vegas somewhere, appearing nightly in "an edgy mix of comedy and magic involving knives, guns, fire, a gorilla and a showgirl." The MIT physics group could absorb all those elements with easy *savoir faire* — except for the fire, which would set off our overhead sprinklers.)

Meanwhile, I noticed on the Penn and Teller Showtime webpage for "Talking to the Dead" that the "experts" are shown all together with links to their homepages. That is, the "Your mother is speaking with my voice" medium (Penn called her, if I remember this right, a "pigdog") is on just the same basis as the respectfully photographed debunking psychologist...

Meanwhile, I am so tempted to stand outside Frank's closed door and make a very soft and ghostly "Whoooooooooo"...

Posted 3/8/05; Department 'Nobel'

Geek celebration: Our billion-second-iversary

That's 1,000,000,000 seconds, a US billion.

I blogged our funny wedding in NJ traffic court — that happened late in the evening of July 3, 1973.

10 seconds, 100 seconds, and 1,000 seconds later, we were still shell-shocked newlyweds, most likely still lost somewhere out in Dutch Neck, NJ.

10,000 seconds after we tied the knot — a bit less than 3 hours — we were eating cake-mix cake and ice cream to celebrate.

At 100,000 seconds (near the July 4/July 5 borderline) Frank was gone! He spent our "honeymoon" in Sicily at a physics summer school; I spent it in Princeton, doing much duller grad school thingummies, and missing him very much. At 1,000,000 seconds, a long 11 days later, our separate-honeymoon phase was still ongoing.

At 10,000,000 seconds, we get to October 26, 1973. We were happily sharing our married-grad-student-housing with two white mice. We had, in those long ago days, our own separate slide rules. (Computers were room-sized, $10,000,000 thingies.)

At 100,000,000 seconds, we're all the way up to September 5, 1976. Frank was all the way up to being an assistant professor, our beautiful daughter Mickey had just turned two, and I was thinking about getting back to grad school.

Now, 1,000,000,000 seconds of marriage takes us to Friday night, March 11, 2005, roughly 11 p.m.

Now, if you think it's geeky to celebrate powers-of-ten-iversaries, remember this: if we were truly geeky we'd celebrating powers-of-two-or-eight-or-sixteen-iversaries.

I'm sure looking forward to our next billion seconds!

Posted 3/10/05; Department 'Life, the Universe, and Everything'

On to the next billion...

In case you young folks are wondering what equipment it takes to celebrate a one-billion-second-iversary, the answer is lots of chocolates and two flashing-lighted rave rings.

No, you don't have to keep on wearing the rave rings for the next billion seconds.

Posted 3/12/05; Department 'Life, the Universe, and Everything'

Enjoyable load of Taurus

Don't tell George Bush, but Frank Wilczek now seems to be famous in France, according to Google. A French astrological website specializing in famous people (including not only "scientifiques" but also "sex symboles") has posted Frank's astrological birth chart.

People who share Frank's May 15 birthday, according to their list of notable Tauri, include twins Madeleine Albright and Trini Lopez, both born in 1937 — and, among many others, Socrates (466 BCE)!

Back to earth for a moment — Wikipedia has Socrates's date of birth as June 4, 470 BCE, and the Encyclopedia Britannica isn't even sure about the year, but what do I care?

It's more fun to picture Socrates as Frank's astrological alter ego. And helping people believe what they want to believe has kept astrologers in business for many, many cycles of Sun and Moon.

Posted 3/24/05;Department 'Life, the Universe, and Everything'

My email from a physics conference in Delft

Our hotel is an old Delft canal house, and our sunny room overlooks an old Delft canal, appropriately called the Oude Delft Canal. I strolled to the big town square called the Markt today, while Frank tried to replace some jet-lagged sleep.

No actual rain fell on my mithril raincoat, but cloudy skies and a wet breeze made me glad I'd brought warm clothing for these few days. It is a lovely, classical Dutch civic outdoor space, with fake wooden shoes now on sale from every ground floor.

Then, after seeing your dad off to his conference, I hopped on a train to Den Haag, to revisit the lovely Vermeers in the Mauritshuis (most notably "Girl with Pearl Earring" and "View of Delft.")

Of course, much of the fun was also riding Dutch trains and trams again, not to mention wandering cheerfully over the bricks and cobbles of many small side streets. Also ordering a lunch of mixed appetizers and being very surprised that it included bits of raw hamburger (?) — which I did not eat, by the way. Now I'll take a much-needed nap, but not before forcing this email into double duty as a traveler's blogpost.

Love you a lot and xxx to all,

Mom

Posted 4/8/05; Department 'Pilgrimages'

Happiness is a Dutch bicycle

On page 32 of his new book, *Happiness: Lessons from a New Science*, Richard Layard charts happiness versus per capita income of various countries. The country with the highest percent who are "Happy" or "Satisfied"?

The Netherlands, by a good bit.

Why? Dutch bicycles, if you want my opinion.

Holland has no monopoly on Layard's big seven — family relationships, financial situation, work, community and friends, health, personal freedom, and personal values — factors he claims can account for much of happiness.

Here's why the Dutch norm of bicycling everywhere creates more happiness:

- ❖ Exercise makes people's bodies feel good.
- ❖ Exercise lifts people's spirits.
- ❖ Bikers are not anonymous the way drivers are; hence their traffic interactions are much more civil.
- ❖ Riding a bike instead of a Hummer to work is just one example of the general Dutch aversion to flaunting wealth — the struggle to keep up with (or better) your neighbors creates much unhappiness in many cultures.
- ❖ Almost running over clueless American tourists who will go home and blog about you gets your pulses racing.

In that last area, Dutch bicycles have just made me happy as well.

Posted 4/7/05; Department 'Pilgrimages'

First glimpses of Riyadh

We got to the Al-Faisaliah Hotel about midnight last night — it's a striking sight against the Riyadh skyline, somehow more welcoming than the Kingdom Tower.

We're here because the King Faisal Foundation very kindly awarded Frank a King Faisal International Prize in Science.

Other prizewinners got here several days ago — we missed camel rides and a picnic in the desert — but Frank had a previous promise to Delft that could not be broken. The KFF was understanding about our delay, and gave us a royal welcome even at midnight.

After meeting a number of charming dignitaries, drinking watermelon juice, eating cotton candy with salted almonds (and not eating even more tempting things we were offered), and then unpacking, we finally got to bed around 2 a.m.

Then we got up at 7 for today's very full schedule. So if I'm incoherent, that's my excuse!

Sandalwood, myrrh, pink rose petals in a glass bowl

This morning, we visited the Riyadh suq, an outdoor market with many different displays of sandalwood, buckets of antique jewelry, and enough inlaid daggers to set off ten kinds of alarms at airport security.

The King Faisal Foundation is taking wonderful care of all their winners — and I'm loving our rooms in the Al-Faisaliyah Hotel, whose luxuries range from fresh fruit and flowers (including floating pink rose petals) to ultra-modern touch pads at each bedside that let you open or close three levels of curtains.

But even outside this VIP cocoon, I've found Saudi Arabia more welcoming than I expected. In Dammam airport, as Frank and I tried to find our way to our next airplane — picking up somebody else's suitcase by mistake, and leaving one of our own suitcases behind — the many Saudis we spoke with went out of their way to help us. I'm told that hospitality to guests is one of the first social skills a Saudi child learns — I can well believe it. Even our suitcase gaffe was met with great tact.

Living in any country on this divided planet, it's easy to pick up some caricature ideas of any other country you can mention. A recent eloquent speech by Saudi foreign minister HRH Prince Saud Al-Faisal addresses a lot of western caricatures of this part of the world.

Another perspective comes from British scholar Carole Hillenbrand, who just won a King Faisal International Prize for her book on the Crusades as seen by Islamic eyes. By the way, Carole would like to make it clear that she is not "fluent" in 11 languages other than English — 11 is merely the number of languages she's studied.

Such modesty is admirable, but…I clearly remember once trying to play a dictionary game in a group that included Carole. We had to give up because none of us could find, in the big dictionary, even one English word that Carole didn't know…

Posted 4/12/05; Department 'Pilgrimages'

In Zurich Airport, thinking about abeyas

Women in Riyadh wear long black abeyas and cover their heads. Some women — fewer than half — also cover their faces. Mia brought glamorous abeyas for both of us from Cairo. I'm sorry this photo doesn't show the opulent gold-patterned sleeve and scarf-ends on mine, but generalize from her fuchsia band and you get the idea.

Many indoor spaces, including our hotel, were abaya-optional zones — full of dynamic, well-educated Saudi women, some of them wearing headscarves and some of them not.

Outdoors, their black color makes them hot when the sun comes pouring down. I predict that the inventive Saudis will soon invent a new, improved tradition of wearing white abeyas in summer.

Now it's Frank's turn with our ethernet cable!

Posted 4/13/05; Department 'Pilgrimages'

Betsy Devine's "roast" of Frank Wilczek and David Gross, at Princeton, April 30, 2005

It's wonderful to see so many friends here tonight — not that I can recognize any of you, all dressed up for a black tie affair! That's a shocker — as we celebrate work done by two guys wearing blue jeans, in 1973.

So I'd like to take us all back to those magical years, when Princeton was still very new to me. I'd rolled into town in my beat-up VW camper, after a lifetime spent in New England girls' schools — where our architectural model is a white-shingled Main-Street house.

So I walked into Princeton's beautiful Graduate College — and the opulent Princeton campus — and what I saw there simply blew me away And I thought, "All this gorgeous architecture, the statues, these gardens of flowers, were put there by people who really cared about learning. And they wanted to inspire people like me — well, okay, maybe not really people exactly like me…" because Princeton had only just started admitting woman graduate students… "okay, maybe I'm not the person they were imagining, but this is my opportunity too, and I'm going to grab it."

Still, I have to admit that the opportunity I was to grab with the most enthusiasm was a cute third-year grad student whose name was Frank Wilczek — a very young third-year grad student, because when I met Frank in June of 1972 he was only about a month past his 21st birthday.

Now, fortunately for my genetic material, 1972 was the summer of the Fischer-Spassky chess matches. And the grad college had only one television — and its position between New York and Philadelphia meant that the antenna could pull in some very large number of channels, it may have been 7 or 8 different channels! — so all the grad students would watch Fischer-Spassky together. And I couldn't help noticing, as we sat there heckling the chess players, whenever Frank Wilczek would shout out a suggestion — "Pawn to king six!" — Boris Spassky or Bobby Fischer would do what he said. And even if Frank disagreed with the rest of the room — if we were all hollering "Take the bishop!" but Frank hollered "Take the knight" — the real chessplayers did what Frank said and not what we said. So I said to myself, "This is a very smart person and I would like to get to know him better."

I had, in fact, already been introduced to Frank. I blush to say, I was introduced to him by my boyfriend at that time. But it wasn't long before Frank and I were an item, and soon our record players — remember those? — were in one apartment.

Perhaps it was fate that somehow brought us together — he with his trusty slide rule and I with my separate but equal slide rule. He with his copy of the CRC Handbook and I with my own copy of the CRC Handbook. For you young folks who don't know why we had CRC handbooks, I'll just say that looking up logarithms was something we often felt obliged to do in 1973.

One of the things I learned when I met Frank Wilczek was that somewhere in Princeton there lived a mighty genius named David Gross. And I also learned that Frank did not want to hear me make jokes about David's last name. Eventually I got to meet this mighty genius and was duly impressed. But David reminded me he gets to speak last and I don't, so perhaps I should say no more…

Jumping ahead for a minute, I was thrilled today when the portraits of Frank and David were added to the gallery of Princeton's Nobel Prize winners in Jadwin Hall. Frank and I used to admire those pictures together, on our many long late-night strolls through the bowels of Princeton, when we would often end up eating bagels and lox in the old Colonial Diner and doing the next day's *New York Times* crossword puzzle, which arrived at the newsstand about 4 a.m.

Jadwin Hall basement had that wonderful gallery, and it also had blackboards full of wonderful gnomic writings by our fellow-midnight-wanderer John Nash. And if I think back to the younger self I was then, I would have been very pleased but not too surprised to know Frank would end up with his picture in Jadwin Hall. But I would have been darn surprised that John Nash got a Nobel Prize before Frank did!

I can remember, in 1972 and 1973, how much it meant to us both to be welcomed into the Princeton physics community. I was so excited the first time David and Shula Gross invited us over to visit them at their house! But that was just the first of many happy times spent together. In fact, if our daughter Amity could be said to have a third parent, that third parent would be Elisheva Gross. Elisheva was just a baby when Frank and I married, but she was such a beautiful and smart and charming baby that she got us both thinking that we'd like our own Elisheva.

Sam and Joan Treiman were also a very important part of bringing young physicists into the physics community. I remember so many physics parties at their house — we would all gather in the living room for food and chat — then the men would disappear down cellar where Sam would whup them one and all at ping pong. Joan made her part of this look so effortless that it would be easy to forget to say "Thank you, Joan," but I don't ever want to forget to say it. Thank you, Joan! I wish Sam were also here to be part of this moment.

What a year 2005 has turned out to be! The World Year of Physics, some say, or the International Year of Physics according to other groups. In England, they're calling it the Year of Einstein. Yes, physics is really in the news these days.

Why even our government is talking now about "The Nuclear Option." Or, in the case of our president, "the nucular option." So, what a great time to be a physicist!

I guess my talk hasn't really been much of a roast — I'd have to say it's really more of a toast. A toast to physics. A toast to the physics community. And here's to Frank and David — congratulations.

Jetlaggedly yours from beautiful Barcelona

The Hotel Neri is beautiful — I love the rooftop garden and even the Flash-y website.

And I'm sure Keats would have been very inspired by the magical, laughing groups of twenty-somethings who keep passing just under our window as they head for what will no doubt be a wonderful party.

On the other hand, it's only 3 p.m., so I can't very well wish they'd all gone home to bed…

Less dozey and more informative blogging later.

Posted 5/2/05; Department 'Pilgrimages'

"The only flooded Amazon forest in Europe"

Frank and I got into the elevator and pushed a button for floor "-5".

Barcelona's science museum CosmoCaixa — which just reopened after a huge expansion — is celebrating the World Year of Physics with lectures in praise of its most beautiful equations — that should give some idea of their energy and ambition.

And how about…

- ❖ A flooded Amazon jungle, on display from its rain-forest canopy to fish swimming in between tree-roots.
- ❖ "Over 90 tons of rock in a single exhibition" (to quote their English website).
- ❖ Six iguanodon skeletons here on a road-trip from Belgium.
- ❖ An enormous Foucault pendulum that knocks things over with satisfying clunks, which was built by the very kind man who also solved 97 computer complications for me.
- ❖ And 50,000 square meters of more of the same…

It's more than enough to wake anyone up from jetlag!

Posted 5/3/05; Department 'Pilgrimages'

10 beans for a rabbit, 100 for a slave

Cocoa beans, that is, and that's just one of the things you can learn at Barcelona's Museum of Chocolate:

- ❖ The Aztec word *xocoatl* means "bitter water" — lily-livered Europeans were the first to flavor it with milk and sugar.
- ❖ The scientific name *Theobroma* means "food of the gods" — an editorial comment by Linnaeus with which I thoroughly agree.
- ❖ In my childhood, chocolate was one of the treats young Catholics were urged to give up for Lent. But medieval monks were allowed plenty of chocolate during their many religious fasts — IMO, a classic early example of the way technology creates new temptations much faster than bureaucrats can create new sins.

Time to get off this smoky cafe computer, so hasta la vista[*]! Oh, one more discovery today — Barcelona hot chocolate is thick and black, like melted Hershey bars or hot chocolate pudding. I drank half a demitasse and I won't need any lunch — not that I'm complaining!

[*]Also *"fins aviat"* — email from Josep Perarnau reminds me that Barcelona is in Catalonia as well as Spain, so the Catalan "see you later" is also on topic…

Posted 5/5/05; Department 'Pilgrimages'

Frank in La Vanguardia...

The things you learn about your husband by reading pull quotes from newspaper interviews...these from *La Vanguardia*:

Tengo 53 años: menos memoria, pero mejor utilizada. Naci in Long Island y soy orgulloso fruto de la escuela pública de Nueva York. Casado, dos hijas...A veces me escuchan. No sé si Dios es; el de los humanos no lo he visto. No juzguen a EE.UU. por sus gubernantes: son mucho más mediocres que el pais.

Via Google language tools: I am 53 years old: less memory, but better used. I was born in Long Island and I am the proud fruit of the public schools of New York. Married, two daughters... Sometimes they listen to me. I do not know if God exists; the one of the humans I have not seen it. Do not judge the U.S.A. by its politicians: they are much more mediocre than the country.

I also discovered that the Spanish for "gluons" is "gluones."

Posted 5/6/05; Department 'Pilgrimages'

Soft watches and traveling Saturdays

Inspired by the theory of general relativity, the iconic painting of melting watches by Catalan painter Salvador Dali, inspires reflection on this traveling Saturday.

Time slip...

> You pack up a tiny part of your normal life and leave all the rest behind.

Time slip...

> An airplane holds you in limbo for many hours.

Time slip...

> Sunrise and sunset in Barcelona are six hours earlier than in Cambridge; you re-set your watch.

Time slip...

But local mealtimes also differ — lunch is at 2 p.m.; dinner at 10 or 11.

So what time is it? Is it time to eat? to sleep? or (of course) to blog!

Posted 5/7/05; Department 'Pilgrimages'

Gaudi nights and days in Barcelona

Heading for chilly Oxford this morning (must pack!)

Just quick miniblog on Barcelona —

- ❖ Wonderful science museum; we spent 6 hours exploring and could have spent more.
- ❖ Wonderful Gaudi buildings and magnificent buildings created by his rivals.
- ❖ Wonderful narrow cobbled streets near the cathedral get cleaned every single day.
- ❖ Wonderful friendly people, even the extremely trendy-looking young staff of the wonderful Hotel Neri.
- ❖ Wonderful food, especially crema Catalana, an improved (if you can believe it) creme brulee with the custard less heavy and the sugar top thicker.

My favorite tourist snapshot: Gaudi's hugely ambitious Sagrada Familia, still under construction, with a busker dressed as Barcelona's Columbus statue, and two tourists (Frank and Betsy) enjoying it all.

Posted 5/10/05; Department 'Pilgrimages'

Word from a lover (and hater) of science museums

Too often modern "museology" is a smudge on the nose of a fine old museum collection. A few years ago, the big museology fad was that old-fashioned science museums were boring. Therefore every science museum, everywhere, cleared some big chunks of space for a show about "Sharks are scary!" or "Snakes are dangerous!"*

I understand that not everybody loves, as I do, the massed ranks of whitened skeletons that Paris displays (probably, just as Georges Cuvier organized them).

But it is possible for museum displays to be fresh instead of stale without losing scientific value. The Barcelona science museum, under the direction of Jorge Wagensberg, is proof of this.

The display of six giant iguanodon skeletons amusingly shows the sequence of scientists' ideas about how to pose them. Did they walk on four legs or two? The huge thumb-claw was depicted as a nose-horn for many years, until intact skeletons made it clear where it went. Another sequence of signs describes scientists' changing theories about the finding of 23 skeletons so close together, and talks about how evidence supports or contradicts each one.

In addition to big displays to ooh and ahhh over — let me re-mention the sunken Amazon rainforest — the interactive exhibits are clever, relevant, and in good repair. (Frank loved the Coriolis force machine and was delighted to find its twin in Madrid. Both museums are funded by Cosmo-Caixa Foundation, so they share inspirations back and forth.)

The CosmoCaixa Science Museums of Barcelona — it's up near Tibidabo — and Madrid — go see both of them if you get a chance!

*You can see exactly the same effect when school curricula get re-written by people who hated school and thought all their subjects were boring or too hard. I was lucky enough to attend public school when some of the smartest and most ambitious career women in my hometown were proud to teach kids the subjects they had learned and loved in those very same schools.

Posted 5/11/05; Department 'Pilgrimages'

Gabriel Garcia Marquez, Frank Wilczek, and Iker Casillas

If you ever go to Madrid, eat at Casa Lucio — everyone else does, including the Spanish king.

Gabriel Garcia Marquez had eaten lunch there on May 10, so when Frank was dragooned to sign their "golden book" that night, his page was right after another Nobel laureate. All the staff made much of us — the owner and his daughter came over and got their photographs taken with Frank — it's no wonder that celebrities like to go there!

After we finished our dinner (about half-past midnight) and were making our way to the door, with Frank being introduced to various patrons and shaking their hands, we met a second entourage of young jet-set types arriving for dinner.

Frank and young Mr. Jet-Set were urged to shake hands with each other, which they duly did, both smiling with baffled but friendly amiability. As the two entourages drew them apart in their different directions, one of our Spanish friends asked, "Do you know who that was? That was the goalkeeper for Real Madrid!"

Real Madrid is the local soccer team, and I have used Google to guess that the name right after Frank's in the golden book will be Iker Casillas. And I'm guessing his bafflement about being introduced to Frank was at least as great as Frank's at being introduced to him.

Of course, food is the real reason to go to Casa Lucio — and the English menu dutifully and honestly translates its local specialties … "gills of hake" … "fat capon" … and so on. "Gills of hake" is a kind of gray soup with fish gills floating in it — but since four out of the nine people at our table ordered it, it probably tastes better than it sounds…

Posted 5/17/05; Department 'Pilgrimages'

Golden Plate: Yogi Berra aka Lawrence Peter Berra…

… made a great funny short speech that ended with "I'm hungry" — the best kind of before-dinner speech — before getting his medal from four-minute miler Roger Bannister. Bill Clinton made an incredibly fascinating and incredibly long speech, but who would have expected him to do anything different?

I'm always thrilled when Frank gets a new award, but I'd never heard of the Golden Plate Achievement Awards. (Thank you, Mr. Google.) It seems the Golden Plate people like it that way — the press is not invited to their parties, but (in my capacity as Mr. Jennifer Lopez) this blogger is.

So here we are, in yet another hotel suite, this one the St. Regis in NYC, and as I head off for the gym I'm hoping that Katie Couric or Sally Field will be on the next treadmill over, because they're here too for the next few days (among many others), doing symposia, meeting some fascinatingly brilliant grad students (200 I think), and getting their own Golden Plates.

More later, but one other big star-struck experience so far was dinner last night in the Temple of Dendur at the Metropolitan Museum. I kept hoping to see Big Bird or Snuffleupagus. Anybody else remember *Don't Eat the Pictures*?

More, from the press office of the President of Botswana… The press office of Botswana has much more information

Posted 6/2/05; Department 'Nobel'

Golden Plate: Naomi Judd heart Leon Lederman

"Do you know Leon Lederman*?" asked a tiny, pretty woman, impeccably made up and wearing a pink cowboy outfit. "Leon is one of my favorite people."

Naomi Judd is a country singer, songwriter, motivational speaker, mother of two famous daughters (singer Wynnona and actress Ashley) — and just all-around full of surprises.

She also has her own website (Get a blog, Naomi!), and a lot of zingy one-liners in the book Amazon let me search so I know that I want to buy it. Amazon doesn't seem to have the CD with "Big Bang Boogie", which she wrote for Leon, and to which he danced with her onstage at a concert.

If I had to make a bet, I would bet that Naomi Judd is one of Leon Lederman's favorite people too.

*Leon Lederman, Nobel 1988, is one of my favorite physicists too, and one of the best joketellers I've ever met. BTW, in physics circles, you meet some very good joketellers.

Posted 6/3/05; Department 'Nobel'

Golden Plate: 'Frank' advice to Rhodes Scholars and etc.

At the American Academy for Achievement's 2005 International Achievement Summit, prizewinners each give a short speech to student delegates — about 200 Rhodes Scholars, etc. from 50 countries. So we all heard some darn good speeches but of course the best was given by Frank Wilczek — short, pithy, informative, and very funny.

I especially like his illustration of how to choose problems wisely:

It's easy to give vague advice, but I will break new ground, and give you an algorithm. Many of you are probably thinking about getting married, and naturally you would like to maximize your chance of finding the best possible mate. I'll give you an algorithm for that...

Wait, I hope this doesn't mean I'm a problem...

If you want Frank's algorithm, and to understand why my new nickname is "n over e plus one", see Frank's 5-minute talk including as a bonus Einstein's favorite joke ("Advice to Students").

Posted 6/5/05; Department 'Nobel'

Golden Plate: Thanks to Wayne and Catherine from n/e + 1

What fun — and now I'm so tired. Partying until 1:30 a.m. and then up at 6 to pack.

What an incredible party.

So here we all were, riding on a bus to Golden Plate Banquet. Frank and I turned around in our seats to ask George Lucas and Dorothy Hamill, who were sitting behind us, if there were any tricky parts to the gold medal ceremony. (They told us there weren't.) Then we started talking about the different things that go wrong at Oscar and Nobel ceremonies.

At the banquet, we sat with Frank's Nobel littermates Aaron Ciechanover, Linda Buck, and David Gross — and John Fogerty of Creedence Clearwater Revival.

Flashback to 1970 or so when I was struggling to convince a classical music fan that there was really good music in rock and roll. Mrs. Krasner challenged me to play her something convincing. Because she loved Beethoven (crash, jangle, thump!) I picked Creedence Clearwater's "Proud Mary." I don't think I really convinced her to switch from Beethoven based on that playing, but I really enjoyed telling John Fogerty the story. He's redhot on stage (and touring now!) but very sweet in person.

So why is this blogger laughing so hard? Just as Linda Buck got ready to snap the picture, Sam Donaldson popped up behind her and started giving her devil's horns.

Isaac Asimov told Marvin Minsky, "Don't ask any questions, just go."

That was good advice.

If this sounds as if I'm bragging — what can I say? Blogging about it is as close as I can come to inviting all my blogfriends to have this fun with me.

Posted 6/4/05; Department 'Life, the Universe, and Everything'

Getting in touch with his roots, and vice versa

Frank's Polish grandmother, aka Grandma Wilczek, compressed a powerful primal force somewhere in her 5-foot-tall frame. Her wedding portrait, lugging a fat sheaf of lilies as long as her arm, dominates one corner of our dining room.

At 19, in Poland's disastrous post-World-War-I years, she fled her hometown, Galician Babice. On Long Island, she met Frank's Polish grandfather — a six-feet-tall blacksmith from Warsaw, who gave us all the last name that means "little wolf."

Frank and I just got back from a great party at the Polish Institute of Arts and Sciences of America — emigre scholars who maintain a wonderful archive of Polish lore here in Manhattan.

I've enjoyed watching Frank get in touch with his Polish roots (and vice versa), but today's PIASA party takes the cake (or, as Grandma Wilczek might say, the chrusciki and paczki).

Posted 6/12/05; Department 'Life, the Universe, and Everything'

I have the nicest in-laws on the planet...

...and that goes for Uncle Walter and Aunt Billie (who came with us to the Stockholm ceremonies) and cousins Cheri and Patti (and many more!) and cousin-in-law Jim too.

But this special picture shows Frank with his mom and dad after the Nobel Monument ceremony in New York City yesterday. It's the first of all the many "praise Frank!" occasions they've been able to be part of. And we are so glad they came.

Posted 6/14/05; Department 'Life, the Universe, and Everything'

In the CliffNotes version of Frank's Nobel adventures...

...we have reached a classic comic subplot. Frank has been invited to address graduation ceremonies at his high school — and at his elementary school.

Talk about dreams of childhood, (or nightmares of childhood)...

And, for another rich Shakespearean irony, *The New York Times* decided to cover last week's Nobel Monument ceremonies in its Sunday Style section, with weddings and glitzy fundraisers. This photo was not online, but (thanks to the sharp eyes of my friend Roberta) I now have a sheet of newsprint featuring Frank and NYC's teenage Nobel essay contest winners sharing a page with party animals David Rockefeller and Henry Kissinger.

Those student essayists will enjoy all the Nobel festivities in December, so I doubt they much care where the NYT put their photos.

Posted 6/22/05; Department 'Nobel'

Pomp and circumstance in Queens

In the background, Ray Charles sang "America the Beautiful." Solemn fifth graders trooped down to the front of the school auditorium, climbed up on stage, and led us in the Pledge.

Which brings me to some very useful advice: when your husband is invited to address his old elementary school graduation — do not wear any eye makeup.

Now we are home for exactly one full day before taking off for a Nobel-laden island in Lake Constance, followed by a lepton-photon conference in Uppsala. I still love traveling, but I am getting tired of packing and unpacking…

Posted 6/24/05; Department 'Pilgrimages'

Sex and physics and Dennis Overbye of the New York Times

Blogging owes some 84.3%* of its success to our pleasure in learning more about people we already know, even slightly or virtually.

I just had the pleasure of trailing along when Dennis Overbye gave Frank a tour of *The New York Times* backstage. (Much more impressive than backstage at CNN, BTW.) So of course I had to read more about Dennis and I came across an excellent interview (in the online zine *Edge*) about his book *Einstein in Love*, whose summation I can't resist sharing:

*All such statistics are, and deserve to be, made up.

I know lots of people like Albert. I might be like him myself. He was a hopeless romantic, he lived on anticipation. He was always yearning for the next thing. He was always envisioning some wonderful life with somebody else, while grimly enduring life with the woman he was with. If I think about it, I would say that that was kind of the key to his psychology, that he had the lure of the perfect situation, the perfect person. Of course if you're Einstein, you want everything that you want your way and then you want to be left alone. So you want love, and you want affection, you want a good meal, but then you don't want any interference outside of that, so you don't want any obligations interfering with your life, with your work. Which is a difficult stance to maintain in an adult relationship; it doesn't work ... If he was around I'd love to buy him a beer ... but I don't know if I'd introduce him to my sister.

Now you know more about Einstein, perhaps, than you wanted to. He could tell us his own side, if he just had a blog.

Posted 6/30/05; Department 'Pilgrimages'

Galactic strawberries and DMZ birding

Last night's amazing dessert at the 55th annual Lindau Nobel event: Topped by a hemisphere of sugar strands, a tower of alternating sorbet-and-ice cream suggests a tiny astronomy radar structure.

I am now in Lindau, Germany, five floors up from the beautiful Bodensee (aka Lake Constance) in a lovely 19th century hotel-with-beach-and-boats. Frank Wilczek and 46 other Nobel laureates and more than 700 students from around the world are on Lindau Island doing scientist stuff.

Before jet-lag sleepiness floors me, I got permission to share some info from my last-night dinner partner Hans Jornvall, who is head of the committee that picks Nobel laureates in medicine as well as a serious nature-lover and birder.

Hans told me something remarkable about Korea's DMZ — a tiny strip of land with an unbroken row of North Korean soldiers pointing big guns into it from the north of it and South Korean soldiers pointing big guns into it from the south. Nobody lives there, nobody farms there, nobody hunts there — so it has gone back to wilderness, full of rare and endangered species. Nice to know there's an upside to world non-peace.

Do not, however, go birding in Korea and then just stroll into the DMZ with your little Bushnells and Peterson's. You will be summarily shot, by one side or the other.

Posted 6/27/05; Department 'Pilgrimages'

How many Nobel laureates can dance on the head of a pin?

Or, never mind the head of a pin, how many Nobel laureates can dance the Polonaise on a tiny dance floor in Lindau? About 47, I'd guess, after last night's party.

Special kudos for stylish moves to Alan Heeger (2000 Chemistry) and of course Frank Wilczek (2004 Physics).

I love the part of traveling where you run into people you already know and like. Since December, 2004, Nobel prizewinners are about 99 44/100% of those people for me. And high on said list is Aaron Ciechanover (2004 Chemistry) who just taught me how to Skype.

Aaron, who won for studying ubiquitin, the useful protein-murderer in our cells, gave a talk yesterday about how cells do quality control by breaking down defective proteins. Just tag the bad protein with some ubiquitin, and a huge scribbly proteasome will find and destroy it. (I am simplifying quite a lot here.)

Cells that can't recycle protein get damaged — old useless proteins get in the way of more healthy processes — so there are a lot of medical possibilities.

Metaphorical possibilities too. I wish all of us had some mental "ubiquitin" — a mechanism to get rid of bad ideas and outmoded beliefs, after testing them carefully to see which ones are defective…

Of course, my ideas are all perfect — including this one — but several of yours, well, they really do need some rethinking.

Posted 6/28/05; Department 'Nobel'

Nineteenth century, with four bars of wireless

The Meeting of Nobel Laureates in Lindau houses its laureates in nineteenth century summer splendor — lawns, wooded allees, lots of roses, afternoons on the terrace looking out over Lake Constance while women in long skirts bring chilled white wine or Eischocolade.

And meanwhile , at the Hotel Bad-Schachen, in our gauze-curtained room, or on the stone terrace, or under the sycamore tree on its green bench , there were between two and four bars of wifi Internet.

If you plan to visit Lake Constance — it's in the leisurely fruit-growing "Grüss Gott" part of Germany, tucked up against Austria and Switzerland — and if you are not too wedded to air conditioning, I recommend the lovely Hotel Bad-Schachen, formerly the White Swan Hotel.

Posted 7/2/05; Department 'Pilgrimages'

Blasts from a mutual past

Thirty years ago...

...on July 3, 1973, I was a kid in my twenties about to get married for the second time, and boy was I nervous. Just about everything went wrong that day.

- ❖ Because we "eloped" with just 4 grad student pals as our witnesses, our families assumed a baby was coming. In fact, we were just too impatient with all the postponements due to "Aunt Agatha can't come June 7, how about the next weekend — no, Uncle Ed can't make the next weekend..." Well, a baby did come, about 15 months later.

❖ I baked us a chocolate wedding cake from a mix, but assumed I wouldn't have to grease and flour my brand-new miracle cake pans lined with Teflon. Wrong! Our guests ate bowls of ice cream with huge hunks of scraped-from-the-pan cake on top.

❖ We got lost driving to Dutch Neck traffic court, so we missed the time of appointment with the judge. We and our pals had to sit through an hour of testimony about stop-sign violations and DUI before being called into the back room for our "ceremony."

One thing went right — I married a wonderful man, who hasn't once stopped surprising me in all the 30 years afterward.

Feel free to throw rice anytime.

Posted 7/3/03; Department 'My Back Pages'

Email! I love it!

Where was email in 1973? The day after our wedding, Frank left for a three-week summer school in Sicily — and the first paleblue air letter didn't arrive until he'd been gone a whole week.

Let me spell that out for you:

a

WHOLE

WEEEEEEEEEEEEEK!!!!!

But I still have all Frank's low-tech letters, thirty years later. Meanwhile, our high-tech early memories, Super 8 film of Mickey learning to walk and talk, have been declared obsolete over and over again. First, we were supposed to get them converted to VHS tape. Then, we were supposed to get the VHS converted to DVDs. Now HDTV is in the offing and who knows what the next format will be.

Email. I love it not least because I can print it out and then it is *mine*.

Posted 2/7/04; Department 'Pilgrimages'

Ladies and gentlemen — the yes-or-no question!

If a man says 'yes' he means maybe.
If he says 'maybe' he means no.
If he says 'no' — he is no gentleman.
If a woman says 'no' she means maybe.
If she says 'maybe' she means yes.
If she says 'yes' — she is no lady.

This was a favorite joke of Paul Dirac (1902–1984), a brilliant physicist so shy that his Cambridge colleagues coined a unit, the dirac, to stand for the smallest measurable quantum of speech. Dirac was what I call an Aleph male — distracted from Alpha ambitions by his private obsessions.

I met Dirac and his wonderful wife Moncie in 1981, the summer before his 80th birthday. There is a famous summer school for physics in the little hilltown of Sicily known as Erice. I was a young physics wife, quite visibly pregnant. (Just one year before, my bottom was pinched black-and-blue by Sicilian men, but my little round belly got me treated me like a queen.)

I have a nerdy fondness for learning languages, so I served as an amateur guide to physicists who didn't speak any Italian. I was delighted to be the chosen companion of a Nobel laureate for an hour or two, even if our mutual goal was to buy him some shoes. Professor Dirac was courteous but shy — and when he smiled, he had a wonderfully naughty twinkle in his eye.

I thought Moncie was a very lucky woman.

Posted 6/5/03; Department 'Learn to Write Funny'

1983: R. P. Feynman and my Valentine

Speaking of Valentines and love (and weren't we?), I love this 1983 picture of Frank Wilczek with legendary physics bad boy R. P. Feynman.

Yes, that's the *Surely you're joking, Mr. Feynman* Feynman, the Challenger disaster frozen O-Rings Feynman, safe-cracking, bongo-playing, Tuva-loving, 1965-Nobel-Prize-winning, 1918–1988 Richard P. Feynman.

About this photograph — first, roll your mental cameras back even farther, to 1972, the year I met Frank Wilczek. When I think back to those intense happy days, I can remember earnestly playing Bob Dylan records and moving the needle to tracks I thought Frank "should" hear. And I'll never forget his introducing me to Feynman, by which he meant the tattered red three-volume *Lectures in Physics* he'd read by himself in high school to learn about physics.

He sat me down next to him on some seedy Grad College sofa, put the first volume into my hands, and waited to enjoy the delight I would surely feel as soon as I started reading. These books were an intellectual treasure that he couldn't wait to share — only later did I find out they were also a part of his personal odyssey — if I hadn't been in love with him before, I would surely have fallen in love with him right then.

Now — ten-plus years, lots of physics, and two great kids later — we get to 1983. Murph Goldberger sent us this photo of Frank with Feynman at his (Murph's) 60th birthday bash. No wonder Frank looks happy — even if Feynman is teasing him and our friend Sam Treiman is giving him "devil horn" fingers…

I've been hard at work collecting Frank Wilczek photos, which have to go out to a whole bunch of publications. So that and some chocolate will have to make up for my Valentine failure to come up with the really perfect love sonnet…

Posted 2/14/05; Department 'Nobel'

1984: *Year of our first Mac*

We got our first Mac in 1984, and Lawrence Krauss (*Physics of Star Trek*) was the Johnny Appleseed who planted the first Macintosh seed.

In 1984, Frank and I were trying to collaborate on a book by using the arcane word-processing stuff on the VAX/VMS at UCSB's Institute for Theoretical Physics.

Then one day, a young physicist named Lawrence Krauss rode into town, carrying with him everywhere a big 15-pound plastic box he called a Mac.[*] We had never imagined lugging our good old Atari 800 on our trips. (We used it to play simple games on an old TV set, and to write simpler games in Atari BASIC.) What did this "Mac" have that our computer didn't?

Lawrence was eager to demo MacWrite and MacPaint. (MacWrite was similar to Microsoft Word, only simpler and much more reliable.) We were instant converts, and so were our kids! I still have pages and pages and pages of dot-matrix printouts of their MacPaint creations to prove it.

The Mac was just what we needed to collaborate.

- ❖ Never mind that you had to save everything on a floppy.[†]
- ❖ Never mind that no single chapter could be more than about 10 pages long.
- ❖ Never mind that the "illustrations" I created with the dot matrix printer were black and white and lumpy with pixels all over.

---

[*] Oops — the Mac plus keyboard plus etc. plus bag weighed 22 lbs! Thanks to Kuba Tatarkiewicz of MIT for links to the original Mac spec.

[†] Thanks to Kuba again, for pinging me that the size of the floppy *"was 400 kB, while RAM of the original Mac was 128 kB, hence the so called Macintosh elbow when you tried to copy a floppy on a single-drive Mac (I once did 27 inserts — it was very painful!)."*

That first Mac was a wonderful revelation of just how much better computers could make people's lives, and we're still grateful to Lawrence Krauss (now here at the APS meeting with us, where he gave 3 (3!) fine talks!)

Long before Lawrence wrote *The Physics of Star Trek*, he helped us boldly go where no Wilczek or Devine had gone before.

Posted 4/19/05; Department 'My Back Pages'

At least our visitors didn't have bolts in their necks…

> MARY SHELLEY is to be commemorated by a blue English Heritage plaque on the London house where she died — an honour proposed in 1975 but resisted by the vicar who then lived there. He objected to the words "author of Frankenstein," presumably for fear of crowds of peasants with torches, and felt that "author(ess) and wife of the poet'" would suffice.

from this month's *Ansible*, Dave Langford's wonderful British sf/fan newsletter.

I have some sympathy with the vicar, because we lived for years in the Princeton house once owned by Albert Einstein. Einstein was adamant he didn't want the house to be a museum — and it wasn't — which didn't stop tourists from ringing the front doorbell.

The most persistent in wanting to come in anyway were the Germans. In less charitable moments, I thought about pointing out that Einstein would happily have grown old in Germany if their ancestors hadn't pitched him out on his ear.

I never thought of the English vicar's solution — "Here lived an author, the husband to two Mrs. Einsteins."

Posted 10/2/03; Department 'Life, the Universe, and Everything'

Return of the King: The absolute best moment

It's hard to pick one best moment in ROTK, but here are my top five (major spoilers ahead!):

Number 5: When Legolas fights a huge elephant plus all its riders.

Number 4: When Gimli tells Legolas, "That only counts as one."

Number 3: When Eowyn and Merry defeat the Witch King.

Number 2: When Sam jumps up to go hit on Rosie Cotton.

Number 1 and absolutely best: Outside the movie, afterward, when my husband says, "Poor Aragorn. He really picked the wrong girl."

Posted 1/11/04; Department 'Heroes and Funny Folks'

Dramatic theory with mosquito continuo

I heard about the Gary Wills review of Bill Clinton's book from my husband, as we went for a humid twilight stroll.

Frank's one-sentence synopsis of Gary Wills: "Wills says Bill Clinton is a tragic hero."

Frank one-sentence improvement of Gary Wills: "In other words, Bill Clinton's a great big jerk."

Betsy's commentary on Frank comment: I think that's a really deep interpretation not only of Bill Clinton but also of tragedy. Think about it — Oedipus–Antigone–Macbeth–Hamlet — royal pains in the butt, every one of them.

Posted 7/23/04; Department 'Heroes and Funny Folks'

Frank Wilczek: The difference between math and physics…

There are certain problems with having a wife who blogs. For example, if you say something funny to a bunch of your high school classmates, your wife is likely to scribble it down on a paper napkin so that she can blog it later.

Then, four months later, she re-finds the paper napkin. So today (ta da!) I'm blogging two fine Frank Wilczek quotes!

> *I went off to college planning to major in math or philosophy —*
> *of course, both those ideas are really the same idea.*
>
> *In physics, your solution should convince a reasonable person.*
> *In math, you have to convince a person who's trying to make trouble.*
> *Ultimately, in physics, you're hoping to convince Nature.*
> *And I've found Nature to be pretty reasonable.*

Now I'm headed to the blogless backwoods of NH — while I'm gone, read the folks on my blogroll, assuming they don't all decide to take time off now too.

Posted 7/9/05; Department 'Heroes and Funny Folks'

Nobel full circle

Nobel full circle: October 5, 2005

One year ago, Frank got the Nobel phone call at 5:30 a.m. This year, one of those phone calls went to Roy Glauber, a Cambridge neighbor. It's full circle time — and, what a wonderful, funny (peculiar and ha-ha) year this has been.

- ❖ Fans of J Lo have besieged my husband for autographs.
- ❖ HRH Prince Carl Phillip of Sweden told me which kind of long underwear is best for the Ice Hotel. (Middleweight, he said, and he was right.)
- ❖ I got to see northern lights at last (wavy and green).
- ❖ Princeton added Frank's picture to its Nobel gallery, a place we loved to visit, way back when we were courting.
- ❖ Skipping over a lot more stuff (Lech Walesa, George Lucas, those aliens, etc. etc. etc. etc… .) I'm racing to sort out, by October 15, the rest of this year's adventures for publication in a book with Frank!

Frank's been invited to far more places than he could possibly go — and we've gone to far more of them than we probably should have. But how could we regret, despite battered suitcases now, and sometimes sore feet, all the places we've seen, the new things we've discovered, and wonderful people we've met?

Thank you, Nobel Foundation, for making this possible.

And thank you, dear readers all, for sharing this journey.

Posted 10/5/05; Department 'Nobel'

Acknowledgments

The author and publisher would like to thank

American Academy of Arts and Sciences
American Association for the Advancement of Science
American Institute of Physics
American Physical Society
Granta Books
MIT Department of Physics
Nature Publishing Group
New York Academy of Sciences
The Nobel Foundation
W. W. Norton
Reed Business Information Ltd
Springer Science and Business Media
Thomson-Gale Group
University of Chicago Press

for their kind permission to include the following materials found in this book:

1. The World's Numerical Recipe, *Daedalus* 131, 142–147 (2001). © American Academy of Arts and Sciences

2. Analysis and Synthesis 1: What Matters for Matter, *Physics Today*, May 2003, pp. 10–11. © American Institute of Physics

3. Analysis and Synthesis 2: Universal Characteristics, *Physics Today*, Jul 2003, pp. 10–11. © American Institute of Physics

4. Analysis and Synthesis 3: Cosmic Groundwork, *Physics Today*, Oct 2003, pp. 10–11. © American Institute of Physics

5. Analysis and Synthesis 4: Limits and Supplements, *Physics Today*, Jan 2004, pp. 10–11. © American Institute of Physics

6. Whence the Force of $F = ma$? 1: Culture Shock, *Physics Today*, Oct 2004, pp. 11–12. © American Institute of Physics

7. Whence the Force of $F = ma$? 2: Rationalization, *Physics Today*, Dec 2004, pp. 10–11. © American Institute of Physics

8. Whence the Force of $F = ma$? 3: Cultural Diversity, *Physics Today*, Jul 2005, pp. 10–11. © American Institute of Physics

9. The Origin of Mass, *Annual Physics @ MIT*, 2003, pp. 24–35. © MIT Department of Physics

10. Mass without Mass 1: Most of Matter, *Physics Today*, Nov 1999, pp. 11–12. © American Institute of Physics

11. Mass without Mass 2: The Medium is the Mass-age, *Physics Today*, Jan 2000, pp. 13–14. © American Institute of Physics

12. QCD Made Simple, *Physics Today*, Aug 2000, pp. 22–28. © American Institute of Physics

13. 10^{12} Degrees in the Shade, *The Sciences*, Jan/Feb 1994, pp. 22–30. © New York Academy of Sciences

14. Back to Basics at Ultrahigh Temperatures, *Physics Today*, April 1998, pp. 11, 13. © American Institute of Physics

15. Scaling Mount Planck 1: A View from the Bottom, *Physics Today*, Jun 2001, pp. 12–13. © American Institute of Physics

16. Scaling Mount Planck 2: Base Camp, *Physics Today*, Nov 2001, pp. 12–13. © American Institute of Physics

17. Scaling Mount Planck 3: Is That All There Is?, *Physics Today*, Aug 2002, pp. 10–11. © American Institute of Physics

18. What Is Quantum Theory?, *Physics Today*, Jun 2000, pp. 11–12. © American Institute of Physics

19. Total Relativity: Mach 2004, *Physics Today*, Apr 2004, pp. 10–11. © American Institute of Physics

20. Life's Parameters, *Physics Today*, Feb 2003, pp. 10–11. © American Institute of Physics

21. The Dirac Equation, from *It Must Be Beautiful: Great Equations of Modern Science*, ed. G. Farmelo, Granta Books, 2002, pp. 102–130. © Granta Books

22. Fermi and the Elucidation of Matter, from *Fermi Remembered*, ed. James W. Cronin, Univ. Chicago Press, 2004, pp. 34–51. © Univ. Chicago Press

23. The Standard Model Transcended, *Nature*, 394, 13–15 (1998). © Nature Publishing Group

24. Masses and Molasses, *New Scientist*, April 1999, pp. 32–37. © Reed Business Information Ltd

25. In Search of Symmetry Lost, *Nature*, 433, 239–247 (2005). © Nature Publishing Group

26. From 'Not Wrong' to (Maybe) Right, *Nature*, 428, 261 (2004). © Nature Publishing Group

27. Unification of Couplings (with S. Dimopoulos and S. A. Raby), *Physics Today*, Oct 1991, pp. 25–33. © American Institute of Physics

28. The Social Benefit of High Energy Physics, from *Building Blocks of Matter*, ed. John S. Rigden, Macmillan Reference USA, 2003. Used by permission of the Gale Group.

29. When Words Fail, *Nature*, 410, 149 (2001). © Nature Publishing Group

30. Why Are There Analogies between Condensed Matter and Particle Theory?, *Physics Today*, Jan 1998, pp. 11, 13. © American Institute of Physics

31. The Persistence of Ether, *Physics Today*, Jan 1999, pp. 11, 13. © American Institute of Physics

32. Reaching Bottom, Laying Foundations, *Nature*, Apr 1999, pp. 4–5 (special issue "A Celebration of Physics" for American Physical Society's 100th Anniversary). © Nature Publishing Group

33. What Did Bohr Do? (Book Review), *Science*, 225, 345 (1991). © American Association for the Advancement of Science

34. Dreams of a Final Theory (Book Review), *Physics Today*, Apr 1993, pp. 59–60. © American Institute of Physics

35. Shadows of the Mind (Book Review), *Science*, 266, 1737–1738 (1994). © American Association for the Advancement of Science

36. The Inflationary Universe (Book Review), *Science*, 276, 1087–1088 (1997). © American Association for the Advancement of Science

37. Is the Sky Made from Pi? (Book Review), *Nature*, 403, 247–248 (2000). © Nature Publishing Group

38. Quantum Field Theory, from the American Physical Society centenary issue of *Rev. Mod. Phys.* 71, S85–S95 (1999); *More Things in Heaven and Earth: A Celebration of Physics at the Millennium*, ed. B. Bederson, Springer Verlag, New York, 1999. © American Physical Society

40. From Concept to Reality to Vision, *Eur. Phys. J.*, C33, S1–S4 (2004). © Springer Science and Business Media

41. Nobel Biography, from *Les Prix Nobel 2004*, pp. 96–99. © The Nobel Foundation

42. Asymptotic Freedom: From Paradox to Paradigm, from *Les Prix Nobel 2004*, pp. 100–124. © The Nobel Foundation

44. Virtual Particles (poem), in *Longing for the Harmonies: Themes and Variations from Modern Physics*, by Frank Wilczek and Betsy Devine, W. W. Norton, 1988. Used by permission of W. W. Norton & Co., Inc.

45. Gluon Rap (poem), in *Longing for the Harmonies: Themes and Variations from Modern Physics*, by Frank Wilczek and Betsy Devine, W. W. Norton, 1988. Used by permission of W. W. Norton & Co., Inc.

46. Reply in Sonnet Form (poem), *Physics Today*, Mar 1999, p. 113. © American Institute of Physics

Betsy Devine would like to thank Sverker Fredriksson, Odd Minde, Jonas Förare, Elizabeth Thomson, and Reuters for permission granted to include photos and text from them. Betsy would also like to thank all the people mentioned in her blog, most of them too briefly — most especially her in-laws Frank and Mary Wilczek, as well as Amity, Mira, Colin, Walter, Billie, Marie, Kim, Bill, Mary, Jeff, Naomi, KC, Lawrence, Kate, Antti, Nadya, and Cecilia, collectively the best Nobel guests ever! Last but not least, Betsy would like to thank Frank Wilczek for making so many fantastic realities possible, only one of which was his winning the Nobel Prize.

Frank and Betsy would also like to express warm thanks to Dorian Devins, Kim Tan, Thecla Teo, Jimmy Low, and the many other talented, hardworking people at World Scientific who helped to make this book a reality!